Pitman Research Notes in Mathematics Series

Submission of proposals for consideration
Suggestions for publication, in the form of outlines and representative samples, are invited by the Editorial Board for assessment. Intending authors should approach one of the main editors or another member of the Editorial Board, citing the relevant AMS subject classifications. Alternatively, outlines may be sent directly to the publisher's offices. Refereeing is by members of the board and other mathematical authorities in the topic concerned, throughout the world.

Preparation of accepted manuscripts
On acceptance of a proposal, the publisher will supply full instructions for the preparation of manuscripts in a form suitable for direct photo-lithographic reproduction. Specially printed grid sheets can be provided and a contribution is offered by the publisher towards the cost of typing. Word processor output, subject to the publisher's approval, is also acceptable.

Illustrations should be prepared by the authors, ready for direct reproduction without further improvement. The use of hand-drawn symbols should be avoided wherever possible, in order to maintain maximum clarity of the text.

The publisher will be pleased to give any guidance necessary during the preparation of a typescript, and will be happy to answer any queries.

Important note
In order to avoid later retyping, intending authors are strongly urged not to begin final preparation of a typescript before receiving the publisher's guidelines. In this way it is hoped to preserve the uniform appearance of the series.

Addison Wesley Longman Ltd
Edinburgh Gate
Harlow, Essex, CM20 2JE
UK
(Telephone (0) 1279 623623)

Titles in this series. A full list is available from the publisher on request.

350 Progress in elliptic and parabolic partial differential equations
A Alvino, P Buonocore, V Ferone, E Giarrusso, S Matarasso, R Toscano and G Trombetti

351 Integral representations for spatial models of mathematical physics
V V Kravchenko and M V Shapiro

352 Dynamics of nonlinear waves in dissipative systems: reduction, bifurcation and stability
G Dangelmayr, B Fiedler, K Kirchgässner and A Mielke

353 Singularities of solutions of second order quasilinear equations
L Véron

354 Mathematical theory in fluid mechanics
G P Galdi, J Málek and J Necas

355 Eigenfunction expansions, operator algebras and symmetric spaces
R M Kauffman

356 Lectures on bifurcations, dynamics and symmetry
M Field

357 Noncoercive variational problems and related results
D Goeleven

358 Generalised optimal stopping problems and financial markets
D Wong

359 Topics in pseudo-differential operators
S Zaidman

360 The Dirichlet problem for the Laplacian in bounded and unbounded domains
C G Simader and H Sohr

361 Direct and inverse electromagnetic scattering
A H Serbest and S R Cloude

362 International conference on dynamical systems
F Ledrappier, J Lewowicz and S Newhouse

363 Free boundary problems, theory and applications
M Niezgódka and P Strzelecki

364 Backward stochastic differential equations
N El Karoui and L Mazliak

365 Topological and variational methods for nonlinear boundary value problems
P Drábek

366 Complex analysis and geometry
V Ancona, E Ballico, R M Mirò-Roig and A Silva

367 Integral expansions related to Mehler–Fock type transforms
B N Mandal and N Mandal

368 Elliptic boundary value problems with indefinite weights: variational formulations of the principal eigenvalue and applications
F Belgacem

369 Integral transforms, reproducing kernels and their applications
S Saitoh

370 Ordinary and partial differential equations. Volume V
P D Smith and R J Jarvis

371 Numerical methods in mechanics
C Conca and G N Gatica

372 Generalized manifolds
K-G Schlesinger

373 Independent axioms for Minkowski space-time
J W Schutz

374 Integral methods in science and engineering Volume one: analytic methods
C Constanda, J Saranen and S Seikkala

375 Integral methods in science and engineering Volume two: approximation methods
C Constanda, J Saranen and S Seikkala

376 Inner product spaces and applications
T M Rassias

377 Functional analysis with current applications in science, technology and industry
M Brokate and A H Siddiqi

378 Classical and quantic periodic motions of multiply polarized spin-particles
A Bahri

379 Analysis, numerics and applications of differential and integral equations
M Bach, C Constanda, G C Hsiao, A-M Sändig and P Werner

380 Numerical analysis 1997
D F Griffiths, D J Higham and G A Watson

381 Real analytic and algebraic singularities
T Fukuda, T Fukui, S Izumiya and S Koike

382 Boundary value problems with equivalued surface and resistivity well-logging
T Li, S Zheng, Y Tan and W Shen

383 Progress in partial differential equations Pont-à-Mousson 1997 Volume 1
H Amann, C Bandle, M Chipot, F Conrad and I Shafrir

384 Progress in partial differential equations Pont-à-Mousson 1997 Volume 2
H Amann, C Bandle, M Chipot, F Conrad and I Shafrir

385 The linear theory of Colombeau generalized functions
M Nedeljkov, S Pilipovic and D Scarpalézos

386 Recent advances in differential equations
H-H Dai and P L Sachdev

387 Progress in holomorphic dynamics
H Kriete

388 Navier–Stokes equations: theory and numerical methods
R Salvi

389 Strongly irreducible operators on Hilbert space
C Jiang and Z Wang

Chunlan Jiang

He Bei University of Technology

Zongyao Wang

East China University of Science and Technology

Strongly irreducible operators on Hilbert space

LONGMAN

Addison Wesley Longman Limited
Edinburgh Gate, Harlow
Essex CM20 2JE, England
and Associated Companies throughout the world.

*Published in the United States of America
by Addison Wesley Longman Inc.*

First published 1998

AMS Subject Classifications (Main) 47A, 47B, 47C
(Subsidiary) 46B, 46C, 46H

ISSN 0269-3674

ISBN 0 582 30594 2

Visit Addison Wesley Longman on the world wide web at
http://www.awl-he.com

British Library Cataloguing in Publication Data

A catalogue record for this book is
available from the British Library

Printed in Great Britain by Henry Ling Ltd.,
at the Dorset Press, Dorchester, Dorset.

To Wei and Yimin

Contents

Preface

Owing to the set-up of the perfect representation theories of normal operators and commutative C^*-algebras, researchers are interested in and working on the structure of non-self-adjoint operators on Hilbert spaces and general C^*-algebras. For example, since P.R. Halmos introduced the notion of irreducibility of operators in 1968, fruitful results in the exploration of the structure of C^*-algebras and operators have been achieved. One of the most typical results is D. Voiculescu's non-commutative Weyl–von Neumann theorem. On the other hand, research on similarity invariants is very important to the study of operator structures, which was shown not only by the work on triangular algebras due to J.R. Ringrose, W.B. Arvison and K.R. Davidson, etc. but also by the systematic research on non-abelian approximation theory by C. Apostol, D.A. Herrero and D. Voiculescu, etc. Irreducibility, however, is only a unitary invariant, and it seems that it cannot essentially reveal and reflect the internal structure of operators.

The purpose of this monograph is to explore the structure of non-self-adjoint operators by analysing and studying the strong irreducibility of operators, which is a similarity invariant. The notion of strong irreducibility of operators was introduced by F. Gilfeather in 1972. Since strongly irreducible operators, or briefly, (SI) operators, and Jordan blocks on a finite dimensional space have many features in common, Z.J. Jiang considers the (SI) operators as the approximate replacement of Jordan blocks on infinite dimensional Hilbert space and hopes that a theorem similar to the Jordan canonical form theorem in matrix theory can be set up with this replacement. In order to enrich the structure of (SI) operators, D.A. Herrero gave the notion of "nice property". Owing to the introduction to these ideas and notions, the theorems on the approximately uniquely (in a certain sense) strongly irreducible decomposition of operators, the spectral pictures, the closures of the similarity orbits and the compact perturbations of (SI) operators were set up.

From the beginning of our research on (SI) operators up until the preparation of this monograph, D.A. Herrero and Z.J. Jiang have contributed a great deal to our work. Herrero raised many intriguing questions on the angle of operator structure.

They involve the uniqueness of the (SI) decomposition of operators, the spectral pictures and compact perturbations of (SI) operators and the applications of the theory of (SI) operators to stable invariant subspaces of operators, etc. Desiring to obtain an approximate Jordan canonical form theorem, Z.J. Jiang asked some questions concerning the description of the norm-closures of (SI) operators and of the direct sum of (SI) operators and quasisimilarity of (SI) operators, etc. It is the raising of these questions that motivates our research with enthusiasm and enjoyment. We should point out here that many of Herrero's results, especially his book Approximation of Hilbert Space operators I supply our research with a strong technical foundation. Without this foundation, our research would have been impossible. In addition, we thank K.R. Davidson, who suggested that the authors consider the (SI) operators in nest algebras. His suggestions add colour not only to our research but also to this monograph.

This monograph consists of seven chapters plus a Chapter 0, which constitutes the prerequisites for this book. It is interesting to observe that for a large number of results concerning (SI) operators, the proofs are mainly constructive. The "main tools" for these construction are contained in Chapter 1 while the "parts" and some other "tools" are developed in Chapters 2 and 3. Most of the material in Chapter 1 is from Chapters 3 and 4 of Herrero's book with a retreatment to a certain extent. Some results in Chapter 1 are our own work. Chapter 2 discusses irreducible operators and gives the definition, examples and fundamental properties of (SI) operators. Chapter 3 considers the (SI) operators in different classes of operators, and constructs (SI) operators according to different requirements which are needed in later chapters. Chapter 4 is about the closure of the direct sum of (SI) operators, the uniqueness of the (SI) decomposition of operators and the realization of (SI) operators according to a prescriptive spectral picture. The following problems are considered in Chapter 5: Given an operator T in $\mathcal{L}(\mathcal{H})$, can we find a compact operator K such that $T+K$ is an (SI) operator? If we can, can the norm of K be arbitrary small?

Chapter 6 treats the (SI) operators in nest algebras and the closure of their unitary orbits, and Chapter 7 consists of some relatively "small" but interesting results (for example, it is proved that each operator is a sum of three (SI) operators) and some applications of (SI) operators.

The authors are deeply indebted to Professors Raul E.Curto, Kenneth R. Davidson and Steven C. Power for their encouragement, suggestions and, above all, for their friendly support. We express our thanks to our colleagues Professors Jiagang

Wang, and Wenci Yu at the East China University of Science and Technology and Guochen Feng, Ronghua Li, Zhengpeng Wang and Zhuoqun Wu at the Jilin University for their interest and support. We also wish to thank miss Hongying Li for the technical expertise with which she typed the manuscript of this monograph.

Chunlan Jiang
Zongyao Wang

0. Background

In this chapter, we review briefly some of the facts about Hilbert space operators and Sobolev space theory which will be needed to read this book. Most of the material can be found in books such as R.A. Adams [1], J.B. Conway [1], K.R. Davidson [1], R.G. Douglas [1] or W. Rudin [1].

0.1 Banach algebras

A *Banach algebra* is an algebra $\mathcal{A}$ over $\mathbb{C}$ that has a norm $\|\cdot\|$ relative to which $\mathcal{A}$ is a Banach space and such that for all a, b in $\mathcal{A}$, $\|ab\| \leq \|a\|\|b\|$. If $\mathcal{A}$ has an identity, e, then it is assumed that $\|e\| = 1$. If $\mathcal{A}$ is a Banach algebra with identity and $a \in \mathcal{A}$, the *spectrum* $\sigma(a)$ of a is defined by $\sigma(a) = \{\lambda \in \mathbb{C} : \lambda - a \text{ is not invertible}\}$. The *left spectrum* $\sigma_l(a)$ is the set $\{\lambda \in \mathbb{C} : \lambda - a \text{ is not left invertible}\}$; the *right spectrum* $\sigma_r(a)$ is defined similarly. The *resolvent set* of a is defined by $\rho(a) = \mathbb{C}\backslash\sigma(a)$. The *left* and *right resolvents* of a are $\rho_l(a) = \mathbb{C}\backslash\sigma_l(a)$ and $\rho_r(a) = \mathbb{C}\backslash\sigma_r(a)$. $\sigma(a)$ is a non-empty compact subset of $\mathbb{C}$ and $\sigma(a) = \sigma_l(a) \cup \sigma_r(a)$.

Let f be holomorphic in a neighbourhood Ω of $\sigma(a)$, and let C be a finite union of Jordan curves such that $\text{ind}_C(\lambda) = 1$ for every λ in $\sigma(a)$. Define

$$f(a) = \frac{1}{2\pi i} \int_C f(z)(ze - a)^{-1}\, dz.$$

Let $\text{Hol}\,(\sigma(a))$ denote the set of all functions that are holomorphic in a neighbourhood of $\sigma(a)$.

Riesz Functional Calculus. *Let a be an element of a Banach algebra $\mathcal{A}$ with identity. Then for every f in $\text{Hol}\,(\sigma(a))$, $f(a)$ is well defined independent of the curve C. The map taking f to $f(a)$ is an algebra homomorphism such that each polynomial $p(z) = \sum_{k=0}^{n} c_k z^k$ is taken to $c_0 e + \sum_{k=1}^{n} c_k a^k$.*

Spectral Mapping Theorem. *For f in $\text{Hol}\,(\sigma(a))$, $\sigma(f(a)) = f(\sigma(a))$.*

Upper Semi-continuity of the Spectrum. *Let a be an element of a Banach algebra $\mathcal{A}$ with identity. Given a bounded open set Ω such that $\sigma(a) \subset \Omega$, there exists $\delta > 0$ such that $\sigma(b) \subset \Omega$, provided $\|a - b\| < \delta$.*

Let $\mathcal{H}$ be a complex, separable, infinite dimensional Hilbert space and let $\mathcal{L}(\mathcal{H})$ denote the algebra of linear, bounded operators acting on $\mathcal{H}$. For each $T \in \mathcal{L}(\mathcal{H})$, $\sigma(T)$, $\sigma_l(T)$, $\sigma_r(T)$, $\rho(T)$, $\rho_l(T)$, $\rho_r(T)$ and $f(T)$ are defined above if $\mathcal{L}(\mathcal{H})$ is chosen as the Banach algebra $\mathcal{A}$, where $f \in \mathrm{Hol}\,(\sigma(T))$.

Riesz Decomposition Theorem. *Assume that $\sigma(T) = \sigma_1 \cup \sigma_2$ where σ_1 and σ_2 are disjoint non-empty compact sets. Then $\mathcal{H}$ is the algebraic direct sum of two invariant subspaces $\mathcal{H}_1$ and $\mathcal{H}_2$ of T, such that $\sigma(T|_{\mathcal{H}_i}) = \sigma_i$ and $\mathcal{H}_i$ is the range of the Riesz idempotent corresponding to σ_i $(i = 1, 2)$, where $T|_{\mathcal{H}_i}$ is the restriction of T on $\mathcal{H}_i$.*

In the Riesz Decomposition Theorem, if $\sigma_1 = \{\lambda\}$ and the dimension of $\mathcal{H}_1$ is finite, then λ is called a *normal eigenvalue* of T. The set of all normal eigenvalues of T will be denoted by $\sigma_0(T)$. Clearly, $\sigma_0(T)$ is contained in the *point spectrum* $\sigma_p(T)$ of T (i.e., the set of all eigenvalues of T).

Let $\mathcal{A}$ be an abelian Banach algebra with identity. A multiplicative linear functional ϕ is an algebra homomorphism: $\mathcal{A} \to \mathcal{C}$ with $\|\phi\| = 1$. They are in a one–to–one correspondence with the maximal ideals of $\mathcal{A}$. The collection Σ of all maximal ideals of $\mathcal{A}$ is called the *maximal ideal space* of $\mathcal{A}$ if it is given the weak * topology induced as a subset of the dual space $\mathcal{A}^*$. It is easy to verify that Σ is a compact Hausdorff space. The *Gelfand Transform* $\hat{a}$ of a is the function $\hat{a}$: $\Sigma \to \mathcal{C}$ defined by $\hat{a}(\phi) = \phi(a)$.

Gelfand's Theorem. *If $\mathcal{A}$ is an abelian Banach algebra with maximal ideal space Σ and $a \in \mathcal{A}$, then the Gelfand transform $\hat{a}$ of a belongs to $C(\Sigma)$ and $\sigma(a) = \{\hat{a}(\phi);\ \phi \in \Sigma\}$. The map $a \to \hat{a}$ of $\mathcal{A}$ into $C(\Sigma)$ is a continuous homomorphism of $\mathcal{A}$ into $C(\Sigma)$ of norm 1 and the kernel of this map is*

$$\begin{aligned} \mathrm{rad}\,\mathcal{A} &= \bigcap\{\mu :\ \mu \textit{ is a maximal ideal of } \mathcal{A}\} \\ &= \big\{a \in \mathcal{A} :\ \sigma(a) = \{0\}\big\}. \end{aligned}$$

The homomorphism $a \to \hat{a}$ of $\mathcal{A}$ into $C(\Sigma)$ is called the *Gelfand transform* of $\mathcal{A}$. The kernel $\mathrm{rad}\,\mathcal{A}$ of the Gelfand transform is called the *radical* of $\mathcal{A}$.

A C^*-algebra is a Banach algebra $\mathcal{C}$ with an operation $*: \mathcal{C} \to \mathcal{C}$ satisfying the conditions:

(i) $(a^*)^* = a, \quad a \in \mathcal{C}$;

(ii) $(ab)^* = b^*a^*, \quad a, b \in \mathcal{C}$;

(iii) $(\alpha a + \beta b)^* = \alpha^* a^* + \beta^* b^*, \quad a, b \in \mathcal{C};\ \alpha, \beta \in \mathbb{C}$; and

(iv) $\|a^*a\| = \|a\|^2, \quad a \in \mathcal{C}$.

(An operation $*: \mathcal{C} \to \mathcal{C}$ satisfying (i)–(iii) is called an involution.)

A *$*$-representation* ρ of a C^*-algebra $\mathcal{C}$ on a Hilbert space $\mathcal{H}_\rho$ is an operator ρ from $\mathcal{C}$ into $\mathcal{L}(\mathcal{H}_\rho)$ which preserves products and involutions (i.e., $\rho(ab) = \rho(a)\rho(b)$ and $\rho(a^*) = [\rho(a)]^*$). If $\mathcal{C}$ has an identity e and $\rho(e) = 1$, then ρ is called *unital*; ρ is *faithful* if it has a trivial kernel. It is well-known that a $*$-representation ρ is faithful if and only if it is an isometric mapping from $\mathcal{C}$ onto $\rho(\mathcal{C}) \subset \mathcal{L}(\mathcal{H}_\rho)$.

Gelfand–Naimark–Segal Theorem. *Every abstract C^*-algebra $\mathcal{C}$ with identity admits a faithful unital $*$-representation ρ in $\mathcal{L}(\mathcal{H}_\rho)$ for a suitable Hilbert space $\mathcal{H}_\rho$, i.e., $\mathcal{C}$ is isometrically $*$-isomorphic to a C^*-algebra of operators. Furthermore, if $\mathcal{C}$ is separable, then $\mathcal{H}_\rho$ can be chosen to be separable.*

von Neumann Double Commutant Theorem. *Let $\mathcal{A}$ be a unital C^*-subalgebra of $\mathcal{L}(\mathcal{H})$. Then the closure of $\mathcal{A}$ in any of the weak operator, strong operator, and weak $*$ topologies is the double commutant $\mathcal{A}''$, where $\mathcal{A}' = \{T \in \mathcal{L}(\mathcal{H});\ AT = TA$ for all A in $\mathcal{A}\}$ and $\mathcal{A}'' = \{\mathcal{A}'\}'$.*

0.2 Compact operators and the compact perturbations of operators

An operator $T \in \mathcal{L}(\mathcal{H})$ is *compact* if the image of the unit ball of $\mathcal{H}$ under T has compact closure. If $\mathcal{K}(\mathcal{H})$ denotes the ideal of all compact operators acting on $\mathcal{H}$ and $\pi: \mathcal{L}(\mathcal{H}) \to \mathcal{A}(\mathcal{H}) := \mathcal{L}(\mathcal{H})/\mathcal{K}(\mathcal{H})$ is the canonical projection of $\mathcal{L}(\mathcal{H})$ onto the (quotient) *Calkin algebra*, then $\sigma_e(T)$, the *essential spectrum* of T, is the spectrum of $\pi(T)$ in $\mathcal{A}(\mathcal{H})$ and $\mathbb{C}\backslash\sigma_e(T)$ is called the *Fredholm domain* of T and is denoted by $\rho_F(T)$. Thus, $\sigma_e(T) = \sigma_{le}(T) \cup \sigma_{re}(T)$, where $\sigma_{le}(T) = \sigma_l(\pi(T))$ (*left essential spectrum* of T) and $\sigma_{re}(T) = \sigma_r(\pi(T))$ (*right essential spectrum* of T).

T is a *semi-Fredholm* operator if the range of T, $\operatorname{ran} T$, is closed and either $\operatorname{nul} T := \dim\ker T$ or $\operatorname{nul} T^* = \dim\ker T^*$ is finite. In this case, the *index* of T is defined by

$$\operatorname{ind} T = \operatorname{nul} T - \operatorname{nul} T^*.$$

If T is a semi-Fredholm operator, then T^*, T^k $(k \geq 1)$ are also semi-Fredholm operators and $\operatorname{ind} T^* = -\operatorname{ind} T$ and $\operatorname{ind} T^k = k \cdot \operatorname{ind} T$. If $K \in \mathcal{K}(\mathcal{H})$, then $T+K$ is semi-Fredholm and $\operatorname{ind}(T+K) = \operatorname{ind} T$. If $\{A_n\}_{n=1}^{\infty}$ is a sequence of operators such that $\|T - A_n\| \to 0$ $(n \to \infty)$, then A_n is semi-Fredholm, $\operatorname{nul} T \geq \limsup_{n\to\infty} \operatorname{nul} A_n$, $\operatorname{nul} T^* \geq \limsup_{n\to\infty} \operatorname{nul} A_n^*$ and $\operatorname{ind} T_n = \operatorname{ind} T$ for all $n \geq n_0$ for some $n_0 \in \mathbb{N}$.

The intersection $\sigma_{lre}(T) := \sigma_{le}(T) \cap \sigma_{re}(T)$ is called *Wolf spectrum* and it includes the boundary $\partial\sigma_e(T)$ of $\sigma_e(T)$. Therefore, it is a non-empty compact subset of $\mathbb{C}$. Its complement $\mathbb{C} \backslash \sigma_{lre}(T)$ coincides with $\rho_{s-F}(T) := \{\lambda \in \mathbb{C} :\ \lambda - T \text{ is semi-Fredholm}\}$, the *semi-Fredholm domain* of T. $\rho_{s-F}(T)$ is the disjoint union of the (possibly empty) open sets $\{\rho_{s-F}^n(T);\ -\infty \leq n \leq \infty\}$, where

$$\rho_{s-F}^n(T) = \{\lambda \in \mathbb{C} :\ \lambda - T \text{ is semi-Fredholm with } \operatorname{ind}(\lambda - T) = n\}.$$

The semi-Fredholm operator T is a *Fredholm operator* if $-\infty < \operatorname{ind} T < \infty$. The well-known Atkinson's theorem asserts that T is Fredholm if and only if $\pi(T)$ is invertible in $\mathcal{A}(\mathcal{H})$. Hence $\rho_F(T) = \{\lambda \in \mathbb{C} :\ \lambda - T \text{ is Fredholm}\}$.

If the *minimal index* of $\lambda - T$, $\lambda \in \rho_{s-F}(T)$, is defined by

$$\min \cdot \operatorname{ind}(\lambda - T) = \min\{\operatorname{nul}(\lambda - T),\ \operatorname{nul}(\lambda - T)^*\},$$

then the function $\lambda \to \min \cdot \operatorname{ind}(\lambda - T)$ is a constant on every component of $\rho_{s-F}(T)$ except for an at most denumerable subset $\rho_{s-F}^s(T)$ without limit points in $\rho_{s-F}(T)$. Furthermore, if $\mu \in \rho_{s-F}^s(T)$ and λ is a point of $\rho_{s-F}(T)$ in the same component as μ but $\lambda \notin \rho_{s-F}^s(T)$, then

$$\min \cdot \operatorname{ind}(\mu - T) > \min \cdot \operatorname{ind}(\lambda - T).$$

$\rho_{s-F}^s(T)$ is the set of *singular points* of the semi-Fredholm domain $\rho_{s-F}(T)$ of T; $\rho_{s-F}^r(T) := \rho_{s-F}(T) \backslash \rho_{s-F}^s(T)$ is the set of *regular points.* It is apparent that $\rho_{s-F}^r(T)$ is open and includes $\rho(T)$. On the other hand, $\rho_{s-F}^s(T)$ includes $\sigma_0(T)$.

Proposition. *Let $T \in \mathcal{L}(\mathcal{H})$. Then $\lambda \in \rho_{s-F}^r(T)$ if and only if*

$$\ker(\lambda - T) \subset \bigcap_{n=1}^{\infty} \operatorname{ran}(\lambda - T)^n \quad \textit{when} \quad \operatorname{nul}(\lambda - T) < \infty$$

or

$$\ker(\lambda - T)^* \subset \bigcap \operatorname{ran}(\lambda - T)^{*n} \quad \textit{when} \quad \operatorname{nul}(\lambda - T)^* < \infty.$$

The *Weyl spectrum* of T is defined by

$$\sigma_W(T) = \bigcap\{\sigma(T+K) : \ K \in \mathcal{K}(\mathcal{H})\}.$$

It is well-known that

$$\sigma_W(T) = \{\lambda \in \mathbb{C} : \ \lambda - T \text{ is not a Fredholm operator of index } 0\},$$

$$\sigma(T) \supset \sigma_W(T) \supset \sigma_e(T) \supset \sigma_{lre}(T)$$

and

$$\varnothing \neq \partial\sigma_W(T) \subset \partial\sigma_e(T) \subset \partial\sigma_{lre}(T).$$

Recall that if K is compact, then $(K^*K)^{1/2}$ can be written as $(K^*K)^{1/2} = \sum_{n=1}^{\infty} \lambda_n e_n \otimes e_n$ with respect to some orthonormal basis $\{e_n\}_{n=1}^{\infty}$ of $\mathcal{H}$, where $\lambda_1 = \|K\| \geq \lambda_2 \geq \lambda_3 \geq \cdots \geq 0$ and $\lambda_n \to 0$ $(n \to \infty)$.

The *Schatten p-class* $C^p(\mathcal{H})$ of compact operators on $\mathcal{H}$ is defined by

$$C^p(\mathcal{H}) := \left\{K \in \mathcal{K}(\mathcal{H}) : \ \sum_{n=1}^{\infty} \lambda_n^p < \infty\right\}, \quad 1 \leq p < \infty,$$

and $C^{\infty}(\mathcal{H}) := \mathcal{K}(\mathcal{H})$.

Theorem. (i) *$C^p(\mathcal{H})$ $(1 \leq p \leq \infty)$ is a self-adjoint ideal in $\mathcal{L}(\mathcal{H})$;*

(ii) *$C^p(\mathcal{H})$ $(1 \leq p < \infty)$ $(C^{\infty}(\mathcal{H}))$ is a Banach space under the norm*

$$|K|_p = \left(\sum_{n=1}^{\infty} \lambda_n^p\right)^{1/p} \qquad (|K|_{\infty} = \|K\|, \ \textit{resp.});$$

(iii) *If $K \in C^1(\mathcal{H})$, then it is possible to define the trace of K by*

$$\operatorname{tr}(K) = \sum_{n=1}^{\infty} (Kf_n, f_n)$$

with respect to some ONB $\{f_n\}_{n=1}^{\infty}$ of $\mathcal{H}$; the result is independent of the particular ONB;

(iv) *$C^1(\mathcal{H})$ is isometrically isomorphic to the dual $\mathcal{K}(\mathcal{H})^*$ of $\mathcal{K}(\mathcal{H})$. This isomorphism is defined by $C^1(\mathcal{H}) \ni K \leftrightarrow \phi_K \in \mathcal{K}(\mathcal{H})^*$, where*

$$\phi_K(X) = \operatorname{tr}(KX), \quad X \in \mathcal{K}(\mathcal{H});$$

(v) $C^1(\mathcal{H})^* = \mathcal{K}(\mathcal{H})^{**}$ *is isometrically isomorphic to* $\mathcal{L}(\mathcal{H})$. *This isomorphism is defined by* $\mathcal{L}(\mathcal{H}) \ni A \leftrightarrow \phi_A \in C^1(\mathcal{H})^*$, *where*

$$\phi_A(K) = \mathrm{tr}\,(AK), \quad K \in C^1(\mathcal{H}).$$

An operator T is called *essentially normal* if the self-commutator

$$[T^*, T] := T^*T - TT^* \in \mathcal{K}(\mathcal{H}).$$

It is apparent that if T is essentially normal, then $\sigma_{le}(T) = \sigma_{re}(T)$ and therefore $-\infty < \mathrm{ind}\,(\lambda - T) < \infty$ for all $\lambda \in \rho_{s-F}(T)$, i.e. $\rho_{s-F}(T) = \rho_F(T)$ and $\sigma_{lre}(T) = \sigma_e(T)$.

There are two classes of essentially normal operators which will often be used in this book.

Let Ω be a non-empty bounded open connected subset of $\mathbb{C}$ such that $(\overline{\Omega})^\circ = \Omega$, where $(\overline{\Omega})^\circ$ denotes the interior of the closure $\overline{\Omega}$ of Ω, and let z be a point of Ω. Then there exists a probability measure μ supported by $\Gamma := \partial\Omega$, such that

$$f(z) = \int_\Gamma f\, d\mu$$

for every function f analytic on some neighbourhood of $\overline{\Omega}$ (D.A. Herrero [1]). Let $M(\Gamma)$ = "multiplication by λ" on $L^2(\Gamma, \mu)$. The subspace $H^2(\Gamma)$ spanned by the functions analytic on some neighbourhood of $\overline{\Omega}$ is obviously invariant under $M(\Gamma)$. Hence

$$M(\Gamma) = \begin{bmatrix} M_+(\Gamma) & Z \\ 0 & M_-(\Gamma) \end{bmatrix} \begin{matrix} H^2(\Gamma) \\ L^2(\Gamma, \mu) \ominus H^2(\Gamma). \end{matrix}$$

Theorem. (i) *$M(\Gamma)$ is normal and $M_+(\Gamma)$, $M_-(\Gamma)$ are essentially normal;*

(ii) $\sigma(M(\Gamma)) = \sigma_e(M(\Gamma)) = \sigma_e(M_+(\Gamma)) = \sigma_e(M_-(\Gamma)) = \Gamma$, $\sigma(M_+(\Gamma)) = \sigma(M_+(\Gamma)) = \overline{\Omega}$ *and* $\mathrm{nul}\,(\lambda - M_+(\Gamma)) = \mathrm{nul}\,(\lambda - M_-(\Gamma))^* = 0$ *and* $\mathrm{ind}\,(\lambda - M_+(\Gamma)) = \mathrm{ind}\,(\lambda - M_-(\Gamma))^* = -1$ *for all* $\lambda \in \Omega$;

(iii) *Z is a compact operator satisfying $\|Z\| \le m(\Omega)/\pi$, where m denotes the planar Lebesgue measure;*

(iv) *If Ω is simply connected, then $M_+(\Gamma)$ and $M_-(\Gamma^*)^*$ are analytic Toeplitz operators with symbol ϕ, where ϕ is the conformal mapping from the unit disc D*

onto Ω *such that* $\phi(D) = \Omega$ *and* $\phi(\partial D) = \partial\Omega$ *(the reader can find the definition of Toeplitz operators in 0.3).*

A different kind of essentially normal operators can be constructed as follows. Let Ω be a non-empty bounded open subset of $\mathcal{C}$ with $(\overline{\Omega})^\circ = \Omega$, and let $N(\Omega)$ be the "multiplication by λ" operator acting on $L^2(\Omega, dm)$. The subspace $A^2(\Omega)$ spanned by the rational functions with poles outside $\overline{\Omega}$ is invariant under $N(\Omega)$. By $N_+(\Omega)$ and $N_-(\Omega)$ we shall denote the restriction of $N(\Omega)$ to $A^2(\Omega)$ and its compression to $L^2(\Omega, dm) \ominus A^2(\Omega)$, respectively. i.e.,

$$N(\Omega) = \begin{bmatrix} N_+(\Omega) & G \\ 0 & N_-(\Omega) \end{bmatrix} \begin{matrix} A^2(\Omega) \\ L^2(\Omega, dA) \ominus A^2(\Omega), \end{matrix}$$

where $N_+(\Omega)$ is called the *Bergman operator.*

Theorem. (i) *$N(\Omega)$ is normal and $N_+(\Omega)$, $N_-(\Omega)$ are essentially normal;*
(ii) $\sigma(N(\Omega)) = \sigma_e(N(\Omega)) = \sigma(N_+(\Omega)) = \sigma(N_-(\Omega)) = \sigma_e(N_-(\Omega)) = \overline{\Omega}$,
$\sigma_e(N_+(\Omega)) = \partial\Omega$, $\operatorname{nul}(\mu - N_+(\Omega)) = 0$ *and* $\operatorname{ind}(\mu - N_+(\Omega)) = -1$ *for all* $\mu \in \Omega$.

Brown–Douglas–Fillmore Theorem. *If T_1 and T_2 are essentially normal operators on $\mathcal{H}$, then a necessary and sufficient condition that T_1 be unitarily equivalent to some compact perturbation of T_2 is that $\sigma_e(T_1) = \sigma_e(T_2)$ and* $\operatorname{ind}(\lambda - T_1) = \operatorname{ind}(\lambda - T_2)$ *for all $\lambda \notin \sigma_e(T_1)$.*

Voiculescu Theorem. *Let $T \in \mathcal{L}(\mathcal{H})$ and ρ be a unital faithful *-representation of a separable C^*-subalgebra of the Calkin algebra $\mathcal{A}(\mathcal{H})$ containing the canonical images $\pi(T)$ and $\pi(I)$, on a separable space $\mathcal{H}_\rho$. Let $A = \rho(\pi(T))$ and k be a positive integer. Given $\epsilon > 0$, there exists $K \in \mathcal{K}(\mathcal{H})$, with $\|K\| < \epsilon$, such that*

$$T - K \simeq T \oplus A^{(\infty)} \simeq T \oplus A^{(k)}.$$

0.3 Toeplitz operators

Let D and C be the open unit disc and unit circle in the complex plane respectively, i.e., $D = \{z \in \mathbb{C} : |z| < 1\}$ and $C = \{z \in \mathbb{C} : |z| = 1\}$, and let μ be Lebesgue measure (the extension of arc length) on the Borel set of C, normalized so that $\mu(C) = 1$. If $e_n(z) = z^n$ for $|z| = 1$ $(n = 0, \pm1, \pm2, \cdots)$, then $\{e_n : n = 0, \pm1, \pm2, \cdots\}$ is an orthonormal basis for $L^2 := L^2(C, \mu)$. The space H^2 is, by definition, the subspace of

L^2 spanned by the e_n's with $n \geq 0$. The space H^∞ is a subspace of $L^\infty := L^\infty(C, \mu)$ consisting of all those functions f in L^∞ for which $\int_C f e_n \, d\mu = 0$ $(n = 1, 2, \cdots)$, i.e. $H^\infty = L^\infty \cap H^2$. The functions in H^2 have analytic extensions in D. In fact, H^2 can be considered as a space consisting of analytic functions in the unit disc D whose series of Taylor coefficients is square-summable; and H^∞ consists of bounded analytic functions in D.

If P is the projection from L^2 onto H^2, and if $\phi \in L^\infty$, then the *Toeplitz operator* T_ϕ induced by ϕ is defined by

$$T_\phi f = P(\phi \cdot f)$$

for all f in H^2. If $\phi \in H^\infty$, T_ϕ is called an *analytic Toeplitz operator.* A matrix $(\lambda_{ij})_{i,j=0}^\infty$ is called a Toeplitz matrix if $\lambda_{i+1,j+1} = \lambda_{ij}$ for all i and j $(= 0, 1, 2, \cdots)$. The simplest non-trivial example of an analytic Toeplitz operator is the unilateral shift $S(= T_z)$.

Theorem. *A necessary and sufficient condition that an operator on H^2 be a Toeplitz operator T_ϕ is that its matrix (λ_{ij}) with respect to the basis $\{e_n : n = 0, 1, 2, \cdots\}$ be a Toeplitz matrix; if that condition is satisfied, then $\lambda_{ij} = \alpha_{i-j}$, where $\phi = \sum_n \alpha_n e_n$ is the Fourier expansion of ϕ.*

An *inner function* is a measurable function $M \in H^2$ for which the radial limits of $|M(z)|$ equal 1 a.e. or as a function on C, $|M(z)| = 1$ a.e. If ϕ is a positive measurable function on C such that $\log \phi \in L^1(C)$, and if

$$Q(z) = c \exp\left\{ \int_C \frac{\omega + z}{\omega - z} \log \phi(\omega) \, d\mu(\omega) \right\}$$

for $z \in D$, then Q is called an *outer function.*

The *Blaschke product* $B(z)$ is a kind of inner function. It has the form

$$B(z) = cz^k \prod_{j=1}^\infty \frac{\overline{\lambda}_j}{|\lambda_j|} \cdot \frac{\lambda_j - z}{1 - \overline{\lambda}_j z}$$

where $k \geq 0$, $c \in \mathbb{C}$ and $\{\lambda_j\}$ is a sequence of non-zero numbers in D and $\sum_{j=1}^\infty (1 - |\lambda_j|) < \infty$. (The last condition guarantees the convergence of the infinite product.)

Theorem. *Suppose c is a constant with $|c| = 1$, B is a Blaschke product, ν is a finite positive Borel measure on C which is singular with respect to Lebesgue measure,* *and*

$$M(z) = cB(z)\exp\left\{-\int \frac{\omega+z}{\omega-z}\,d\nu(\omega)\right\} \qquad (z \in D).$$

Then M is an inner function, and every inner function is of this form.

Factorization Theorem. *If $f \in H^2$ and f is not identically 0, then f is the product of an outer function Q and an inner function M : $f = MQ$, where*

$$Q(z) = \exp\left\{\int \frac{\omega+z}{\omega-z}\log|f(\omega)|\,d\mu(\omega)\right\}.$$

Beurling Theorem. *A subspace H_1 of H is invariant under the unilateral shift S if and only if H_1 is of the form*

$$H_1 = \phi H^2 := \{\phi f : \ f \in H^2\}$$

for some inner function ϕ.

0.4 Nest algebras

A *nest* is a chain $\mathcal{N}$ of subspaces of a Hilbert space $\mathcal{H}$ containing $\{0\}$ and $\mathcal{H}$ which is closed under intersection and closed span. It is well-known that for a nest, there is a spectral measure $E(t)$ on $[0,1]$ such that $\mathcal{N} = \{E[0,t]\mathcal{H} : \ t \in [0,1]\}$ and the compact subset $\operatorname{supp} E$ of $[0,1]$ is ordering isomorphic to and topologically homeomorphic to $\mathcal{N}$, where $\mathcal{N}$ is given the order topology, which is equivalent to the strong operator topology on $\mathcal{N}$. Here $\operatorname{supp} E$ has the order and the related topology induced on it by the usual topology of the real line. Furthermore, E is uniquely determined by $\mathcal{N}$ up to order-preserving homeomorphisms of $[0,1]$ onto itself, that is, if F is another spectral measure associated to $\mathcal{N}$ (as indicated above), then there exists a strictly increasing continuous function ϕ mapping $[0,1]$ onto itself such that $E = \phi(F)$ and $F = \phi^{-1}(E)$. We will not distinguish a nest $\mathcal{N}$ from the support of its spectral measure.

For each $M \in \mathcal{N}$, let $M_- = \bigvee\{M' \in \mathcal{N} : \ M' \subsetneqq M\}$. If $M_- \neq M$, $M \ominus M_-$ is called an *atom* of $\mathcal{N}$ and the cardinal number $\dim M \ominus M_-$ is called the *dimension of the atom.* $\lambda \in \mathcal{N}$ corresponds to an atom if and only if λ is the right end point of an interval in the complementary open set of $\mathcal{N}$ in $[0,1]$. A nest is called *continuous* if it has no atoms. If the atoms of $\mathcal{N}$ span the space $\mathcal{H}$, then $\mathcal{N}$ is *atomic.* A nest is *maximal* if it is not contained in any larger nest. It is easy to show that a nest is maximal if and only if all its atoms are one dimensional.

The *nest algebra* $\operatorname{alg}\mathcal{N}$ associated with $\mathcal{N}$ is the family of operators defined by

$$\operatorname{alg}\mathcal{N} := \{T \in \mathcal{L}(\mathcal{H}) : \; TM \subset M \text{ for all } M \in \mathcal{N}\}.$$

Proposition. *The nest algebra* $\operatorname{alg}\mathcal{N}$ *is a closed subalgebra of* $\mathcal{L}(\mathcal{H})$ *in the weak operator topology.*

Given a subalgebra $\mathcal{A}$ of $\mathcal{L}(\mathcal{H})$, the lattice of all invariant subspaces of $\mathcal{A}$ is denoted by $\operatorname{Lat}\mathcal{A}$. Similarly, if $\mathcal{L}$ is a lattice of subspaces of $\mathcal{H}$, then $\operatorname{alg}\mathcal{L}$ denotes the algebra of all operators leaving each element of $\mathcal{L}$ invariant. A lattice is *reflexive* if $\mathcal{L} = \operatorname{Lat}\operatorname{alg}\mathcal{L}$; an algebra $\mathcal{A}$ is *reflexive* if $\mathcal{A} = \operatorname{alg}\operatorname{Lat}\mathcal{A}$.

Theorem. *Every nest* $\mathcal{N}$ *and nest algebra* $\operatorname{alg}\mathcal{N}$ *is reflexive.*

0.5 The Sobolev space $W^{2,2}(\Omega)$

A non-empty bounded open subset Ω of the complex plane C is a *Cauchy domain* if the following conditions are satisfied: (i) Ω has finitely many components, the closures of any two of which are disjoint, and (ii) the boundary $\partial\Omega$ of Ω is composed of a finite positive number of closed rectifiable Jordan curves, no two of which intersect. If all the curves of $\partial\Omega$ are regular analytic Jordan curves, we shall say that Ω is an *analytic Cauchy domain*. A connected Cauchy domain is called a *Cauchy region*.

Let Ω denote a Cauchy region and let $W^{2,2}(\Omega)$ denote the Sobolev space:

$$W^{2,2}(\Omega) := \left\{ f \in L^2(\Omega, dA) : \begin{array}{c} \text{the distributional partial derivatives of first} \\ \text{and second order of } f \text{ belong to } L^2(\Omega, dA) \end{array} \right\},$$

where dA denotes the planar Lebesgue measure.

It is well-known that $W^{2,2}(\Omega)$ is a Hilbert space under the norm

$$\|f\| = \left(\int_\Omega \sum_{|\alpha| \leq 2} \left| D^\alpha f \right|^2 dA \right)^{1/2},$$

where α is a 2-index and D^α is one of the differential operators of order $|\alpha|$.

Sobolev Embedding Theorem. *If* Ω *is an analytic Cauchy domain, then* $W^{2,2}(\Omega)$ *is embedded continuously in* $C(\overline{\Omega})$.

Theorem. *Let* Ω *be an analytic Cauchy region. Then* $W^{2,2}(\Omega)$ *is an abelian Banach algebra with identity under pointwise multiplication and an equivalent norm whose maximal ideal space can be naturally identified with* $\overline{\Omega}$ *via "point evaluation",* $\mathcal{M} \subset W^{2,2}(\Omega)$ *is a maximal ideal of* $W^{2,2}(\Omega)$ *if and only if* $\mathcal{M}$ *is of the form*

$$\mathcal{M} = \{f \in W^{2,2}(\Omega);\ f(x_0) = 0\} \quad \textit{for some } x_0 \in \overline{\Omega}.$$

Extension Theorem. *If* Ω *is an analytic Cauchy domain, then there is a bounded linear operator* E *from* $W^{22}(\Omega)$ *into* $W^{22}(\mathbb{R}^2)$ *such that for each* $f \in W^{22}(\Omega)$, $Ef|_\Omega = f$.

1. The main tools

In this chapter we shall analyse the main tools needed in this book. Most of them are developed within the framework of operator theory: The Rosenblum's equation $AX - XB = C$, the triangular representation of C. Apostol and the theorem of C. Apostol, C. Foiaş and D. Voiculescu on the normal restriction of small compact perturbations of operators, the theory of Cowen–Douglas operators, and the similarity orbit theorem. Some material presented in this chapter is from D.A. Herrero [1] and C. Apostol, L.A. Fialkow, D.A. Herrero and D. Voiculescu [1], and some of the other results are our own work. Further analysis on the theory of Cowen–Douglas operators will be developed in Chapters 2, 3 and 4.

1.1 The Rosenblum operator: $X \mapsto AX - XB$

Let $\mathcal{A}$ be a Banach algebra with identity. For $a \in \mathcal{A}$, $L_a \in \mathcal{L}(\mathcal{A})$ is defined by $L_a(b) = ab$ ($b \in \mathcal{A}$). Similarly, $R_a \in \mathcal{L}(\mathcal{A})$ is given by $R_a(b) = ba$. It is obvious that $\sigma(L_a) = \sigma(R_a) = \sigma(a)$ for all a in $\mathcal{A}$.

Given $a, b \in \mathcal{A}$, $\tau_{ab} \in \mathcal{L}(\mathcal{A})$ is defined by $\tau_{ab}(c) = ac - cb$, i.e., $\tau_{ab} = L_a - R_b$. It is easily seen that L_a and R_b commute and that

$$\tau_{(a-\lambda)(b-\mu)} = \tau_{ab} - (\lambda - \mu),$$

$(\lambda, \mu \in \mathbb{C})$, so that

$$\begin{aligned} \|\tau_{ab}\| = \|\tau_{(a-\lambda)(b-\lambda)}\| &\leq \inf_{\mu \in \mathbb{C}} \{\|L_{(a-\mu)}\| + \|R_{(b-\mu)}\|\} \\ &= \inf_{\mu \in \mathbb{C}} \{\|a - \mu\| + \|b - \mu\|\}. \end{aligned}$$

Proposition 1.1. $\sigma(\tau_{ab}) \subset \sigma(a) - \sigma(b)$.

Proof. Let $\mathcal{B} \subset \mathcal{L}(\mathcal{A})$ be a maximal abelian algebra containing L_a and R_b. Clearly, $1 \in \mathcal{B}$ and $(\lambda - L_a)^{-1}$, $(\mu - R_b)^{-1}$ belong to $\mathcal{B}$ for all $\lambda \notin \sigma(L_a)$ and $\mu \notin \sigma(R_b)$. Let

$\Sigma_{\mathcal{B}}$ be the maximal ideal space of $\mathcal{B}$. Then the Gelfand theory implies that

$$\begin{aligned}\sigma(\tau_{ab}) &= \sigma(L_a - R_b) = \{\phi(L_a - R_b) : \ \phi \in \Sigma_{\mathcal{B}}\} \\ &= \{\phi(L_a) - \phi(R_b) : \ \phi \in \Sigma_{\mathcal{B}}\} \\ &\subset \{\phi(L_a) : \ \phi \in \Sigma_{\mathcal{B}}\} - \{\phi(R_b) : \ \phi \in \Sigma_{\mathcal{B}}\} \\ &= \sigma(L_a) - \sigma(R_b) = \sigma(a) - \sigma(b).\end{aligned}$$

□

As a matter of fact when $\mathcal{A} = \mathcal{L}(\mathcal{H})$, this inclusion is actually an equality.

Theorem 1.2. *Let $\mathcal{H}$ be a separable Hilbert space and $A, B \in \mathcal{L}(\mathcal{H})$. Then $\sigma(\tau_{AB}) = \sigma(A) - \sigma(B)$.*

In order to prove this theorem, we need some concepts and lemmas.

Let $\mathcal{X}$ be a complex Banach space. The *approximate point spectrum* of $A \in \mathcal{L}(\mathcal{X})$ is the set

$$\sigma_\pi(A) := \{\lambda \in \mathbb{C} : \ \lambda - A \text{ is not bounded below}\},$$

and

$$\sigma_\delta(A) := \{\lambda \in \mathbb{C} : \ \lambda - A \text{ is not onto}\}$$

is the *approximate defect spectrum* of A.

If $\mathcal{X}$ is a Hilbert space, $\sigma_\pi(A) = \sigma_l(A) = \sigma_r(A^*)^*$ and $\sigma_\delta(A) = \sigma_r(A) = \sigma_l(A^*)^*$, where $\Omega^* = \{\lambda^* : \lambda \in \Omega\}$ for each $\Omega \subset \mathbb{C}$.

Lemma 1.3. *Given a Banach space $\mathcal{X}$, there is an isometric imbedding of $\mathcal{X}$ into a larger Banach space $\mathcal{X}'$, and a mapping $A \to A'$ of $\mathcal{L}(\mathcal{X})$ into $\mathcal{L}(\mathcal{X}')$ which is an isometric isomorphism such that every A' is an extension of A and $\sigma_p(A') = \sigma_\pi(A') = \sigma_\pi(A)$.*

Proof. Let $l^\infty(\mathcal{X})$ be the Banach space of all bounded sequences of elements of $\mathcal{X}$ with the norm $\|\{x_n\}_{n=1}^\infty\| = \sup_n \|x_n\|_{\mathcal{X}}$ and let $c_0(\mathcal{X}) = \{\{x_n\} \in l^\infty(\mathcal{X}) : \|x_n\| \to 0 \ (n \to \infty)\}$. It is easily seen that $c_0(\mathcal{X})$ is a subspace of $l^\infty(\mathcal{X})$.

Define $\mathcal{X}' = l^\infty(\mathcal{X})/c_0(\mathcal{X})$ and the imbedding of $\mathcal{X}$ into $\mathcal{X}'$ by $x \mapsto [\{x_n\}]$ (=the coset of $\{x_n\}$), where $x_n = x$ for all $n = 1, 2, \cdots$. Clearly, this mapping is an isometric isomorphism of $\mathcal{X}$ into $\mathcal{X}'$.

Similarly, given A in $\mathcal{L}(\mathcal{X})$, define $A' \in \mathcal{L}(\mathcal{X}')$ by $A'[\{x_n\}] = [\{Ax_n\}]$; then $A \mapsto A'$ defines an isometric isomorphism from $\mathcal{L}(\mathcal{X})$ into $\mathcal{L}(\mathcal{X}')$.

If $\lambda \in \sigma_\pi(A)$, then there exists a sequence $\{x_n\}_{n=1}^\infty$ of unit vectors in $\mathcal{X}$ such that $\|(\lambda - A)x_n\| \to 0$ $(n \to \infty)$. Thus $A'[\{x_n\}] = [\{Ax_n\}] = [\lambda\{x_n\}] = \lambda[\{x_n\}]$, whence it follows that $\sigma_\pi(A) \subset \sigma_p(A') \subset \sigma_\pi(A')$.

On the other hand, if $\|(\lambda - A)x\| \geq c\|x\|$ for some $c > 0$ and for all x in $\mathcal{X}$, then it can be easily seen that $\|(\lambda - A')[\{x_n\}]\| = \|[\{(\lambda - A)x_n\}]\| \geq c\|[\{x_n\}]\|$ for all $[\{x_n\}] \in \mathcal{X}'$. Hence $\sigma_\pi(A') \subset \sigma_\pi(A)$ and therefore

$$\sigma_p(A') = \sigma_\pi(A') = \sigma_\pi(A). \qquad \square$$

Lemma 1.4. *Let $\mathcal{H}$ be a Hilbert space and $A, B \in \mathcal{L}(\mathcal{H})$. Then*
(i) $\sigma_\pi(\tau_{AB}) \subset \sigma_l(A) - \sigma_r(B)$;
(ii) $\sigma_\delta(\tau_{AB}) \subset \sigma_r(A) - \sigma_l(B)$.

Proof. Let $\lambda \in \sigma_\pi(L_A - R_B)$; then by Lemma 1.3, $\lambda \in \sigma_p((L_A - R_B)') = \sigma_p(L'_A - R'_B)$ and $L'_A R'_B = R'_B L'_A$. It readily follows that the subspace

$$\mathcal{M} = \ker(L'_A - R'_B - \lambda)$$

is invariant under L'_A and R'_B, and $R'_B|_{\mathcal{M}} = (L'_A - \lambda)|_{\mathcal{M}}$. Thus, if $\alpha \in \sigma_\pi(L'_A|_{\mathcal{M}})$ then $\alpha - \lambda \in \sigma_\pi(R'_B|_{\mathcal{M}})$ and therefore

$$\begin{aligned} \lambda = \alpha - (\alpha - \lambda) &\in \sigma_\pi(L'_A|_{\mathcal{M}}) - \sigma_\pi(R'_B|_{\mathcal{M}}) \\ &\subset \sigma_\pi(L'_A) - \sigma_\pi(R'_B) \\ &= \sigma_\pi(L_A) - \sigma_\pi(R_B) \end{aligned}$$

and $\sigma_\pi(\tau_{AB}) \subset \sigma_\pi(L_A) - \sigma_\pi(R_B)$.

Let $\lambda \in \sigma_l(A)$; then there exists a sequence $\{x_n\}_{n=1}^\infty$ of unit vectors such that $\|(\lambda - A)x_n\| \to 0$ $(n \to \infty)$. It follows that

$$\|(\lambda - L_A)(x_n \otimes x_n)\| = \|(\lambda - A)x_n \otimes x_n\| = \|(\lambda - A)x_n\| \to 0,$$

so that $\lambda - L_A$ cannot be bounded below. Hence $\sigma_l(A) \subset \sigma_\pi(L_A)$.

Conversely, if $\lambda \in \sigma_\pi(L_A)$, then there exists a sequence $\{X_n\}_{n=1}^\infty$ in $\mathcal{L}(\mathcal{H})$ such that $\|X_n\| = 1$ for all $n = 1, 2, \cdots$, and $\|(\lambda - A)X_n\| \to 0$ $(n \to \infty)$. For each n, choose $y_n \in \mathcal{H}$ such that $1 \leq \|y_n\| \leq 2$ and $\|X_n y_n\| = \|X_n\| = 1$, then $\|(\lambda - A)(X_n y_n)\| \to 0$ $(n \to \infty)$, whence we conclude that $\lambda \in \sigma_\pi(A)$. It follows that $\sigma_l(A) = \sigma_\pi(L_A)$. Similarly, we can prove that $\sigma_\pi(R_B) = \sigma_r(B)$. Therefore $\sigma_\pi(\tau_{AB}) \subset \sigma_l(A) - \sigma_r(B)$.

The second statement follows by similar arguments. □

Let $T \in \mathcal{L}(\mathcal{H})$ and let Ω be an open subset of $\rho_r(T)$. A continuous function $F : \Omega \to \mathcal{L}(\mathcal{H})$ satisfying the equations

$$(\lambda - T)F(\lambda) \equiv 1 \quad (\lambda \in \Omega)$$

and

$$F(\lambda) - F(\mu) = (\mu - \lambda)F(\lambda)F(\mu) \quad (\lambda, \mu \in \Omega)$$

will be called a *right resolvent* for T defined on Ω. It is completely apparent that such a function F is an analytic function defined on Ω with derivative $F'(\lambda) = -F^2(\lambda)$.

Similarly, if Φ is an open subset of $\rho_l(T)$ and $G : \Phi \to \mathcal{L}(\mathcal{H})$ is a continuous function satisfying

$$G(\lambda)(\lambda - T) \equiv 1 \quad (\lambda \in \Phi)$$

and

$$G(\lambda) - G(\mu) = (\mu - \lambda)G(\lambda)G(\mu) \quad (\lambda, \mu \in \Phi),$$

then G is called a *left resolvent* for T defined on Φ. Clearly, G is analytic on Φ. Moreover, if $F(\lambda)$ is a right resolvent for T^* (defined on $\Omega = \Phi^*$), then $G(\lambda) = F(\lambda^*)^*$ is a left resolvent for T on Φ.

For any $\lambda \in \rho_r(T)$, $\mu \in \rho_l(T)$, we shall put

$$\begin{aligned} R_r(\lambda, T) &= (\lambda - T)^*[(\lambda - T)(\lambda - T)^*]^{-1} \\ R_l(\mu, T) &= [(\mu - T)^*(\mu - T)]^{-1}(\mu - T)^*. \end{aligned} \tag{1.1}$$

It is easily seen that $R_r(\cdot, T) : \rho_r(T) \to \mathcal{L}(\mathcal{H})$ $(R_l(\cdot, T) : \rho_l(T) \to \mathcal{L}(\mathcal{H}))$ is continuous on this domain and satisfies the equation

$$(\lambda - T)\,R_r(\lambda, T) \equiv 1,\ \lambda \in \rho_r(T) \quad (R_l(\mu, T)\,(\mu - T) \equiv 1,\ \mu \in \rho_l(T),\ \text{resp.}).$$

However, $R_r(\lambda, T)$ $(R_l(\lambda, T))$ does not satisfy, in general, the resolvent equation, i.e., it is not a right (resp. left) resolvent for T. In fact, there is the following stronger result in G.R. Allan [1] [2] and M.A. Shubin [1].

Lemma 1.5. *Given $T \in \mathcal{L}(\mathcal{H})$, there exist analytic functions $R_T(\lambda)$ and $L_T(\lambda)$ defined on $\rho_r(T)$ and, respectively, $\rho_l(T)$ such that*

$$\begin{aligned} (\lambda - T)R_T(\lambda) &\equiv 1 \quad \lambda \in \rho_r(T) \\ L_T(\lambda)(\lambda - T) &\equiv 1 \quad \lambda \in \rho_l(T). \end{aligned}$$

Proof. Here, we give only a sketch of the proof. The reader is referred to D. H. Herrero [1] for more details. Let Ω be a bounded connected component of $\rho_r(T)$, $\lambda_0 \in \Omega$ and let B be the right inverse of $(\lambda_0 - T)$. Choose $r > 0$ such that

$$D_r = \{\lambda \in \mathbb{C} : \ |\lambda - \lambda_0| < r\} \subset \Omega$$

and

$$f(\lambda) := \sum_{n=0}^{\infty} (-1)^n (\lambda - \lambda_0)^n B^{n+1}$$

is analytic in D_r. Thus $(\lambda - T)f(\lambda) \equiv 1$ $(\lambda \in D_r)$. By analytic continuation, we get an analytic function on the whole of Ω satisfying

$$(\lambda - T)f(\lambda) \equiv 1 \quad (\lambda \in \Omega).$$

Therefore we have an analytic function $R_T(\lambda)$ on $\rho_r(T)$ such that

$$(\lambda - T)R_T(\lambda) \equiv 1 \quad (\lambda \in \rho_r(T).$$

Applying the first conclusion to T^*, we get the second conclusion. $\square$

Lemma 1.6. *Assume $A, B \in \mathcal{L}(\mathcal{H})$. Then*
(i) $\sigma_\delta(\tau_{AB}) = \sigma_r(\tau_{AB}) = \sigma_r(A) - \sigma_l(B)$;
(ii) $\sigma_\pi(\tau_{AB}) = \sigma_l(\tau_{AB}) = \sigma_l(A) - \sigma_r(B)$.

Proof. Since $\sigma_r(\tau_{AB}) := \{\lambda \in \mathbb{C} : (\lambda - \tau_{AB})$ has no right inverse in $\mathcal{L}(\mathcal{L}(\mathcal{H})).\}$, $\sigma_\delta(\tau_{AB}) \subset \sigma_r(\tau_{AB})$. By Lemma 1.4 $\sigma_\delta(\tau_{AB}) \subset \sigma_r(A) - \sigma_l(B)$. Let $\mu = \alpha - \beta$, where $\alpha \in \sigma_r(A)$ and $\beta \in \sigma_l(B)$, and $\tau_{AB} - \mu$ is onto. Since $\operatorname{ran}(\tau_{AB} - \mu)$ is closed, there is a constant $m > 0$ such that $\|(\tau_{AB} - \mu)(T)\| \geq m \operatorname{dist}(T, \ker(\tau_{AB} - \mu))$, for all T in $\mathcal{L}(\mathcal{H})$. Since $\alpha \in \sigma_r(A)$ and $\beta \in \sigma_r(B)$, we can find unit vectors x, y in $\mathcal{H}$ such that $\|(A - \alpha)^* x\| < m/4$ and $\|(B - \beta)y\| < m/4$. Set $C = \frac{m}{2} x \otimes y$, then $\|C\| = (Cy, x) = m/2$. Since $\tau_{AB} - \mu$ is onto, there exists $X \in \mathcal{L}(\mathcal{H})$ such that

$$(\tau_{AB} - \mu)(X) = \tau_{A-\alpha, B-\beta}(X) = (A - \alpha)X - X(B - \beta) = C.$$

In particular, X can be chosen so that $\operatorname{dist}(X, \ker(\tau_{AB} - \mu)) \geq \|X\|/2$, or $\frac{m}{2}\|X\| \leq \|C\|$, Thus, we have

$$\begin{aligned} 0 < \|C\| &= (Cy, x) = ((\tau_{AB} - \mu)(X)y, x) \\ &= ([(A - \alpha)X - X(B - \beta)]y, x) \\ &\leq |(Xy, (A - \alpha)^* x)| + |((B - \beta)y, X^* x)| \\ &\leq \frac{2}{m} \|C\| \{\|(A - \alpha)^* x\| + \|(B - \beta)y\|\} < \|C\|. \end{aligned}$$

The contradiction implies that $\sigma_\delta(\tau_{AB}) = \sigma_r(A) - \sigma_l(B) \subset \sigma_r(\tau_{AB})$.

Assume that $\mu \notin \sigma_r(A) - \sigma_l(B)$; equivalently, $\sigma_r(A-\mu) \cap \sigma_l(B) = \varnothing$. Since $\sigma_r(A-\mu)$ and $\sigma_l(B)$ are non-empty, compact, and disjoint, there exist Cauchy domains Ω_A, Ω_B and Ω such that $\sigma_r(A-\mu) \subset \Omega_A \subset \overline{\Omega}_A \subset \Omega$, $\sigma_l(B) \subset \Omega_B$ and $\overline{\Omega}_B \cap \overline{\Omega} = \varnothing$. It follows from Lemma 1.5 that there exists an analytic function $R_{A-\mu}(\lambda)$ defined on $\rho_r(A-\mu)$ such that $(A-\mu-\lambda)R_{A-\mu}(\lambda) \equiv 1$ on this domain. Similarly, there exists an analytic function $L_B(\lambda)$ defined on $\rho_l(B)$ such that $L_B(\lambda)(B-\lambda) \equiv 1$ on this domain.

Clearly, $R_{A-\mu}$ and L_B are analytic on a neighbourhood of $\partial\Omega$. Let Ω_0 be a component of Ω. Since $\sigma_l(B) \cap \overline{\Omega} = \varnothing$, it follows from Cauchy's theorem that $\int_{\partial\Omega_0} L_B(\lambda)\,d\lambda = 0$. Therefore

$$\frac{1}{2\pi i}\int_{\partial\Omega} L_B(\lambda)\,d\lambda = 0.$$

Let $\partial'\Omega = \partial\Omega \cap \rho(A-\mu)$ and $\partial''\Omega = \partial\Omega \backslash \rho(A-\mu)$. It is not difficult to check that $\partial'\Omega$ is the (positively oriented) boundary of a Cauchy domain containing $\sigma(A-\mu)$, while $\partial''\Omega$ is the (negatively oriented) boundary of a Cauchy domain $\Omega'' \subset \rho_r(A-\mu) \cap \sigma(A-\mu)$ such that $(\Omega'')^- \cap \overline{\Omega} = \partial''\Omega$. Hence, by Cauchy's theorem and the properties of the analytic right inverse, we have

$$-\int_{\partial''\Omega} R_{A-\mu}(\lambda)\,d\lambda = 0$$

and

$$\begin{aligned}\frac{1}{2\pi i}\int_{\partial\Omega} R_{A-\mu}(\lambda)\,d\lambda &= \frac{1}{2\pi i}\int_{\partial'\Omega} R_{A-\mu}(\lambda)\,d\lambda \\ &= \frac{1}{2\pi i}\int_{\partial'\Omega} [\lambda-(A-\mu)]^{-1}\,d\lambda = 1.\end{aligned}$$

Given Y in $\mathcal{L}(\mathcal{H})$, let

$$X = \phi(Y) = -\frac{1}{2\pi i}\int_{\partial\Omega} R_{A-\mu}(\lambda) Y L_B(\lambda)\,d\lambda.$$

It is completely apparent that ϕ defines a bounded mapping from $\mathcal{L}(\mathcal{H})$ into itself.

We shall verify that ϕ is a right inverse of $\tau_{AB} - \mu$. Indeed,

$$\begin{aligned}
&(\tau_{AB} - \mu)\phi(Y) = (A-\mu)X - XB \\
&= -\frac{1}{2\pi i}\int_{\partial\Omega} (A-\mu)R_{A-\mu}(\lambda)YL_B(\lambda)\,d\lambda + \frac{1}{2\pi i}\int_{\partial\Omega} R_{A-\mu}(\lambda)YL_B(\lambda)B\,d\lambda \\
&= -\frac{1}{2\pi i}\int_{\partial\Omega} (A-\mu-\lambda)R_{A-\mu}(\lambda)YL_B(\lambda)\,d\lambda + \frac{1}{2\pi i}\int_{\partial\Omega} R_{A-\mu}(\lambda)YL_B(\lambda)(B-\lambda)\,d\lambda \\
&= -\frac{1}{2\pi i}\int_{\partial\Omega} YL_B(\lambda)\,d\lambda + \frac{1}{2\pi i}\Big(\int_{\partial\Omega} R_{A-\mu}(\lambda)\,d\lambda\Big)Y = Y,
\end{aligned}$$

i.e., $(\tau_{AB} - \mu)\phi = 1$ on $\mathcal{L}(\mathcal{H})$ and $\mu \notin \sigma_r(\tau_{AB})$.

Hence, $\sigma_r(\tau_{AB}) \subset \sigma_r(A) - \sigma_l(B)$ and the proof of (i) is complete.

Since $\sigma_l(\tau_{AB}) := \{\lambda \in \mathcal{C} : \lambda - \tau_{AB}$ has no left inverse in $\mathcal{L}(\mathcal{L}(\mathcal{H})).\}$, $\sigma_\pi(\tau_{AB}) \subset \sigma_l(\tau_{AB})$. By Lemma 1.4 $\sigma_\pi(\tau_{AB}) \subset \sigma_l(A) - \sigma_r(B)$.

Assume that $\mu \in \sigma_l(A) - \sigma_r(B)$, i.e., $\mu = \alpha - \beta$ for some $\alpha \in \sigma_l(A)$ and some $\beta \in \sigma_r(B)$. Then given $\epsilon > 0$, we can find unit vectors $x, y \in \mathcal{H}$ such that $\|(A-\alpha)x\| < \epsilon$ and $\|(B-\beta)^* y\| < \epsilon$. Thus

$$\begin{aligned}
\|(\tau_{AB} - \mu)(x \otimes y)\| &= \|\tau_{A-\alpha, B-\beta}(x \otimes y)\| \\
&= \|(A-\alpha)(x \otimes y) - (x \otimes y)(B-\beta)\| \\
&= \|(A-\alpha)x \otimes y - x \otimes (B-\beta)^* y\| \\
&\le \|(A-\alpha)x\| + \|(B-\beta)^* y\| < 2\epsilon.
\end{aligned}$$

Since $\|x \otimes y\| = 1$ and ϵ can be chosen arbitrarily small, $\tau_{AB} - \mu$ is not bounded below, i.e., $\mu \in \sigma_\pi(\tau_{AB})$ and $\sigma_\pi(\tau_{AB}) = \sigma_l(A) - \sigma_r(B) \subset \sigma_l(\tau_{AB})$.

Assume that $\mu \notin \sigma_l(A) - \sigma_r(B)$ or $\sigma_l(A-\mu) \cap \sigma_r(B) = \varnothing$ and let Φ_A, Φ_B and Φ be Cauchy domains such that $\sigma_l(A-\mu) \subset \Phi_A \subset \overline{\Phi}_A \subset \Phi$, $\sigma_r(B) \subset \Phi_B$ and $\overline{\Phi}_B \cap \overline{\Phi} = \varnothing$. By Lemma 1.5 there exist an analytic function $R_B(\lambda)$ defined on $\rho_r(B)$ and an analytic function $L_{A-\mu}(\lambda)$ defined on $\rho_l(A-\mu)$, such that $(B-\lambda)R_B(\lambda) \equiv 1$ for all $\lambda \in \rho_r(B)$ and $L_{A-\mu}(\lambda)(A-\mu-\lambda) \equiv 1$ for all $\lambda \in \rho_l(A-\mu)$. By the same arguments used in the proof of (i), we can easily check that

$$\psi(Y) = \frac{1}{2\pi i}\int_{\partial\Phi} L_{A-\mu}(\lambda)YR_B(\lambda)\,d\lambda$$

defines a left inverse of $\tau_{AB} - \mu$. It readily follows that $\sigma_l(A) - \sigma_r(B) \supset \sigma_l(\tau_{AB})$.

□

Corollary 1.7. *Let $\mathcal{H}_1$ and $\mathcal{H}_2$ be two Hilbert spaces, let $A \in \mathcal{L}(\mathcal{H}_1)$, $B \in \mathcal{L}(\mathcal{H}_2)$ and assume that $\sigma_r(A) \cap \sigma_l(B) = \varnothing$. Then τ_{AB} is surjective.*

Proof. Assume that $\mathcal{H}_1$ and $\mathcal{H}_2$ are infinite dimensional; then we can identify them via a unitary mapping of $\mathcal{H}_1$ onto $\mathcal{H}_2$, i.e., we can directly assume that $\mathcal{H}_1 = \mathcal{H}_2 = \mathcal{H}$. By Lemma 1.6, τ_{AB} is surjective.

If $\dim \mathcal{H}_1 < \infty$ and $\dim \mathcal{H}_2 = \infty$, define $\alpha = \|A\| + \|B\| + 1$ and consider the operator $\alpha \oplus A \in \mathcal{L}(\mathcal{H}_2 \oplus \mathcal{H}_1)$; then $\sigma_r(\alpha \oplus A) \cap \sigma_l(B) = \varnothing$. It follows from the first part of the proof that $\tau_{\alpha\oplus A,B}$ is surjective. For each $C \in \mathcal{L}(\mathcal{H}_2, \mathcal{H}_1)$, set $C' = \begin{pmatrix} 0 \\ C \end{pmatrix} \in \mathcal{L}(\mathcal{H}_2, \mathcal{H}_2 \oplus \mathcal{H}_1)$. Since $\tau_{\alpha\oplus A,B}$ is onto, there exists a $W = \begin{pmatrix} W_1 \\ W_2 \end{pmatrix} \in \mathcal{L}(\mathcal{H}_2, \mathcal{H}_2 \oplus \mathcal{H}_1)$ such that $\tau_{\alpha\oplus A,B}(W) = C'$ or $\begin{pmatrix} \alpha W_1 \\ AW_2 \end{pmatrix} - \begin{pmatrix} W_1 B \\ W_2 B \end{pmatrix} = \begin{pmatrix} 0 \\ C \end{pmatrix}$. Thus $AW_2 - W_2 B = C$ or $\tau_{AB}(W_2) = C$, i.e., τ_{AB} is onto.

The cases $\dim \mathcal{H}_2 < \infty$, and $\dim \mathcal{H}_1 = \infty$ or both $\mathcal{H}_1$ and $\mathcal{H}_2$ are finite dimensional spaces can be similarly analysed to reach the same conclusion: τ_{AB} is onto. □

By similar arguments we have

Corollary 1.8. *Let $\mathcal{H}_1$ and $\mathcal{H}_2$ be two Hilbert spaces, and let $A \in \mathcal{L}(\mathcal{H}_1)$, $B \in \mathcal{L}(\mathcal{H}_2)$ and assume that $\sigma_l(A) \cap \sigma_r(B) = \varnothing$. Then τ_{AB} is injective.*

Now we are in a position to prove Theorem 1.2.

Proof of Theorem 1.2. By Lemma 1.6,

$$\begin{aligned} \sigma(\tau_{AB}) &= \sigma_r(\tau_{AB}) \cup \sigma_l(\tau_{AB}) \\ &= \left[\sigma_r(A) - \sigma_l(B)\right] \cup \left[\sigma_l(A) - \sigma_r(B)\right] \\ &= \left\{[\sigma_r(A) \cup \sigma_l(A)] - [\sigma_l(B) \cup \sigma_r(B)]\right\} \\ &= \sigma(A) - \sigma(B). \end{aligned}$$

□

The following several results on τ_{AB} are frequently used in this book.

Proposition 1.9. *Let $A, B \in \mathcal{L}(\mathcal{H})$; then τ_{AB} is onto if and only if $\tau_{AB}\big|_{\mathcal{K}(\mathcal{H})}$ is a mapping from $\mathcal{K}(\mathcal{H})$ onto $\mathcal{K}(\mathcal{H})$.*

We need the following lemma.

Lemma 1.10. *Let* $\tau = \tau_{AB}\big|_{\mathcal{K}(\mathcal{H})}$. *Then* $\tau^* = -\tau_{BA}\big|_{C^1(\mathcal{H})}$ *and* $(\tau^*)^* = \tau_{AB}$, *where* τ^* *is the dual operator of* τ.

Proof. Recall that $C^1(\mathcal{H})$ is isometrically isomorphic to the dual $\mathcal{K}(\mathcal{H})^*$ of $\mathcal{K}(\mathcal{H})$. This isomorphism is defined by $C^1(\mathcal{H}) \ni K \leftrightarrow \phi_K \in \mathcal{K}(\mathcal{H})^*$, where

$$\phi_K(X) = \operatorname{tr}(KX), \quad X \in \mathcal{K}(\mathcal{H}).$$

Thus,

$$\begin{aligned}\phi_{\tau^*(K)}(X) = \phi_K(\tau(X)) &= \operatorname{tr}[K(AX - XB)] \\ &= \operatorname{tr}(KAX) - \operatorname{tr}(KXB) \\ &= \operatorname{tr}(KAX) - \operatorname{tr}(BKX) \\ &= \operatorname{tr}(-\tau_{BA}(K)X) \\ &= \phi_{-\tau_{BA}(K)}(X).\end{aligned}$$

Therefore $\tau^* = -\tau_{BA}\big|_{C^1(\mathcal{H})}$.

Since $\mathcal{L}(\mathcal{H})$ is isometrically isomorphic to the dual $C^1(\mathcal{H})^*$ of $C^1(\mathcal{H})$, by similar arguments we can prove that $(\tau^*)^* = \tau_{AB}$. □

Proof of Proposition 1.9. If $\tau = \tau_{AB}\big|_{\mathcal{K}(\mathcal{H})}$ is a mapping from $\mathcal{K}(\mathcal{H})$ onto $\mathcal{K}(\mathcal{H})$, then by Lemma 1.10, $\tau^* \left(= -\tau_{BA}\big|_{C^1(\mathcal{H})}\right)$ is one to one and has closed range. By Lemma 1.10 again $(\tau^*)^* = \tau_{AB}$ is a mapping onto $\mathcal{L}(\mathcal{H})$.

If τ_{AB} is a mapping from $\mathcal{L}(\mathcal{H})$ onto $\mathcal{L}(\mathcal{H})$, then by Lemma 1.10 $-\tau_{BA}\big|_{C^1(\mathcal{H})}$ is one to one and has closed range. Therefore, $\tau_{AB}\big|_{\mathcal{K}(\mathcal{H})}$ is onto. □

More precisely, Fialkow's results read as follows (L.A. Fialkow [1]).

Theorem 1.11. *Let* $A, B \in \mathcal{L}(\mathcal{H})$. *Then the following are equivalent for* τ_{AB} :

(i) τ_{AB} *is surjective;*

(ii) $\sigma_r(A) \cap \sigma_l(B) = \varnothing$;

(iii) $\operatorname{ran}\tau_{AB}$ *contains the set of finite rank operators;*

(iv) $\tau_{AB}\big|_J$ *is surjective for every norm ideal* J.

Corollary 1.12. *The following are equivalent for* $\tau_{\pi(A)\pi(B)}$:

(i) $\tau_{\pi(A)\pi(B)}$ *is surjective;*

(ii) $\sigma_{re}(A) \cap \sigma_{le}(B) = \varnothing$;

where $\tau_{\pi(A)\pi(B)}$ *is the Rosenblum operator defined in the Calkin algebra.*

Theorem 1.13. *Let $A, B \in \mathcal{L}(\mathcal{H})$. Then the following are equivalent:*

(i) $\operatorname{ran} \tau_{AB}$ *is dense in* $\mathcal{L}(\mathcal{H})$;

(ii) (1) $\sigma_{re}(A) \cap \sigma_{le}(B) = \varnothing$, *and*

(2) *there exists no non-zero* $X \in C^1(\mathcal{H})$ *such that* $BX = XA$;

(iii) *Given* $Y \in \mathcal{L}(\mathcal{H})$ *and* $\epsilon > 0$, *there exists* $X \in \mathcal{L}(\mathcal{H})$ *such that* $\tau_{AB}(X) - Y \in \mathcal{K}(\mathcal{H})$ *and* $\|\tau_{AB}(X) - Y\| < \epsilon$.

The proofs of Theorem 1.11 and Theorem 1.13 are omitted.

Proposition 1.14. *Let $A, B \in \mathcal{L}(\mathcal{H})$. Assume that*

$$\mathcal{H} = \bigvee \{\ker (\lambda - B)^k : \lambda \in \Gamma, k \geq 1\}$$

for a certain subset Γ *of the point spectrum* $\sigma_p(B)$ *of* B, *and* $\sigma_p(A) \cap \Gamma = \varnothing$; *then* τ_{AB} *is injective.*

Proof. Let p be a monic polynomial with zeros in Γ and let $x \in \mathcal{H}$ be any vector such that $p(B)x = 0$; then $AX = XB$ implies

$$p(A)Xx = Xp(B)x = 0.$$

Since $p(A)$ is injective, we infer $Xx = 0$. It readily follows that

$$\ker X \supset \bigvee \{\ker (\lambda - B)^k : \lambda \in \Gamma, k \geq 1\} = \mathcal{H}.$$

Hence, $X = 0$. □

1.2 Apostol's triangular representation

An operator $A \in \mathcal{L}(\mathcal{H})$ is a *triangular operator* if it admits an upper triangular matrix; i.e.,

$$A = \begin{bmatrix} a_{11} & a_{12} & a_{13} & \cdots \\ & a_{22} & a_{23} & \cdots \\ & & a_{33} & \cdots \\ & 0 & & \ddots \end{bmatrix} \tag{1.2}$$

with respect to a suitable ONB. In operator theory, triangular operators form a class of operators whose construction is clearer.

Given $T \in \mathcal{L}(\mathcal{H})$, let

$$\mathcal{H}_r(T) = \bigvee \{\ker (\lambda - T) : \lambda \in \rho_{s-F}^r(T)\}$$

and let

$$\mathcal{H}_l(T) = \bigvee \{\ker(\lambda - T)^* : \lambda \in \rho_{s-F}^r(T)\};$$

then $\mathcal{H}_r(T)$ and $\mathcal{H}_l(T)$ are invariant subspaces of T and T^* respectively. We shall show that $T_r = T|_{\mathcal{H}_r(T)}$ and $T_l^* = T^*|_{\mathcal{H}_l(T)}$ are triangular operators. Apostol's triangular representation theorem will indicate that T can be expressed by T_r and T_l in a certain sense and can increase our knowledge about the structure of operators. Before the theorem is established, we need several results related to triangular operators.

Lemma 1.15. *If $B \in \mathcal{L}(\mathcal{H})$, $\lambda \in \sigma(B)$ and $\|(\lambda - B)x\| \geq M\|x\|$ for some $M > 0$ and all $x \in \mathcal{H}$, then $\|A - B\| \geq M$ for every triangular operator A.*

Proof. Assume that A has an upper triangular matrix of the form (1.2) with respect to the $ONB\{e_n\}_{n=1}^{\infty}$. Let P_n be the orthogonal projection of $\mathcal{H}$ onto $\bigvee\{e_1, e_2, \cdots, e_n\}$ and let y be a unit vector orthogonal to $\operatorname{ran}(\lambda - B)$ (so that $(\lambda - B)^* y = 0$); clearly, $P_n A P_n = A P_n$.

Given ϵ, $0 < \epsilon < 1$, there exists $n_0 = n_0(\epsilon)$ such that $\|P_n(\lambda - B)^* P_n y\| < \epsilon$ for all $n \geq n_0$. $P_n(\lambda - B)^* P_n$ can be regarded as an operator acting on the finite dimensional space $\mathcal{M}_n = \operatorname{ran} P_n$. It follows from the finite dimensionality of $\mathcal{M}_n$ that this subspace contains a unit vector z_n such that

$$\|[P_n(\lambda - B)^* P_n]^* z_n\| = \|P_n(\lambda - B)P_n z_n\| < \epsilon.$$

We have $P_n z_n = z_n$, and

$$\begin{aligned}
\|A - B\| &\geq \liminf_{n\to\infty} \|(1 - P_n)(A - B)P_n\| \\
&\geq \liminf_{n\to\infty} \|(1 - P_n)(\lambda - B)P_n\| \\
&\geq \liminf_{n\to\infty} \|(1 - P_n)(\lambda - B)P_n z_n\| \\
&\geq \liminf_{n\to\infty} \|(\lambda - B)z_n\| - \liminf_{n\to\infty} \|P_n(\lambda - B)P_n z_n\| \\
&\geq M - \epsilon.
\end{aligned}$$

Since ϵ can be chosen arbitrarily small, we conclude that $\|A - B\| \geq M$. □

Proposition 1.16. *Let A be a triangular operator with matrix (1.2) with respect to the $ONB\,\{e_n\}_{n=1}^{\infty}$, and let $d(A) = \{a_{nn}\}_{n=1}^{\infty}$ be the diagonal sequence of A; then*

(i) $\sigma(A) = \sigma_l(A) = \sigma_{lre}(A) \cup \sigma_p(A)$;

(ii) *Every clopen subset σ of $\sigma(A)$ meets $d(A)$, and every component of $\sigma(A)$ meets $d(A)^-$. In particular, every isolated point of $\sigma(A)$ belongs to $d(A)$;*

(iii) $\sigma_p(A^*) \subset d(A)^*$.

Proof. (i) The inclusions $\sigma_l(A) \subset \sigma(A)$ and $\sigma_{lre}(A) \cup \sigma_p(A) \subset \sigma(A)$ are obvious. If $\lambda \in \sigma(A) \backslash \sigma_l(A)$, then $(\lambda - A)$ is left invertible and therefore there exists $M > 0$ such that $\|(\lambda - A)x\| \geq M\|x\|$ for all $x \in \mathcal{H}$. By Lemma 1.15, $M \leq 0$, thus $\sigma(A) = \sigma_l(A)$. If $\lambda \in \sigma(A)$ and $\lambda \notin \sigma_{lre}(A)$, i.e., $\lambda \in \sigma(A) \cap \rho_{s-F}(A)$, then $\operatorname{ran}(\lambda - A)$ is closed. Since $\lambda \in \sigma_l(A)$, $\lambda - A$ is not one to one, i.e., $\lambda \in \sigma_p(A)$. Therefore $\sigma(A) = \sigma_{lre}(A) \cup \sigma_p(A)$.

(ii) Let $E(\sigma; A)$ be the Riesz idempotent corresponding to σ and $\mathcal{H}(\sigma; A) = E(\sigma; A)\mathcal{H}$. Thus $\mathcal{H}(\sigma; A)$ is invariant under A and $\sigma(B) = \sigma$, where $B = A|_{\mathcal{H}(\sigma;A)}$.

Note

$$\begin{aligned}\mathcal{H}(\sigma; A) &= E(\sigma; A)\big(\bigvee \{\ker(a_{nn} - A)^k : k \geq 1,\, a_{nn} \in d(A)\}\big)\\ &= \bigvee \{E(\sigma; A)\ker(a_{nn} - A)^k : k \geq 1,\, a_{nn} \in d(A)\}\\ &\subset \bigvee \{\ker(a_{nn} - B)^k : k \geq 1,\, a_{nn} \in d(A)\}\\ &\subset \mathcal{H}(\sigma; A),\end{aligned}$$

i.e., $\mathcal{H}(\sigma; A) = \bigvee\{\ker(a_{nn} - B)^k : k \geq 1, a_{nn} \in d(A)\}$. This implies that B is triangular and $d(A) \cap \sigma(B) = d(A) \cap \sigma \neq \varnothing$.

If τ is a component of $\sigma(A)$, then

$$\tau = \bigcap\{\sigma : \sigma \text{ is clopen in } \sigma(A) \text{ and } \sigma \supset \tau\}.$$

Since each σ meets $d(A)$, $\tau \cap d(A)^- \neq \varnothing$.

(iii) If $(\lambda - A)^* y = 0$ and $\mathcal{M}_n = \bigvee\{e_k : 1 \leq k \leq n\}$, then

$$(y, (\lambda - A)e_n) = ((\lambda - A)^* y, e_n) = 0 \quad (n = 1, 2, \cdots),$$

i.e., $y \in [(\lambda - A)\mathcal{M}_n]^\perp$ $(n = 1, 2, \cdots)$. If $\lambda \notin d(A)$, then $(\lambda - A)\mathcal{M}_n = \mathcal{M}_n$ for all n. Thus $y \in \big(\bigvee\{\mathcal{M}_n : 1 \leq n < \infty\}\big)^\perp = \{0\}$ and $\sigma(A^*)^* \subset d(A)$. □

Proposition 1.17. *Let $T \in \mathcal{L}(\mathcal{H})$ and $\mu \in \rho_{s-F}(T)$; then $Q = \lim_{\lambda \to \mu} P_{\ker(\lambda - T)}$ exists in the norm topology. Q is the orthogonal projection onto*

$$\ker(\mu - T) \cap \left\{\bigcap_{n=1}^{\infty} \operatorname{ran}(\mu - T)^n\right\}.$$

Proof. Assume first that $\mu - T$ is onto. By the stability of the semi-Fredholm operator, $\lambda - T$ is onto for all λ in some neighbourhood Ω of μ. Set $P(\lambda) = 1 - R_r(\lambda, T)(\lambda - T)$ $(\lambda \in \Omega)$, where $R_r(\lambda, T)$ is defined by (1.1). Computations indicate that $P(\lambda)^* = P(\lambda)$ and $P(\lambda)^2 = P(\lambda)$, i.e., $P(\lambda)$ is an orthogonal projection. It is obvious that $\ker(\lambda - T) \subset \operatorname{ran} P(\lambda)$. If $P(\lambda)x = x$, then $x - R_r(\lambda, T)(\lambda - T)x = x$, i.e., $R_r(\lambda, T)(\lambda - T)x = (\lambda - T)^*[(\lambda - T)(\lambda - T)^*]^{-1}(\lambda - T)x = 0$. Since $(\lambda - T)$ is onto, $(\lambda - T)^*$ is injective and $(\lambda - T)x = 0$. Thus $\operatorname{ran} P(\lambda) \subset \ker(\lambda - T)$ and $P(\lambda) = P_{\ker(\lambda - T)} = 1 - R_r(\lambda, T)(\lambda - T)$. It follows that $P_{\ker(\lambda - T)}$ converges in the norm to $1 - R_r(\mu, T)(\mu - T) = P_{\ker(\mu - T)}$ as $\lambda \to \mu$.

If $\mu - T$ is not onto, set $\mathcal{M} = \bigcap_{n=1}^{\infty} \operatorname{ran}(\mu - T)^n$. Because $(\mu - T)$ is semi-Fredholm, $\mathcal{M}$ is closed and $(\mu - T)\mathcal{M} = \mathcal{M}$. Hence $(\mu - T|_{\mathcal{M}})\mathcal{M} = \mathcal{M}$.

For $\lambda \neq \mu$ and $x \in \ker(\lambda - T)$, since $(\lambda - T)(\mu - T)x = 0$ and $x = (\mu - \lambda)^{-1}(\mu - T)x$, $(\mu - T)\ker(\lambda - T) = \ker(\lambda - T)$ and so $\ker(\lambda - T) \subset \mathcal{M}$. It follows that

$$P_{\ker(\lambda - T)} = P_{\ker(\lambda - T|_{\mathcal{M}})} P_{\mathcal{M}}$$

and the first paragraph shows that this is the desired limit. □

Corollary 1.18. *Let $T \in \mathcal{L}(\mathcal{H})$ and $\mu \in \rho_{s-F}(T)$. The following conditions are equivalent:*

(i) *μ is a regular point of $\rho_{s-F}(T)$;*

(ii) *$P_{\ker(\lambda - T)}$ is continuous at $\lambda = \mu$.*

Proof. (i)⇒(ii). If $\operatorname{nul}(\lambda - T)$ is a finite constant in some neighbourhood of μ, then $\ker(\lambda - T) \subset \bigcap_{n=1}^{\infty} \operatorname{ran}(\lambda - T)^n$. By Proposition 1.17, $P_{\ker(\lambda - T)}$ is continuous at $\lambda = \mu$.

If $\operatorname{nul}(\lambda - T)^*$ is a finite constant in a neighbourhood of μ, then

$$\ker(\lambda - T)^* \subset \bigcap_{n=1}^{\infty} \operatorname{ran}(\lambda - T)^{*n}.$$

Note that since $\lambda - T$ is semi-Fredholm,

$$\operatorname{ran}(\lambda - T) = [\ker(\lambda - T)^*]^{\perp} \supset [\operatorname{ran}(\lambda - T)^{*n}]^{\perp} = \ker(\lambda - T)^n, \quad n = 1, 2, \cdots.$$

Thus, if $y \in \ker(\lambda - T)$, then $y = (\lambda - T)y_1$ for some $y_1 \in \mathcal{H}$. Therefore, $\operatorname{ran}(\lambda - T) \supset \ker(\lambda - T)^2 \supset \{z : (\lambda - T)z = y\} \neq \varnothing$, so that $y = (\lambda - T)^2 y_2$ for some $y_2 \in \mathcal{H}$ and $\operatorname{ran}(\lambda - T) \supset \ker(\lambda - T)^3 \supset \{z : (\lambda - T)^2 z = y\} \neq \varnothing$. By induction, there exists

y_n in $\mathcal{H}$ such that $y = (\lambda - T)^n y_n$, $n = 1, 2, \cdots$. Hence $y \in \bigcap_{n=1}^{\infty} \operatorname{ran}(\lambda - T)^n$ and $\ker(\lambda - T) \subset \bigcap_{n=1}^{\infty} \operatorname{ran}(\lambda - T)^n$. By the first paragraph, we conclude that $P_{\ker(\lambda - T)}$ is continuous at $\lambda = \mu$.

(ii)$\Rightarrow$(i). If $P_{\ker(\lambda - T)}$ is continuous at $\lambda = \mu$, Proposition 1.17 implies that $\ker(\mu - T) \subset \bigcap_{n=1}^{\infty} \operatorname{ran}(\mu - T)^n$. By the same argument used in the first part of the proof, we can conclude that $\ker(\mu - T)^* \subset \bigcap_{n=1}^{\infty} \operatorname{ran}(\mu - T)^{*n}$. Therefore $\mu \in \rho_{s-F}^r(T)$. □

Let $\mathcal{H}_r(T)$ and $\mathcal{H}_l(T)$ be given as in the beginning of this section. Let $\mathcal{H}_0(T)$ be the orthogonal complement of $\mathcal{H}_r(T) + \mathcal{H}_l(T)$. Denote the compression of T to $\mathcal{H}_r(T)$, $\mathcal{H}_l(T)$ and $\mathcal{H}_0(T)$ by T_r, T_l and T_0 respectively. Apostol's triangular representation and its basic properties are established in the following

Theorem 1.19. (i) *$\mathcal{H}_r(T)$ is orthogonal to $\mathcal{H}_l(T)$, so that*

$$\mathcal{H} = \mathcal{H}_r(T) \oplus \mathcal{H}_0(T) \oplus \mathcal{H}_l(T).$$

(ii) *$\mathcal{H}_r(T)$ and $\mathcal{H}_r(T) \oplus \mathcal{H}_0(T)$ are invariant for T, and T admits a 3×3 upper triangular operator matrix representation*

$$T = \begin{bmatrix} T_r & * & * \\ 0 & T_0 & * \\ 0 & 0 & T_l \end{bmatrix} \tag{1.3}$$

with respect to the above decomposition.

(iii) *$T_r = T|_{\mathcal{H}_r(T)}$ is a triangular operator,*

$$\sigma(T_r) = \sigma_l(T_r) = \sigma_{lre}(T_r) \cup [\rho_r(T_r) \backslash \rho(T_r)],$$

$\sigma(T_r)$ is a perfect set, every component of $\sigma_{lre}(T_r)$ intersects the set $\sigma_p(T_r)^-$ and $\sigma_p(T_r^*) = \varnothing$, so that

$$\rho_{s-F}^s(T_r) = \varnothing \text{ and } \min\operatorname{ind}(\lambda - T_r) = 0 \text{ for all } \lambda \in \rho_{s-F}(T_r).$$

(iv) *T_l is the adjoint in $\mathcal{L}(\mathcal{H}_l(T))$ of the triangular operator $T^*|_{\mathcal{H}_l(T)}$,*

$$\sigma(T_l) = \sigma_r(T_l) = \sigma_{lre}(T_l) \cup [\rho_l(T_l) \backslash \rho(T_l)],$$

$\sigma(T_l)$ is a perfect set, every component of $\sigma_{lre}(T_l^*)$ intersects the set $\sigma_p(T_l^*)^-$ and $\sigma_p(T_l) = \varnothing$, so that

$$\rho_{s-F}^s(T_l) = \varnothing \text{ and } \min\operatorname{ind}(\lambda - T_l) = 0 \text{ for all } \lambda \in \rho_{s-F}(T_l).$$

(v) $\lambda \to P_{\ker(\lambda - T)}$ ($\lambda \in \rho_{s-F}(T)$) *is continuous at* $\lambda \in \rho_{s-F}^r(T)$ *and discontinuous at* $\lambda \in \rho_{s-F}^s(T)$.

(vi) $\rho_{s-F}(T) \subset \rho_r(T_r) \cap \rho_l(T_l)$.

(vii) $\rho_{s-F}^r(T) \subset \rho(T_0)$.

(viii) $\rho_{s-F}^s(T) \subset \sigma_0(T_0)$.

(ix) $\sigma_0(T) \subset \rho_r(T_r) \cap \rho_l(T_l) \cap \sigma_0(T_0)$.

Proof. (i) Since $\lambda \in \rho_{s-F}^r(T)$, either $\ker(\lambda - T) \subset \operatorname{ran}(\lambda - T)$ or $\ker(\lambda - T)^* \subset \operatorname{ran}(\lambda - T)^*$. If $\mu \in \rho_{s-F}^r(T)$ and $\mu \neq \lambda$, then $(\lambda - T)\ker(\mu - T) = \ker(\mu - T)$ and so $[\ker(\lambda - T)^*]^\perp = \operatorname{ran}(\lambda - T) \supset \ker(\mu - T)$.

By Corollary 1.18, $\|P_{\ker(\mu-T)} - P_{\ker(\lambda-T)}\| \to 0$ ($\mu \to \lambda$), whence it readily follows that $\ker(\lambda - T) \subset [\ker(\lambda - T)^*]^\perp$. Hence,

$$\mathcal{H}_r(T) = \bigvee\{\ker(\lambda - T) : \lambda \in \rho_{s-F}^r(T)\} \perp \bigvee\{\ker(\lambda - T)^* : \lambda \in \rho_{s-F}^r(T)\} = \mathcal{H}_l(T)$$

and, a fortiori, $\mathcal{H} = \mathcal{H}_r(T) \oplus \mathcal{H}_0(T) \oplus \mathcal{H}_l(T)$.

(ii) It is easily seen that for each $\lambda \in \mathbb{C}$, $\ker(\lambda - T)$ is invariant for T. A fortiori, $\mathcal{H}_r(T) = \bigvee\{\ker(\lambda - T) : \lambda \in \rho_{s-F}^r(T)\}$ and $\mathcal{H}_l = \bigvee\{\ker(\lambda - T)^* : \lambda \in \rho_{s-F}^r(T)\}$ are invariant for T and, respectively, for T^*. Hence, $\mathcal{H}_r(T)$ and $\mathcal{H}_r(T) \oplus \mathcal{H}_0(T) = \mathcal{H}_l(T)^\perp$ are invariant for T, and T admits a matrix representation of the form (1.3).

(iii) Let $\{\lambda_n\}_{n=1}^\infty$ be a denumerable dense subset of $\rho_{s-F}^r(T)$. By Corollary 1.18, $\mathcal{H}_r(T) = \bigvee\{\mathcal{M}_n\}_{n=1}^\infty$, where $\mathcal{M}_n = \ker(\lambda_n - T)$. It is apparent that T_r has an upper triangular matrix with respect to an *ONB* obtained by Gram–Schmidt orthonormalization of a union of suitably chosen *ONB*'s of the subspaces $\{\mathcal{M}_n\}_{n=1}^\infty$ and the diagonal $d(T_r) = \{\lambda_n\}_{n=1}^\infty$.

Since the dual of a triangular operator has at most countably many point spectra and since $\{\lambda_n\}_{n=1}^\infty$ can be arbitrarily chosen under the condition $\{\lambda_n\}^- = [\rho_{s-F}^r(T)]^-$, it readily follows that $\sigma_p(T_r^*) = \emptyset$ and a fortiori, that $\rho_{s-F}^s(T_r) = \emptyset$, $\min\operatorname{ind}(\lambda - T_r) = 0$ for all $\lambda \in \rho_{s-F}(T_r)$.

By Proposition 1.16, $\sigma(T_r) = \sigma_l(T_r) = \sigma_{lre}(T_r) \cup \sigma_p(T_r)$, every component of $\sigma(T_r)$ intersects $d(T_r)^- \subset [\rho_{s-F}^+(T_r)]^-$ (observe that $\lambda - T_r$ is onto for each $\lambda \in \rho_{s-F}^r(T_r)$) and $\sigma_p(T_r^*)^* \subset d(T_r) = \{\lambda_n\}_{n=1}^\infty$.

It is also clear that $\sigma(T_r)$ does not have any isolated point. Thus either $\mathcal{H}_r(T) = \{0\}$ and $\sigma(T_r) = \emptyset$, or $\sigma(T_r)$ is a non-empty perfect set and

$$\sigma(T_r) = \sigma_{lre}(T_r) \cup \sigma_p(T_r) = \sigma_{lre}(T_r) \cup [\rho_r(T_r) \backslash \rho(T_r)].$$

Since $\rho^+_{s-F}(T_r) = \rho_r(T_r)\backslash\rho(T_r)$, it follows that every component of $\sigma_{lre}(T_r)$ intersects $[\rho_r(T_r)\backslash\rho(T_r)]^-$. Since $\rho_r(T_r)\backslash\rho(T_r) \subset \sigma_p(T_r)$, every component of $\sigma_{lre}(T_r)$ intersects $\sigma_p(T_r)^-$.

(iv) In order to prove (iv), we only have to apply the arguments of (iii) to $T^*|_{\mathcal{H}_l(T)}$.

(v) is the content of Corollary 1.18.

(vi) If $\lambda \in \rho_{s-F}(T)\backslash\rho_r(T_r)$, then since $\rho^+_{s-F}(T_r) = \rho_r(T_r)\backslash\rho(T_r)$, $\lambda \in \sigma_{lre}(T_r) \subset \sigma_{lre}(T)$, a contradiction. Hence $\rho_{s-F}(T) \subset \rho_r(T_r)$. A similar argument shows that $\rho_{s-F}(T) \subset \rho_l(T_l)$.

(vii) Assume that $\mathcal{H}_r(T) \neq \{0\}$ and $\mathcal{H}_l(T) \neq 0$; then $\mathcal{H}_r(T)$, $\mathcal{H}_l(T)$, $\mathcal{H}_r(T) \oplus \mathcal{H}_0(T)$ and $\mathcal{H}_0(T) \oplus \mathcal{H}_l(T)$ are infinite dimensional spaces.

Suppose that $\lambda \in \rho^r_{s-F}(T)$ and $\operatorname{ind}(\lambda - T) > -\infty$; then $\lambda - \pi(T_l)$ is invertible and $\lambda - \pi(T)$ is right invertible. By decomposing $\mathcal{H}$ as $[\mathcal{H}_r(T) \oplus \mathcal{H}_0(T)] \oplus \mathcal{H}_l(T)$, $\lambda - \pi(T)$ can be written as the 2×2 matrix

$$\lambda - \pi(T) = \begin{bmatrix} \left[\lambda - \begin{pmatrix} \pi(T_r) & * \\ 0 & \pi(T_0) \end{pmatrix}\right] & \pi(R) \\ 0 & \lambda - \pi(T_l) \end{bmatrix} = \begin{bmatrix} p & r \\ 0 & q \end{bmatrix}.$$

Let $\begin{bmatrix} a & b \\ c & d \end{bmatrix}$ be the inverse of $\lambda - \pi(T)$. Since $\lambda - \pi(T_l) = q$ is right invertible and $\begin{bmatrix} p & r \\ 0 & q \end{bmatrix}\begin{bmatrix} a & b \\ c & d \end{bmatrix} = \begin{bmatrix} 1 & 0 \\ 0 & 1 \end{bmatrix}$, $c = 0$. By similar arguments $\lambda - \pi(T_0)$ is right invertible.

Similarly, if $\operatorname{ind}(\lambda - T) < \infty$, we conclude that $\lambda - \pi(T_0)$ is left invertible. Hence $\lambda \in \rho_{s-F}(T_0)$ and therefore either $\lambda - \pi(T_0)$ is invertible, or $\operatorname{nul}(\lambda - T_0) \neq 0$ or $\operatorname{nul}(\lambda - T_0)^* \neq 0$. Thus, in order to complete the proof, it only remains to show that $\lambda \notin \sigma_p(T_0) \cup \sigma_p(T_0^*)^*$.

Assume that $(\lambda - T_0)x = 0$. Then $(\lambda - T)x \perp \mathcal{H}_0(T)$. The invariance of $\mathcal{H}_r(T) \oplus \mathcal{H}_0(T)$ implies that $(\lambda - T)x \in \mathcal{H}_r(T)$. Since $\lambda \in \rho_{s-F}(T)$ and $\sigma_p(T_r^*) = \varnothing$, $\lambda - T_r$ is onto. Therefore there is some $y \in \mathcal{H}_r(T)$ such that $(\lambda - T_r)y = (\lambda - T)x$. Since $y - x \in \ker(\lambda - T) \subset \mathcal{H}_r(T)$, $x = 0$. An analogous argument shows that $\operatorname{nul}(\lambda - T_0)^* = 0$ and so $\lambda - T_0$ is invertible, i.e., $\lambda \in \rho(T_0)$.

If either $\mathcal{H}_r(T) = \{0\}$ or $\mathcal{H}_l(T) = \{0\}$, then the result follows by an even simpler argument.

(viii) Let $\mu \in \rho^s_{s-F}(T)$. If μ is not an eigenvalue of T_0, then (vi) shows that $\ker(\mu - T) = \ker(\mu - T_r)$. Since $\ker(\mu - T) \subset \mathcal{H}_r(T)$ and $\mu - T_r$ is right invertible,

μ cannot be an eigenvalue of $T_{\mathcal{H}_r(T)^\perp}$ where $T_{\mathcal{H}_r(T)^\perp} = \begin{bmatrix} T_0 & * \\ 0 & T_l \end{bmatrix}$. It is easily seen that $\operatorname{ran}(\mu - T_{\mathcal{H}_r(T)^\perp})$ is closed and therefore, $\mu \in \rho_l(T_{\mathcal{H}_r(T)^\perp})$.

For each λ in the open set $\rho_l(T_{\mathcal{H}_r(T)^\perp})$ and $x_\lambda \in \ker(\lambda - T)$, we have

$$0 = P_{\mathcal{H}_r(T)^\perp}(\lambda - T)x_\lambda = (\lambda - T_{\mathcal{H}_r(T)^\perp})P_{\mathcal{H}_r(T)^\perp}x_\lambda.$$

Since $\lambda - T_{\mathcal{H}_r(T)^\perp}$ is injective, it follows that $P_{\mathcal{H}_r(T)^\perp}x_\lambda = 0$, i.e., $x_\lambda \in \mathcal{H}_r(T)$. Thus, we can find a neighbourhood Ω of μ such that

$$\ker(\lambda - T) = \ker(\lambda - T_r), \quad \text{for all } \lambda \in \Omega.$$

This implies that $\mu \in \rho^r_{s-F}(T)$, a contradiction. Hence $\mu \in \sigma_p(T_0)$. On the other hand, the first part of the proof of (vii) indicates that $\mu \in \rho_{s-F}(T_0)$ and by (vii) μ is an isolated point of $\sigma(T_0)$. Therefore, $\mu \in \sigma_0(T_0)$.

(ix) Note that $\sigma_0(T) \subset \rho^s_{s-F}(T)$. The conclusion follows immediately from (vi) and (viii). $\square$

1.3 Correction by compact perturbation of the singular behaviour of operators

An operator $T \in \mathcal{L}(\mathcal{H})$ such that

$$\min \cdot \operatorname{ind}(\lambda - T) = 0 \quad \text{for all } \lambda \in \rho_{s-F}(T)$$

will be called a *smooth operator*. The main aim of this section is to prove that for each $T \in \mathcal{L}(\mathcal{H})$, there is a compact operator K such that $T + K$ is smooth.

In what follows, $P_{\mathcal{N}}$ denotes the orthogonal projection of $\mathcal{H}$ onto the subspace $\mathcal{N}$; if $\mathcal{N}$ is an invariant subspace of T, $T|_{\mathcal{N}}$ and $T_{\mathcal{N}}$ denote the restriction and compression of T onto $\mathcal{N}$ respectively.

Proposition 1.20. *Given $T \in \mathcal{L}(\mathcal{H})$, there exists $K \in \mathcal{K}(\mathcal{H})$ such that $\rho^s_{s-F}(T+K) = \varnothing$ and $\min \cdot \operatorname{ind}(T+K-\lambda) = \min \cdot \operatorname{ind}(T-\lambda)$ for all $\lambda \in \rho^r_{s-F}(T)$; moreover, K can be chosen to be equal to a normal compact operator such that*

$$\|K\| = \max\left\{\operatorname{dist}[\lambda, \partial\rho_{s-F}(T)] : \lambda \in \rho^s_{s-F}(T)\right\}.$$

Proof. Let

$$T = \begin{bmatrix} T_r & T_{12} & T_{13} \\ 0 & T_0 & T_{23} \\ 0 & 0 & T_l \end{bmatrix} \begin{matrix} \mathcal{H}_r(T) \\ \mathcal{H}_0(T) \\ \mathcal{H}_l(T) \end{matrix}$$

be Apostol's triangular representation of T and let $\{\mu_n\}_{n=1}^m$ $(0 \leq m \leq \infty)$ be an enumeration of the eigenvalues of $\sigma_0(T_0) \cap \rho_{s-F}^s(T)$. Then T_0 admits the representation

$$T_0 = \begin{bmatrix} \mu_1 + T_1 & & & & * \\ & \mu_2 + T_2 & & & \\ & & \ddots & & \\ & & & \mu_n + T_n & \\ & 0 & & & \ddots \\ & & & & & T_\infty \end{bmatrix} \begin{matrix} \mathcal{H}_1 \\ \mathcal{H}_2 \\ \vdots \\ \mathcal{H}_n \\ \vdots \\ \mathcal{H}_\infty \end{matrix}$$

where $\mathcal{H}_1, \mathcal{H}_2, \cdots, \mathcal{H}_n, \cdots$ are defined so that $\oplus_{i=1}^n \mathcal{H}_i$ coincides with the Riesz subspace of T_0 corresponding to the clopen subset $\{\mu_1, \mu_2, \cdots, \mu_n\}$ of $\sigma(T_0)$ $(1 \leq n < m)$ and $\mathcal{H}_\infty = \mathcal{H}_0(T) \ominus \{\oplus_{1 \leq n < m} \mathcal{H}_n\}$. It is not difficult to check that $\sigma(T_\infty) \subset \sigma(T_0) \backslash \{\mu_n\}_{n=1}^\infty$.

For each μ_n, let $\lambda_n \in \partial\rho_{s-F}(T)$ be any point such that

$$|\mu_n - \lambda_n| = \text{dist}\,[\mu_n, \sigma(T) \backslash \rho_{s-F}(T)].$$

Thus $\lambda_n \in \sigma_{lre}(T)$. Let T_0' be the operator obtained from T_0 by replacing μ_n by λ_n $(1 \leq n < m)$, and

$$T' = \begin{bmatrix} T_r & T_{12} & T_{13} \\ 0 & T_0' & T_{23} \\ 0 & 0 & T_l \end{bmatrix}.$$

Then computations indicate that $\rho_{s-F}^r(T) \subset \rho_{s-F}^r(T')$ and therefore $\mathcal{H}_r(T) = \mathcal{H}_r(T')$, $\mathcal{H}_l(T) = \mathcal{H}_l(T')$, $\mathcal{H}_0(T) = \mathcal{H}_0(T')$, $T_r = (T')_r$ and $T_l = (T')_l$. It follows from Theorem 1.19 that $\rho_{s-F}^s(T') = \varnothing$. Note that $K = T' - T$ is a diagonal operator with diagonal elements 0 and $\lambda_n - \mu_n$ $(1 \leq n < m)$. It is obvious that K is compact with $\|K\| = \max\{|\lambda_n - \mu_n| : 1 \leq n < \infty\} = \max\{\text{dist}\,[\mu_n, \partial\rho_{s-F}(T)] : 1 \leq n < m\}$. Computations show that $\min \cdot \text{ind}\,(T + K - \lambda) = \min \cdot \text{ind}\,(T - \lambda)$ for all $\lambda \in \rho_{s-F}^r(T)$.

□

Lemma 1.21. *Let $T \in \mathcal{L}(\mathcal{H})$, $\mu \in \rho_{s-F}^s(T) \backslash \sigma_0(T)$ and $\epsilon > 0$. Then there exists a finite rank operator K_μ such that*

(i) $\|K_\mu\| < \epsilon$, $\mu \in \rho^r_{s-F}(T+K_\mu)$;

(ii) $\min \cdot \mathrm{ind}\,(\lambda - (T+K_\mu))^k = \min \cdot \mathrm{ind}\,(\lambda - T)^k$, $\lambda \in \rho_{s-F}(T)\backslash\{\mu\}$, $k = 1, 2, \cdots$;

(iii) *If* $\nu \in \rho^s_{s-F}(T)\backslash\{\mu\}$ *and* $T \sim T_\nu \oplus T'_\nu$, *where* T_ν *acts on a finite dimensional space,* $\sigma(T_\nu) = \{\nu\}$ *and* $\nu \in \rho^r_{s-F}(T'_\nu)$, *then* $\nu \in \rho^s_{s-F}(T+K_\mu)$ *and* $T + K_\mu \sim T_\nu \oplus T''_\nu$, *where* $\nu \in \rho^r_{s-F}(T''_\nu)$.

Proof. By Theorem 1.19, $\mu \in \rho_{s-F}(T) \subset \rho_r(T_r) \cap \rho_l(T_l) \subset \rho_{s-F}(T_r) \cap \rho_{s-F}(T_l)$. Since at least one of $\ker(\lambda - T)$, $\ker(\lambda - T)^*$ is non-trivial when λ is in the same component of $\rho_{s-F}(T)$ as μ but $\lambda \in \rho^r_{s-F}(T)$, then $\mu \in \sigma(T_r) \cup \sigma(T_l)$. Thus $\mu \in \rho_{s-F}(T_r) \cap \sigma(T_r)$ or $\mu \in \rho_{s-F}(T_l) \cap \sigma(T_l)$ i.e. $\mu \in \sigma_p(T_r) \cup \sigma_p((T_l)^*)^*$. Suppose that $\mu \in \sigma_p((T_l)^*)^*$ and denote by $\mathcal{M}$ the Riesz spectral subspace of T_0 corresponding to $\sigma(T_0)\backslash\{\mu\}$. Set $\mathcal{H}_\mu = \mathcal{H}_0(T) \ominus \mathcal{M}$. Then

$$T = \begin{bmatrix} T_r & T_{12} & T_{13} & T_{14} \\ 0 & T_0|_{\mathcal{M}} & T_{23} & T_{24} \\ 0 & 0 & T_\mu & T_{34} \\ 0 & 0 & 0 & T_l \end{bmatrix} \begin{matrix} \mathcal{H}_r(T) \\ \mathcal{M} \\ \mathcal{H}_\mu \\ \mathcal{H}_l(T). \end{matrix}$$

Let $S' = \begin{bmatrix} T_\mu & T_{34} \\ 0 & T_l \end{bmatrix}$. Since $\sigma(T_\mu) = \{\mu\}$ and $\dim \mathcal{H}_\mu < \infty$, we can find a finite rank operator $A \in \mathcal{L}(\mathcal{H}_\mu)$ such that $\|A\| < \epsilon/2$, $\sigma(T_\mu + A) = \{\mu\}$ and $\mathrm{nul}\,(T_\mu + A - \mu) = 1$. Define

$$S'' = \begin{bmatrix} T_\mu + A & T_{34} \\ 0 & T_l \end{bmatrix} \begin{matrix} \mathcal{H}_\mu \\ \mathcal{H}_l(T). \end{matrix}$$

By Theorem 1.19 (vi), $\sigma_r(T_\mu + A) \cap \sigma_l(T_l) = \varnothing$. Therefore Theorem 1.11 indicates that there exists $X \in \mathcal{L}(\mathcal{H}_l(T), \mathcal{H}_\mu)$ such that $(T_\mu + A)X - XT_l = T_{34}$. Set

$$W = \begin{bmatrix} 1_{\mathcal{H}_\mu} & X \\ 0 & 1_{\mathcal{H}_l(T)} \end{bmatrix},$$

then $S_1 = WS''W^{-1} = (T_\mu + A) \oplus T_l$.

Let $f \in \ker(T_\mu + A - \mu)$, $0 < \|f\| < \epsilon/(2\|W\|\,\|W^{-1}\|)$, and let $e \in \ker(T_l - \mu)^*$, $\|e\| = 1$. Set $S_2 = S_1 + e \otimes f \in \mathcal{L}(\mathcal{H}_\mu \oplus \mathcal{H}_l(T))$, i.e.,

$$S_2 = \begin{bmatrix} T_\mu + A & 0 \\ e \otimes f & T_l \end{bmatrix} \begin{matrix} \mathcal{H}_\mu \\ \mathcal{H}_l(T). \end{matrix}$$

If $\sigma_p(S_2) \neq \varnothing$, there exist $\lambda \in \mathbb{C}$, $y \in \mathcal{H}_\mu$ and $z \in \mathcal{H}_l(T)$ such that $(S_2 - \lambda)(y+z) = 0$ or $(T_\mu + A - \lambda)y = 0$ and $(T_l - \lambda)z + (y, f)e = 0$.

If $\lambda \neq \mu$, since $\sigma(T_\mu + A) = \{\mu\}$, $y = 0$. Since $\sigma_p(T_l) = \varnothing$, $z = 0$.

If $\lambda = \mu$, then $y \in \ker(T_\mu + A - \mu)$ and $(T_l - \mu)z = -(y, f)e$. Since $e \in \ker(T_l - \mu)^* = [\operatorname{ran}(T_l - \mu)]^\perp$, $(y, f) = 0$. It follows that $y = 0$ and $z = 0$ from $\operatorname{nul}(T_\mu + A - \mu) = 1$ and $\sigma_p(T_l) = \varnothing$. Therefore, $\sigma_p(S_2) = \varnothing$.

Since $\mathcal{H}_\mu$ is finite dimensional, it is not difficult to check that $\mathcal{H}_\mu \subset \ker(S_2 - \mu)^{*n}$, where $n = \dim \mathcal{H}_\mu$, and $\mathcal{H}_\mu \subset \bigvee\{\ker(S_2 - \lambda)^* : \lambda \in \rho_{s-F}^r(T)\}$, whence we conclude that $\mathcal{H}_l(S_2) = \mathcal{H}_\mu \oplus \mathcal{H}_l(T)$ and $(S_2)_l = S_2$.

Set $S_3 = W^{-1} S_2 W$ and let

$$S = \begin{bmatrix} T_r & T_{12} & S_{13} \\ 0 & T_0|_{\mathcal{M}} & S_{23} \\ 0 & 0 & S_3 \end{bmatrix} \begin{matrix} \mathcal{H}_r(T) \\ \mathcal{M} \\ \mathcal{H}_\mu \oplus \mathcal{H}_l(T), \end{matrix}$$

where $S_{13} = [T_{13}, T_{14}]$, $S_{23} = [T_{23}, T_{24}]$; then $K_\mu = S - T$ is a finite rank operator such that $\|K_\mu\| \le \|A\| + \|W\|\,\|W^{-1}\|\,\|f\| < \epsilon$, $\mathcal{H}_r(S) = \mathcal{H}_r(T)$, $\mathcal{H}_l(S) = \mathcal{H}_\mu \oplus \mathcal{H}_l(T)$ and $S_l = S_3$, and $\mathcal{H}_0(S) = \mathcal{M}$ and $S_0 = T_0|_{\mathcal{M}}$. Now (i), (ii) and (iii) can be easily verified by using Theorem 1.19.

If $\mu \in \sigma_p(T_r)$, then we decompose T as

$$T = \begin{bmatrix} T_r & T_{12} & T_{13} & T_{14} \\ 0 & T_\mu & T_{23} & T_{24} \\ 0 & 0 & (T_0)_{\mathcal{M}} & T_{34} \\ 0 & 0 & 0 & T_l \end{bmatrix} \begin{matrix} \mathcal{H}_r(T) \\ \mathcal{H}_\mu \\ \mathcal{M} \\ \mathcal{H}_l(T), \end{matrix}$$

where $\mathcal{H}_\mu$ is the finite dimensional Riesz subspace of T_0 corresponding to $\{\mu\}$, $\mathcal{M} = \mathcal{H}_0(T) \ominus \mathcal{H}_\mu$, $T_\mu = T|_{\mathcal{H}_\mu}$ and $(T_0)_{\mathcal{M}}$ is the compression of T_0 onto $\mathcal{M}$. Applying the above arguments to $\begin{bmatrix} T_r & T_{12} \\ 0 & T_\mu \end{bmatrix}^*$ instead of S', we can complete the proof of the lemma. □

Proposition 1.22. *Let $T \in \mathcal{L}(\mathcal{H})$ and $\epsilon > 0$; then there exists $K \in \mathcal{K}(\mathcal{H})$ with $\|K\| < \epsilon$, such that*

(i) *$\sigma_0(T+K) = \sigma_0(T)$ and $(T+K)|_{\mathcal{H}(\lambda;\, T+K)} \sim T|_{\mathcal{H}(\lambda;\, T)}$ for all $\lambda \in \sigma_0(T)$, where $\mathcal{H}(\lambda;\, A)$ denotes the Riesz subspace of operator A corresponding to $\lambda \in \sigma_0(A)$;*

(ii) *$\rho_{s-F}^r(T+K) = \rho_{s-F}(T) \backslash \sigma_0(T)$;*

(iii) *$\min \cdot\operatorname{ind}(T + K - \lambda) = \min \cdot\operatorname{ind}(T - \lambda)$ for all $\lambda \in \rho_{s-F}^r(T)$.*

Proof. Let $\Delta = \{\lambda \in \rho_{s-F}(T) \backslash [\sigma_0(T) \cup \rho(T)] : \operatorname{dist}[\lambda, \partial\rho_{s-F}(T)] \ge \epsilon/2\}$. Clearly, Δ is compact and $\Lambda = \Delta \cap [\rho_{s-F}^s(T) \backslash \sigma_0(T)]$ is finite. Applying Lemma 1.21 a finite number of times, once for each point in Λ, we can find a finite rank operator K_1 with $\|K_1\| < \epsilon/2$ such that $T + K_1$ satisfies (i), (iii) and $\rho^s_{s-F}(T + K_1) \cap \Delta = \varnothing$.

Now we can repeat the argument of the proof of Proposition 1.20 in order to "erase" the points of $\rho^s_{s-F}(T + K_1)\backslash\sigma_0(T)$. Since $\operatorname{dist}[\lambda, \partial\rho_{s-F}(T + K_1)] < \epsilon/2$ for all $\lambda \in \rho^s_{s-F}(T + K_1)\backslash\sigma_0(T + K_1)$, there exists a compact operator K_2 with $\|K_2\| < \epsilon/2$ such that $T + K_1 + K_2$ satisfies (i), (ii) and (iii), i.e. $K = K_1 + K_2$ is the required compact operator. □

Theorem 1.23. *Let $T \in \mathcal{L}(\mathcal{H})$ and $\epsilon > 0$; then there exists $K \in \mathcal{K}(\mathcal{H})$ such that*

$$\|K\| < \epsilon + \max\{\operatorname{dist}[\lambda, \partial\rho_{s-F}(T)] : \lambda \in \sigma_0(T)\}$$

and $\min\cdot\operatorname{ind}(T + K - \lambda) = 0$ *for all $\lambda \in \rho_{s-F}(T)$, i.e., T is a smooth operator. In particular,*

$$\sigma(T + K) = \{\lambda \in \mathbb{C} : \lambda - T \textit{ is not a Fredholm operator of index } 0\} = \sigma_W(T).$$

Proof. By Propositions 1.20 and 1.22, we can find $K_1 \in \mathcal{K}(\mathcal{H})$ such that

$$\|K_1\| < \frac{\epsilon}{2} + \max\{\operatorname{dist}[\lambda, \partial\rho_{s-F}(T)] : \lambda \in \sigma_0(T)\},$$

$\rho^r_{s-F}(A) = \rho_{s-F}(A)$ and $\min\cdot\operatorname{ind}(\lambda - A) = \min\cdot\operatorname{ind}(\lambda - T)$ for all $\lambda \in \rho^r_{s-F}(T)$, where $A = T + K_1$.

Thus, in order to complete the proof it suffices to show that there exists $K_2 \in \mathcal{K}(\mathcal{H})$ with $\|K_2\| < \epsilon/2$, such that $\min\cdot\operatorname{ind}(A + K_2 - \lambda) = 0$ for all $\lambda \in \rho_{s-F}(A) = \rho_{s-F}(T)$.

Let $\Delta = \{\lambda \in \rho_{s-F}(A) : \operatorname{dist}[\lambda, \partial\rho_{s-F}(A)] \geq \epsilon/6\}$ and $\Delta_1 = \{\lambda \in \Delta : \min\cdot\operatorname{ind}(\lambda - A) \neq 0\}$.

Claim 1. There exist y, $z \in \mathcal{H}$ such that

$$P_{\ker(\lambda-A)}y \neq 0, \ P_{\ker(\lambda-A)^*}z \neq 0, \quad \text{for all } \lambda \in \Delta_1.$$

Proof of Claim 1. Set $\mathcal{M}_1 = \bigvee\{\ker(\lambda - A) : \lambda \in \Delta_1\}$ and $\mathcal{M}_2 = \bigvee\{\ker(\lambda - A)^* : \lambda \in \Delta_1\}$. Then by Theorem 1.19 $\mathcal{M}_1 \subset \mathcal{H}_r(A)$, $\mathcal{M}_2 \subset \mathcal{H}_l(A)$ and $\Delta_1 \subset \rho_r(A_r) \cap [\rho_r(A_l^*)]^*$. Since $\ker(\lambda - A) = \ker(\lambda - A_r) \oplus 0$ and $\ker(\lambda - A)^* = \ker(\lambda - A_l)^* \oplus 0$, $P_{\ker(\lambda-A)} = P_{\ker(\lambda-A_r)}$ and $P_{\ker(\lambda-A)^*} = P_{\ker(\lambda-A_l)^*}$. Thus it suffices to show that there exist $y \in \mathcal{H}_r(A)$ and $z \in \mathcal{H}_l(A)$ such that

$$P_{\ker(\lambda-A_r)}y \neq 0, \ P_{\ker(\lambda-A_l)^*}z \neq 0 \quad \text{for all } \lambda \in \Delta_1.$$

Applying Proposition 1.40 of Section 1.5, we can find such y, z.

Define $C_\alpha = \alpha z \otimes y$, $\alpha > 0$.

Claim 2. $\min\cdot\text{ind}\,(A + C_\alpha - \lambda) = \min\cdot\text{ind}\,(A - \lambda) - 1$, for all $\lambda \in \Delta_1$, $\alpha > 0$.

Proof of Claim 2. Since $\lambda \in \Delta_1$, $\text{nul}\,(\lambda - A) < \infty$ or $\text{nul}\,(\lambda - A)^* < \infty$. Without loss of generality, we assume that $\text{nul}\,(\lambda - A) < \infty$. Let $x \in \ker(A + C_\alpha - \lambda)$, i.e.,

$$(A - \lambda)x = -\alpha(x, y)z.$$

Decomposing $z = z_1 + z_2$, where $0 \neq z_1 = P_{\ker(\lambda - A)^*}z \in [\text{ran}\,(\lambda - A)]^\perp$, $z_2 \in \text{ran}\,(\lambda - A)$, we have

$$(A - \lambda)x = -\alpha(x, y)(z_1 + z_2).$$

Since $z_1 \in [\text{ran}\,(\lambda - A)]^\perp$ and $\alpha > 0$, $(x, y) = 0$ and $x \in \ker(\lambda - A)$. Decomposing $y = y_1 + y_2$, where $0 \neq y_1 = P_{\ker(\lambda - A)}y$, $y_2 \in [\ker(\lambda - A)]^\perp$. Since $(x, y) = 0$, $x \perp y_1$, i.e., we have

$$\ker(A + C_\alpha - \lambda) = \{x : \ x \in \ker(\lambda - A) \text{ and } x \perp y_1, \ y_1 = P_{\ker(\lambda - A)}y\}.$$

This implies that

$$\text{nul}\,(A + C_\alpha - \lambda) = \text{nul}\,(A - \lambda) - 1, \text{ for all } \lambda \in \Delta_1.$$

Since C_α is compact, $\text{ind}\,(A + C_\alpha - \lambda) = \text{ind}\,(A - \lambda)$. Therefore,

$$\min\cdot\text{ind}\,(A + C_\alpha - \lambda) = \min\cdot\text{ind}\,(A - \lambda) - 1, \text{ for all } \lambda \in \Delta_1.$$

Since $\min\cdot\text{ind}\,(\lambda - A) = 0$ for all $\lambda \in \Delta\backslash\Delta_1$, $\Delta\backslash\Delta_1 \subset \rho_l(A) \cup \rho_r(A)$. Thus there exists $\eta_1 > 0$ such that for $x \in \mathcal{H}$

$$\|(\lambda - A)x\| \geq \eta_1\|x\|, \quad \lambda \in \rho_l(A) \cap (\Delta\backslash\Delta_1)$$

and

$$\|(\lambda - A)^*x\| \geq \eta_1\|x\|, \quad \lambda \in \rho_r(A) \cap (\Delta\backslash\Delta_1).$$

Therefore, when $0 < \alpha < \min\{\epsilon/6, \eta_1/2\}$ we have $\|C_\alpha\| < \epsilon/6$ and

$$\|(A + C_\alpha - \lambda)x\| \geq \frac{\eta_1}{2}\|x\|, \quad \lambda \in \rho_l(A) \cap (\Delta\backslash\Delta_1)$$

and

$$\|(A+C_\alpha-\lambda)^*x\| \geq \frac{\eta_1}{2}\|x\|, \quad \lambda \in \rho_r(A)\cap(\Delta\backslash\Delta_1).$$

Thus $\min\cdot\mathrm{ind}\,(A+C_\alpha-\lambda)=0$ for all $\lambda\in\Delta\backslash\Delta_1$.

Now we can apply the same argument to $A+C_\alpha$, etc. After finitely many steps, we obtain a finite rank operator C_1, with $\|C_1\|<\epsilon/6$, such that

$$\min\cdot\mathrm{ind}\,(A+C_1-\lambda)=0 \quad \text{for all } \lambda\in\Delta \text{ and } \sigma(A+C_1)\subset[\sigma(A)_{\epsilon/6}]^\circ,$$

the interior of the $\epsilon/6$-neighbourhood of $\sigma(A)$.

Let $\{\Omega_j\}_{1\leq j<m}$ $(0\leq m\leq\infty)$ be an enumeration of the components of $\sigma(A+C_1)\cap\rho_{s-F}(A)$ such that $\min\cdot\mathrm{ind}(\lambda-A)\neq 0$ for all $\lambda\in\Omega_j$. Clearly, $\bigcup_j\Omega_j\subset[\partial\rho_{s-F}(A)]_{\epsilon/6}$. A repetition of our previous argument shows that there exists a positive constant $\eta<\min\{\epsilon/6,\,\eta_1/2\}$ such that $\min\cdot\mathrm{ind}\,(A+C_1+B-\lambda)=0$ for all $\lambda\in\Delta$ and $\sigma(A+C_1+B)\subset[\sigma(A)_{\epsilon/6}]^\circ$ for any B in $\mathcal{L}(\mathcal{H})$ such that $\|B\|\leq\eta$.

If $\lambda_1\in\rho^r_{s-F}(A+C_1)\cap\Omega_1$, then the arguments used for the construction of C_α can be repeated here in order to find a finite rank operator $C_{2,1}$ with $\|C_{2,1}\|<\eta/2$ such that $\min\cdot\mathrm{ind}\,(A+C_1+C_{2,1}-\lambda_1)=0$. By induction, we can construct finite rank operators $C_{2,1}, C_{2,2},\cdots,C_{2,n}\cdots$, $\|C_{2,n}\|<\epsilon/2^n$ $(n=1,2,\cdots)$ such that $C_2=\sum_{n=1}^\infty C_{2,n}\in\mathcal{K}(\mathcal{H})$, $\|C_2\|<\eta$, $\min\cdot\mathrm{ind}\,(A+C_1+C_2-\lambda)=0$ for all $\lambda\in\rho^r_{s-F}(A+C_1+C_2)$,

$$\Delta\subset\rho^r_{s-F}(A+C_1+C_2) \text{ and } \sigma(A+C_1+C_2)\subset[\sigma(A)_{\epsilon/6}]^\circ.$$

Since

$$\max\{\mathrm{dist}\,[\lambda,\,\partial\rho_{s-F}(A)]:\ \lambda\in\rho^s_{s-F}(A+C_1+C_2)\}<\frac{\epsilon}{6},$$

we can apply the arguments of Proposition 1.20 in order to find a compact normal operator C_3, $\|C_3\|<\epsilon/6$, such that

$$\min\cdot\mathrm{ind}\,(A+C_1+C_2+C_3-\lambda)=0 \quad \text{for all } \lambda\in\rho_{s-F}(A).$$

Set $K_2=C_1+C_2+C_3$; the proof of the theorem is now complete. □

By arguments similar to Theorem 1.23, we can prove

Theorem 1.24. *Let $T\in\mathcal{L}(\mathcal{H})$, $\epsilon>0$ and Ω be an open component of $\sigma(T)\cap\rho_{s-F}(T)$; if $\inf\{\min\cdot\mathrm{ind}\,(\lambda-T):\ \lambda\in\rho^r_{s-F}(T)\cap\sigma(T)\}=\min\cdot\mathrm{ind}\,\{(\lambda-T):\ \lambda\in$*

$\Omega \cap \rho^r_{s-F}(T)\} = \alpha > 0$ *and* $0 \leq k < \alpha$, *then there exists a compact operator* K *such that*

$$\|K\| < \epsilon + \max\{\text{dist}\,[\lambda,\, \partial\rho_{s-F}(T)] :\ \lambda \in \sigma_0(T)\},$$
$$\min \cdot\text{ind}\,(T + K - \lambda) = k, \quad \lambda \in \Omega,$$
$$\rho_{s-F}(T + K) = \rho^r_{s-F}(T + K) = \rho_{s-F}(T)$$

and

$$\min \cdot\text{ind}\,(T + K - \lambda) \geq k \textit{ for all } \lambda \in \rho^r_{s-F}(T)\backslash\Omega.$$

1.4 The compact perturbations of operators and their normal restrictions

Given T in $\mathcal{L}(\mathcal{H})$, we cannot in general expect to find an infinite dimensional invariant subspace $\mathcal{M}$ of T such that $T|_{\mathcal{M}}$ is normal. But an important approximation argument (due to C. Apostol, C. Foiaş and D. Voiculescu) asserts that some arbitrarily small compact perturbation of T always has that property. This result will have important applications in this book.

Theorem 1.25. *Given* $T \in \mathcal{L}(\mathcal{H})$, *a non-empty compact subset* Γ_l *of* $\sigma_{le}(T)$ *and* $\epsilon > 0$, *there exist an infinite dimensional subspace* $\mathcal{H}_\epsilon \subset \mathcal{H}$ *and* $\mathcal{K}_\epsilon \in C^1(\mathcal{H})$ *such that* $|\mathcal{K}_\epsilon|_1 < \epsilon$, $\mathcal{H}_\epsilon$ *is an invariant subspace of* $T - \mathcal{K}_\epsilon$, $T_\epsilon = (T - \mathcal{K}_\epsilon)|_{\mathcal{H}_\epsilon}$ *is a diagonal normal operator of uniform infinite multiplicity and* $\sigma(T_\epsilon) = \sigma_e(T_\epsilon) = \Gamma_l$.

Proof. Let $\{\lambda_n\}_{n=1}^\infty$ be a sequence of complex numbers such that $\{\lambda_n\}^- = \Gamma_l$ and

$$\text{card}\,\{m :\ \lambda_m = \lambda_n\} = \infty$$

for each $n = 1, 2, \cdots$. Since $\Gamma_l \subset \sigma_{le}(T) \subset \sigma_l(T)$, we can find a unit vector $e_1 \in \mathcal{H}$ such that $\|(\lambda_1 - T)e_1\| < \epsilon/4$. Consider the decomposition

$$T = \begin{bmatrix} \lambda_1 + t_{11} & T^1_{12} \\ T^1_{21} & T^1_{22} \end{bmatrix} \begin{matrix} e_1 \\ \{e_1\}^\perp \end{matrix};$$

then $\begin{bmatrix} t_{11} \\ T^1_{12} \end{bmatrix} :\ \bigvee\{e_1\} \to \mathcal{H}$ is a rank-one operator with $\left\| \begin{bmatrix} t_{11} \\ T^1_{21} \end{bmatrix} \right\| < \epsilon/4$ and $T^1_{12} :\ \{e_1\}^\perp \to \bigvee\{e_1\}$ is also a rank-one operator. Thus, we can find a finite dimensional subspace $\mathcal{M}_1 \supset \bigvee\{e_1\}$ such that $\|T^1_{12}|_{\mathcal{M}_1^\perp}\| < \epsilon/4$. Let R_1 be the orthogonal projection of $\mathcal{H}$ onto $\mathcal{M}_1^\perp$ and define C_1 by

$$C_1 = \begin{bmatrix} t_{11} & 0 \\ T^1_{21} & 0 \end{bmatrix} + \begin{bmatrix} 0 & T^1_{12} \\ 0 & 0 \end{bmatrix} R_1.$$

Then $C_1 \in C^1(\mathcal{H})$ and $|C_1|_1 < \left\| \begin{bmatrix} t_{11} \\ T_{21}^1 \end{bmatrix} \right\|_1 + |T_{12}^1|_{\mathcal{M}_1^\perp}|_1 < \epsilon/2$ and

$$T - C_1 = \begin{bmatrix} \lambda_1 & T_{12}^1|_{\mathcal{N}_1} & 0 \\ 0 & A_1 & B_1 \\ 0 & D_1 & T_1 \end{bmatrix} \begin{matrix} e_1 \\ \mathcal{N}_1 \\ \mathcal{M}_1^\perp \end{matrix} = \begin{bmatrix} \lambda_1 & 0 & E_1 \\ 0 & T_1 & D_1 \\ 0 & B_1 & A_1 \end{bmatrix} \begin{matrix} e_1 \\ \mathcal{M}_1^\perp \\ \mathcal{N}_1, \end{matrix}$$

where $\mathcal{N}_1 = \mathcal{M}_1 \cap \{e_1\}^\perp$.

Clearly, $\sigma_{le}(T_1) = \sigma_{le}(T)$. Thus, we can apply the same argument to T_1 in order to obtain a unit vector $e_2 \in \mathcal{M}_1^\perp$, a $C_2 \in C^1(\mathcal{H})$ with $|C_2|_1 < \epsilon/4$ and a finite dimensional subspace $\mathcal{M}_2$ such that $\mathcal{M}_1^\perp \supset \mathcal{M}_2 \supset \{e_2\}$ and

$$T - (C_1 + C_2) = \begin{bmatrix} \begin{bmatrix} \lambda_1 & 0 \\ 0 & \lambda_2 \end{bmatrix} & 0 & E_2 \\ 0 & T_2 & D_2 \\ 0 & B_2 & A_2 \end{bmatrix} \begin{matrix} e_1 \\ e_2 \\ \mathcal{M}_1^\perp \ominus \mathcal{M}_2. \\ \\ \end{matrix}$$

By induction, we can find orthonormal vectors $e_1, e_2, \cdots, e_n, \cdots$ and trace class operators $C_1, C_2, \cdots, C_n, \cdots$ such that $K_\epsilon = \sum_{n=1}^\infty C_n \in C^1(\mathcal{H})$, $|K_\epsilon|_1 \le \sum_{n=1}^\infty |C_n|_1 < \epsilon$ and

$$T - K_\epsilon = \begin{bmatrix} T_\epsilon & E \\ 0 & A \end{bmatrix} \begin{matrix} \mathcal{H}_\epsilon \\ \mathcal{H}_\epsilon^\perp, \end{matrix}$$

where $\mathcal{H}_\epsilon = \bigvee \{e_n\}_{n=1}^\infty$ and $T_\epsilon = \sum_{n=1}^\infty \lambda_n e_n \otimes e_n$. Thus T_ϵ has the desired properties.

□

Corollary 1.26. *Given $T \in \mathcal{L}(\mathcal{H})$, non-empty compact subsets $\Gamma_l \subset \sigma_{le}(T)$ and $\Gamma_r \subset \sigma_{re}(T)$, $\epsilon > 0$ and diagonal normal operators N_l and N_r of uniform infinite multiplicity such that $\Gamma_l \subset \sigma(N_l) = \sigma_e(N_l) \subset (\Gamma_l)_\epsilon$ and $\Gamma_r \subset \sigma(N_r) = \sigma_e(N_r) \subset (\Gamma_r)_\epsilon$, there exists an operator $L \in \mathcal{L}(\mathcal{H})$ such that*

$$L \simeq \begin{bmatrix} N_l & * & * \\ 0 & A & * \\ 0 & 0 & N_r \end{bmatrix}$$

and $\|T - L\| < 2\epsilon$.

Proof. Applying the inductive argument of the proof of Theorem 1.25 alternately to T and T^*, we can find $K \in C^1(\mathcal{H})$, $\|K\| \le |K|_1 < \epsilon/2$ such that

$$T - K = \begin{bmatrix} M_l & * & * \\ 0 & A & * \\ 0 & 0 & M_r \end{bmatrix},$$

where M_l and M_r are diagonal normal operators of uniform infinite multiplicity such that $\sigma(M_l) = \Gamma_l$ and $\sigma(M_r) = \Gamma_r$.

Since $\Gamma_l \subset \sigma(N_l) = \sigma_e(N_l) \subset (\Gamma_l)_\epsilon$ and $\Gamma_r \subset \sigma(N_r) = \sigma_e(N_r) \subset (\Gamma_r)_\epsilon$, and since N_l and N_r are diagonal normal operators of uniform infinite multiplicity, we can "spread" the spectral measures of M_l and M_r (concentrated in $\sigma_p(M_l)$ and $\sigma_p(M_r)$, resp.) in order to get normal operators N_l' and N_r' such that $N_l' \simeq N_l$, $N_r' \simeq N_r$, $\|M_l - N_l'\| < 3\epsilon/2$ and $\|M_r - N_r'\| < 3\epsilon/2$.

The operator L obtained by replacing M_l and M_r by N_l' and N_r', respectively, satisfies all our requirements. □

Theorem 1.27. *Let $T \in \mathcal{L}(\mathcal{H})$, $\epsilon > 0$ and let Φ be an analytic Cauchy domain satisfying $\sigma_{lre}(T) \subset \Phi \subset \left[\sigma_{lre}(T)\right]_\epsilon$; then there exists $T_\epsilon \in \mathcal{L}(\mathcal{H})$ such that*

(i) $\sigma_{lre}(T_\epsilon) = \overline{\Phi}$;

(ii) $\rho_{s-F}^r(T_\epsilon) = \rho_{s-F}(T_\epsilon)\backslash\sigma_0(T_\epsilon)$, $\text{ind}\,(\lambda - T_\epsilon) = \text{ind}\,(\lambda - T)$ *and* $\min\cdot\text{ind}\,(\lambda - T_\epsilon)^k \leq \min\cdot\text{ind}\,(\lambda - T)^k$ *for all* $\lambda \in \rho_{s-F}(T_\epsilon)$ *and* $k = 1, 2, \cdots$;

(iii) $\sigma(T_\epsilon)$ *consists of finitely many connected components; the number of connected components is less than or equal to the number of connected components of* $\sigma(T)$;

(iv) $\|T - T_\epsilon\| < \epsilon$.

Proof. Let $\delta = \text{dist}\,(\overline{\Phi}, \partial[\sigma_{lre}(T)]_\epsilon)$. By Voiculescu's Theorem, there exist $K_1 \in \mathcal{K}(\mathcal{H})$, $\|K_1\| < \delta/8$, and a unitary operator U such that

$$U(T + K_1)U^* = A \oplus T,$$

where $\sigma_e(A) = \sigma(A) = \sigma_e(T)$, $\sigma_{lre}(A) = \sigma_{lre}(T)$, $\text{ind}\,(\lambda - A) = \infty$ or $\text{ind}\,(\lambda - A) = -\infty$. By Theorem 1.25 and Proposition 1.22, there exists $K_2 \in \mathcal{K}(\mathcal{H})$, $\|K_2\| < \delta/8$ such that

$$U(T + K_1 + K_2)U^* = \begin{bmatrix} N & A_{12} \\ 0 & A_\infty \end{bmatrix} \oplus T = B \oplus T,$$

$\sigma(B) = \sigma_e(B) = \sigma_e(T)$, and $\min\cdot\text{ind}\,(\lambda - B) = 0$ for all $\lambda \in \rho_{s-F}(B)$, where $B = \begin{bmatrix} N & A_{12} \\ 0 & A_\infty \end{bmatrix}$ and N is a diagonal normal operator of uniform infinite multiplicity and $\sigma(N) = \sigma_{lre}(T)$. By the argument of Corollary 1.26

$$\|L - U(T + K_1 + K_2)U^*\| < \epsilon - \delta$$

where $L = \begin{bmatrix} M & A_{12} \\ 0 & A_\infty \end{bmatrix} \oplus T$, and M is a diagonal normal operator of uniform infinite multiplicity with $\sigma(M) = \sigma_e(M) = \overline{\Phi}$. Setting $T_\epsilon = U^*LU$, the T_ϵ satisfies all the requirements. □

Let $T \in \mathcal{L}(\mathcal{H})$. λ is an *approximate normal eigenvalue* of T if for each $\epsilon > 0$, there is a unit vector e_ϵ such that $\|(\lambda - T)e_\epsilon\| < \epsilon$ and $\|(\lambda - T)^* e_\epsilon\| < \epsilon$. If $\lambda \in \sigma_{lre}(T) \subset \sigma_{le}(T)$, by Theorem 1.25, there exist a compact operator K and an infinite dimensional subspace $\mathcal{H}_1$ such that

$$T = \begin{bmatrix} \lambda & E \\ 0 & A \end{bmatrix} \begin{matrix} \mathcal{H}_1 \\ \mathcal{H}_1^\perp \end{matrix} + K.$$

Let $\{e_n\}_{n=1}^\infty$ be an ONB of $\mathcal{H}_1$. Since $e_n \to 0$ weakly, $\|Ke_n\| \to 0$ $(n \to \infty)$, i.e. $\|(\lambda - T)e_n\| \to 0$.

If T is a *hyponormal operator*, i.e., $[T^*, T] = T^*T - TT^*$ is a positive semi-definite hermitian operator, then

$$((\lambda - T)^*(\lambda - T)e_n, e_n) \geq ((\lambda - T)(\lambda - T)^* e_n, e_n)$$

or

$$\|(\lambda - T)e_n\| \geq \|(\lambda - T)^* e_n\| \qquad (n = 1, 2, \cdots).$$

Thus λ is an approximate normal eigenvalue of T.

If T is essentially normal, i.e., $[T^*, T] = K_1 \in \mathcal{K}(\mathcal{H})$, then

$$\|(\lambda - T)^*(\lambda - T)e_n - (\lambda - T)(\lambda - T)^* e_n\| = \|K_1 e_n\| \to 0 \qquad (n \to \infty).$$

Therefore, $\|(\lambda - T)e_n\|^2 - \|(\lambda - T)^* e_n\|^2 \to 0$ $(n \to \infty)$. Since $(\lambda - T)e_n \to 0$, $(\lambda - T)^* e_n \to 0$ and λ is an approximate normal eigenvalue.

Using these arguments, we in fact can get the following proposition.

Proposition 1.28. *Assume that T is hyponormal or essentially normal. Then $\sigma_{lre}(T)$ is contained in the set of approximate normal eigenvalues of T. Furthermore, if λ is an approximate normal eigenvalue of the operator $A \in \mathcal{L}(\mathcal{H})$, then λ is an approximate normal eigenvalue of $A + K$ for all $K \in \mathcal{K}(\mathcal{H})$.*

Using Proposition 1.28, we can get

Theorem 1.29. *Given T, $N \in \mathcal{L}(\mathcal{H})$, such that $\sigma_{lre}(T)$ is contained in the set of approximate normal eigenvalues of T, N is a normal operator with $\sigma(N) = \sigma_e(N) = \Gamma$, $\Gamma \subset \sigma_{lre}(T)$ and given $\epsilon > 0$, there exists a compact operator K with $\|K\| < \epsilon$ such that*

$$T - K \simeq \begin{bmatrix} N & 0 \\ 0 & T \end{bmatrix}.$$

Proof. Let $\{\lambda_n\}_{n=1}^{\infty}$ be a sequence of complex numbers such that $\{\lambda_n\}^- = \Gamma$ and $\operatorname{card}\{m : \lambda_m = \lambda_n\} = \infty$ for each $n = 1, 2, \cdots$. By the definition of approximate normal eigenvalue, there exists a unit vector e_1 such that $\|(\lambda_1 - T)e_1\| < \epsilon/16$ and $\|(\lambda_1 - T)^* e_1\| < \epsilon/16$. Thus, we have $\|P_{e_1}(\lambda_1 - T)P_{e_1}\| < \epsilon/16$, $\|P_{e_1}(\lambda_1 - T)P_{e_1^\perp}\| = \|P_{e_1^\perp}(\lambda_1 - T)^* P_{e_1}\| < \epsilon/16$, and $\|P_{e_1^\perp}(\lambda_1 - T)P_{e_1}\| < \epsilon/16$, where P_{e_1} and $P_{e_1^\perp}$ denote the orthogonal projections onto $\bigvee\{e_1\}$ and, respectively, $[\bigvee\{e_1\}]^\perp$. Under the decomposition $\mathcal{H} = \bigvee\{e_1\} \bigoplus [\bigvee\{e_1\}]^\perp$, T admits the representation

$$T = \begin{bmatrix} \lambda_1 + t_{11} & T_{12} \\ T_{21} & T_1 \end{bmatrix}.$$

Set $K_1 = \begin{bmatrix} t_{11} & T_{12} \\ T_{21} & 0 \end{bmatrix}$; then $K_1 \in \mathcal{K}(\mathcal{H})$, $\|K_1\| < \epsilon/8$ and $T - K_1 = \begin{bmatrix} \lambda_1 & 0 \\ 0 & T_1 \end{bmatrix}$. Clearly, $\sigma_{lre}(T_1) = \sigma_{lre}(T)$. Repeating the argument, we can get $K_2 \in \mathcal{K}(\mathcal{H})$, $\|K_2\| < \epsilon/2^4$, such that

$$T - K_1 - K_2 = \begin{bmatrix} \lambda_1 & 0 & 0 \\ 0 & \lambda_2 & 0 \\ 0 & 0 & T_2 \end{bmatrix} \begin{matrix} \bigvee\{e_1\} \\ \bigvee\{e_2\} \\ [\bigvee\{e_1, e_2\}]^\perp. \end{matrix}$$

By induction, we can find an orthonormal sequence $e_1, e_2, \cdots, e_n, \cdots$ in $\mathcal{H}$ and a sequence $K_1, K_2, \cdots, K_n, \cdots$ in $\mathcal{K}(\mathcal{H})$ such that $\|K_n\| < \epsilon/2^{n+2}$ $(n = 1, 2, \cdots)$ and

$$T - \sum_{n=1}^{m} K_n = \begin{bmatrix} \sum_{n=1}^{m} \lambda_n e_n \otimes e_n & 0 \\ 0 & T_{m+1} \end{bmatrix} \begin{matrix} \bigvee\{e_1, \cdots, e_m\} \\ [\bigvee\{e_1, \cdots, e_m\}]^\perp \end{matrix} \qquad m = 1, 2, \cdots.$$

Set $C_1 = \sum_{n=1}^{\infty} K_n$, $N_1 = \sum_{n=1}^{\infty} \lambda_n e_n \otimes e_n$; then

$$T - C_1 = \begin{bmatrix} N_1 & 0 \\ 0 & T_\infty \end{bmatrix}$$

with respect to the decomposition $\mathcal{H} = \bigvee\{e_n : 1 \le n < \infty\} \bigoplus [\bigvee\{e_n\}]^\perp$, and $C_1 \in \mathcal{K}(\mathcal{H})$, $\|C_1\| < \epsilon/4$. Applying Voiculescu's non-commutative Weyl-von Neumann

Theorem to N_1, we can find a compact operator C_2', $\|C_2'\| < \epsilon/8$, such that $N_1 + C_2' \simeq N \oplus N_1$. Therefore,

$$T - C_1 - C_2 \simeq \begin{bmatrix} N & 0 & 0 \\ 0 & N_1 & 0 \\ 0 & 0 & T_\infty \end{bmatrix}$$

where C_2 is compact and $\|C_2\| < \epsilon/8$. Finally, we can find a compact operator C_3 with $\|C_3\| < \epsilon/4$, such that

$$T - C_1 - C_2 - C_3 \simeq \begin{bmatrix} N & 0 \\ 0 & T \end{bmatrix}.$$

Set $K = C_1 + C_2 + C_3$; then K satisfies all the requirements of the theorem. □

If T is essentially normal, $\sigma_{lre}(T) = \sigma_e(T)$, thus we have

Corollary 1.30. *Let $T \in \mathcal{L}(\mathcal{H})$ be essentially normal, $\Gamma \subset \sigma_e(T)$ be a non-empty perfect set, and let N be normal with $\sigma(N) = \Gamma$, $\epsilon > 0$; then there exists $K \in \mathcal{K}(\mathcal{H})$ such that $\|K\| < \epsilon$ and*

$$T - K \simeq \begin{bmatrix} N & 0 \\ 0 & T \end{bmatrix}.$$

1.5 Some results on Cowen–Douglas operators

Definition 1.31. Let Ω be a bounded connected open set in $\mathbb{C}$, and n be a positive integer. The set $\mathcal{B}_n(\Omega)$ of *Cowen–Douglas operators of index* n is the set of operators B in $\mathcal{L}(\mathcal{H})$ satisfying

(i) $\sigma(B) \supset \Omega$;
(ii) $\operatorname{ran}(\lambda - B) = \mathcal{H}$ for all $\lambda \in \Omega$;
(iii) $\operatorname{nul}(\lambda - B) = n$ for all $\lambda \in \Omega$;
(iv) $\bigvee \{\ker(\lambda - B) : \lambda \in \Omega\} = \mathcal{H}$.

Example 1.32. Let $\{e_n\}_{n=1}^{\infty}$ be an *ONB* of $\mathcal{H}$, and $S \in \mathcal{L}(\mathcal{H})$ be given by $Se_n = e_{n+1}$ $(n \geq 1)$, i.e., S is the (forward) shift. Then $S^* \in B_1(D)$, where D is the open unit disc.

Example 1.33. Given $T \in \mathcal{L}(\mathcal{H})$ such that $\rho^n_{s\text{-}F}(T) = \{\lambda \in \rho_{s\text{-}F}(T) : \operatorname{ind}(\lambda - T) = n\} \neq 0$, let Ω be a component of $\rho^n_{s-F}(T)$ and $\mathcal{H}(\Omega) = \bigvee \{\ker(\lambda - T) : \lambda \in \Omega \cap \rho^r_{s-F}(T)\}$; then $\mathcal{H}(\Omega)$ is an invariant subspace of T. By Theorem 1.19, $T(\Omega) = T|_{\mathcal{H}(\Omega)} \in \mathcal{B}_n(\Omega)$.

We need some propositions and lemmas in order to analyse the properties of Cowen–Douglas operators clearly.

Lemma 1.34. *Let $T \in \mathcal{L}(\mathcal{H})$, and $\mathcal{M}$ be an invariant subspace of T with $0 \in \rho_{re}(T|_{\mathcal{M}})$. Then $\operatorname{ran} T_{\mathcal{M}^\perp}$ is closed if and only if $\operatorname{ran} T$ is closed.*

Proof. Since $0 \in \rho_{re}(T|_{\mathcal{M}})$, $P_{\ker(T|_{\mathcal{M}})^*}$ is a finite rank operator. Let

$$T = \begin{bmatrix} T|_{\mathcal{M}} & E \\ 0 & T_{\mathcal{M}^\perp} \end{bmatrix} \begin{matrix} \mathcal{M} \\ \mathcal{M}^\perp \end{matrix}$$

where $E = P_{\mathcal{M}} T|_{\mathcal{M}^\perp}$. Since $\operatorname{ran} P_{\ker(T|_{\mathcal{M}})^*}$ is finite dimensional,

$$\mathcal{N} := \{z \in \mathcal{M}^\perp : \; P_{\ker(T|_{\mathcal{M}})^*} Ez = 0\}$$

is an infinite dimensional subspace, $\mathcal{N} \subset \mathcal{M}^\perp$ and $\mathcal{M}^\perp \ominus \mathcal{N}$ is finite dimensional. Note that $H \ominus (\mathcal{M} \oplus \mathcal{N}) = \mathcal{M}^\perp \ominus \mathcal{N}$ is finite dimensional, thus $\operatorname{ran} T$ is closed if and only if $T(\mathcal{M} \oplus \mathcal{N})$ is closed. Since $\mathcal{M}^\perp \ominus \mathcal{N}$ is finite dimensional, the proof is reduced to proving that $T(\mathcal{M} \oplus \mathcal{N})$ is closed if and only if $T_{\mathcal{M}^\perp}\mathcal{N}$ is closed. Since $P_{\ker(T|_{\mathcal{M}})^*} P_{\mathcal{M}} T\mathcal{N} = 0$ and $0 \in \rho_{re}(T|_{\mathcal{M}})$, $\operatorname{ran} T|_{\mathcal{M}}$ is closed and $P_{\mathcal{M}} T\mathcal{N} \subset [\ker(T|_{\mathcal{M}})^*]^\perp = \operatorname{ran}(T|_{\mathcal{M}}) = T\mathcal{M}$. Therefore $T(\mathcal{M} \oplus \mathcal{N}) = T\mathcal{M} + T\mathcal{N} \subset T\mathcal{M} + P_{\mathcal{M}} T\mathcal{N} + P_{\mathcal{M}^\perp} T\mathcal{N} = T\mathcal{M} + P_{\mathcal{M}^\perp} T\mathcal{N} = T\mathcal{M} + T_{\mathcal{M}^\perp}\mathcal{N}$.

On the other hand, for any $y \in \mathcal{M}$ and $z \in \mathcal{N}$, since $P_{\mathcal{M}} T\mathcal{N} \subset T\mathcal{M}$, there exists $y_0 \in \mathcal{M}$ such that $T|_{\mathcal{M}} y_0 = -P_{\mathcal{M}} Tz$. Thus,

$$\begin{aligned} Ty + T_{\mathcal{M}^\perp} z &= Ty + (Ty_0 + P_{\mathcal{M}} Tz) + T_{\mathcal{M}^\perp} z \\ &= T(y + y_0 + z) \subset T(\mathcal{M} \oplus \mathcal{N}). \end{aligned}$$

This implies that $T(\mathcal{M} \oplus \mathcal{N}) = T\mathcal{M} \oplus T_{\mathcal{M}^\perp}\mathcal{N}$. Since $T\mathcal{M}$ is closed, $T(\mathcal{M} \oplus \mathcal{N})$ is closed if and only if $T_{\mathcal{M}^\perp} N$ is closed. ☐

Corollary 1.35. *Given $T \in \mathcal{L}(\mathcal{H})$ with closed range and given that $\mathcal{M}$ is an invariant subspace of T satisfying $(T\mathcal{M})^- = \mathcal{M}$ and $0 \in \rho_{le}(T_{\mathcal{M}^\perp})$; then $0 \in \rho_r(T|_{\mathcal{M}})$.*

Proof. Since $\operatorname{ran} T^*$ is closed and $0 \in \rho_{re}[(T_{\mathcal{M}^\perp})^*]$, $(T|_{\mathcal{M}})^*$ has closed range by Lemma 1.34. Since $\operatorname{ran} T|_{\mathcal{M}}$ is dense in $\mathcal{M}$, $(T|_{\mathcal{M}})^*$ is injective. Therefore $0 \in \rho_l[(T_{\mathcal{M}})^*]$ and $0 \in \rho_r(T|_{\mathcal{M}})$. ☐

Lemma 1.36. *Given $T \in \mathcal{L}(\mathcal{H})$ with closed range, let $\mathcal{M}$ be an invariant subspace of T satisfying $\ker T \subset \mathcal{M}$ and $0 \in \rho_r(T|_{\mathcal{M}})$; then $0 \in \rho_l(T_{\mathcal{M}^\perp})$ and the set $\{\lambda \in \mathbb{C} : \ker(\lambda - T) \subset \mathcal{M}\}$ contains a neighbourhood of 0.*

Proof. Since $\ker T \subset \mathcal{M}$, $0 \notin \sigma_p(T_{\mathcal{M}^\perp})$ (otherwise there exists $x \in \ker T$ and $x \notin \mathcal{M}$). By Lemma 1.34, $\operatorname{ran} T_{\mathcal{M}^\perp}$ is closed and $0 \in \rho_l(T_{\mathcal{M}^\perp})$. Since $\rho_r(T|_{\mathcal{M}})$ and $\rho_l(T|_{\mathcal{M}^\perp})$ are open sets containing 0, there is a neighbourhood Ω of 0 such that

$$\Omega \subset \rho_r(T|_{\mathcal{M}}) \cap \rho_l(T_{\mathcal{M}^\perp}).$$

Let $\lambda \in \Omega$ and $x \in \ker(\lambda - T)$, where $x = x_1 + x_2$, $x_1 \in \mathcal{M}$ and $x_2 \in \mathcal{M}^\perp$. Since $\lambda \in \rho_l(T_{\mathcal{M}^\perp})$, $x_2 = 0$. Thus $x = x_1 \in \mathcal{M}$, i.e.,

$$\Omega \subset \{\lambda \in \mathbb{C} : \ker(\lambda - T) \subset \mathcal{M}\}. \qquad \square$$

Lemma 1.37. *Let $T \in \mathcal{L}(\mathcal{H})$, $\{\lambda_k\}_{k=1}^n \subset \sigma_p(T)$ be pairwise distinct and let $\{m_k\}_{k=1}^n$ be natural numbers; then*

$$\sum_{k=1}^n \ker(\lambda_k - T)^{m_k} = \ker \prod_{k=1}^n (\lambda_k - T)^{m_k}.$$

Proof. "$\subset$" is obvious.

We only prove the inclusion "$\supset$" of the lemma when $n = 2$. The case of $n > 2$ can be proved similarly.

Let $x \in \ker(\lambda_1 - T)^{m_1}(\lambda_2 - T)^{m_2}$; then $(\lambda_2 - T)^{m_2}x \in \ker(\lambda_1 - T)^{m_1}$. Computation shows that

$$\begin{aligned}\sum_{i=0}^{m_1-1} \frac{(T-\lambda_1)^i}{(\lambda_2-\lambda_1)^i} - \sum_{i=0}^{m_1-1} \frac{(T-\lambda_1)^{i+1}}{(\lambda_2-\lambda_1)^{i+1}} &= 1 - \frac{(T-\lambda_1)^{m_1}}{(\lambda_2-\lambda_1)^{m_1}} \\ &= (\lambda_2 - T)\Big(\sum_{i=0}^{m_1-1} \frac{(T-\lambda_1)^i}{(\lambda_2-\lambda_1)^{i+1}}\Big).\end{aligned}$$

Set $y = (\lambda_2 - T)^{m_2}x$ and $x_1 = \left(\sum_{i=0}^{m_1-1} \frac{(T-\lambda_1)^i}{(\lambda_2-\lambda_1)^{i+1}}\right)^{m_2} y$. Then $x_1 \in \ker(\lambda_1 - T)^{m_1}$.

Let $x_2 = x - x_1$. Calculations indicate that

$$\begin{aligned}(\lambda_2 - T)^{m_2} x_2 &= (\lambda_2 - T)^{m_2} x - (\lambda_2 - T)^{m_2} x_1 \\ &= y - (\lambda_2 - T)^{m_2} \left(\sum_{i=0}^{m_1 - 1} \frac{(T-\lambda_1)^i}{(\lambda_2 - \lambda_1)^{i+1}} \right)^{m_2} y \\ &= y - \left[1 - \frac{(T-\lambda_1)^{m_1}}{(\lambda_2-\lambda_1)^{m_1}} \right]^{m_2} y \\ &= \sum_{k=1}^{m_2} (-1)^{k+1} \binom{m_2}{k} \frac{(T-\lambda_1)^{m_1 k}}{(\lambda_2-\lambda_1)^{m_1 k}} (\lambda_2 - T)^{m_2} x = 0.\end{aligned}$$

Thus $x \in \ker(\lambda_1 - T)^{m_1} + \ker(\lambda_2 - T)^{m_2}$. □

Lemma 1.38. *Given $T \in \mathcal{L}(\mathcal{H})$ and $x \in \mathcal{H}$. If $\xi \in \rho_r(T)$ is a limit point of the set $\{\lambda \in \mathbb{C} : P_{\ker(\lambda - T)} x = 0\}$, then $P_{\ker(\lambda-T)} x = 0$ for all $\lambda \in \Omega_\xi$, where Ω_ξ is the component of $\rho_r(T)$ containing ξ.*

Proof. Set $\Delta = \{\lambda \in \Omega_\xi : P_{\ker(\lambda-T)} x = 0\}$ and assume that $\Delta \neq \Omega_\xi$. Since Ω_ξ contains the limit points of Δ, we can find $\mu \in \Omega_\xi \cap \partial\Delta$ and $\{\mu_k\}_{k=1}^\infty \subset \Delta$ such that $\mu_k \neq \mu$ and $\lim \mu_k = \mu$ $(k \to \infty)$. Since $\min\cdot\mathrm{ind}(\lambda - T) = 0$ for all $\lambda \in \rho_r(T) \subset \rho_{s-F}^r(T)$, by Theorem 1.19 $P_{\ker(\lambda - T)}$ is continuous at $\lambda = \mu$. Therefore $\ker(\mu - T) \subset \mathcal{M} := \bigvee \{\ker(\mu_k - T) : k = 1, 2, \cdots\}$ and T admits the representation

$$T = \begin{bmatrix} T|_{\mathcal{M}} & T_{12} \\ 0 & T_{\mathcal{M}^\perp} \end{bmatrix} \begin{matrix} \mathcal{M} \\ \mathcal{M}^\perp \end{matrix}$$

with respect to the decomposition $\mathcal{H} = \mathcal{M} \oplus \mathcal{M}^\perp$.

Since $\ker(\mu - T) \subset \mathcal{M}$, $\mathrm{ran}(\mu - T|_{\mathcal{M}})$ is closed. But $\mathrm{ran}(\mu - T|_{\mathcal{M}})$ is obviously dense; hence $\mu \in \rho_r(T|_{\mathcal{M}})$. Let $\Omega_0 = \{\lambda \in \Omega_\xi : \ker(\lambda - T) \subset \mathcal{M}\}$. By Lemma 1.36, Ω_0 is a neighbourhood of μ. Since $P_{\ker(\mu_k - T)} x = 0$, $x \in \mathcal{M}^\perp$. Therefore $P_{\ker(\lambda-T)} x = 0$ for all $\lambda \in \Omega_0$. This contradicts $\mu \in \partial\Delta$. Hence $\Delta = \Omega_\xi$. □

Lemma 1.39. *Let $T \in \mathcal{L}(\mathcal{H})$, $x \in \mathcal{H}$ and $\xi \in \rho_r(T)$ such that $P_{\ker(\xi - T)^k} x = 0$ $(k = 1, 2, \cdots)$; then $P_{\ker(\lambda - T)} x = 0$ for all $\lambda \in \Omega_\xi$, where Ω_ξ is the component of $\rho_r(T)$ containing ξ.*

Proof. Let $\mathcal{M} = \bigvee \{\ker(\xi - T)^k : k = 1, 2, \cdots\}$. Since $\xi - T$ is surjective,

$$(\xi - T)\ker(\xi - T)^{k+1} = \ker(\xi - T)^k \quad (k = 0, 1, 2, \cdots).$$

Therefore, $\operatorname{ran}(\xi - T|_{\mathcal{M}})$ is dense in $\mathcal{M}$.

On the other hand, $\ker(\xi - T) \subset \mathcal{M}$. By the same argument of Lemma 1.38, $\xi \in \rho_r(T|_{\mathcal{M}})$. By Lemma 1.36, $\Omega_0 = \{\lambda \in \Omega_\xi : \ker(\lambda - T) \subset \mathcal{M}\}$ is a neighbourhood of ξ. Since $x \in \mathcal{M}^\perp$, $P_{\ker(\lambda-T)}x = 0$ for all $\lambda \in \Omega_0$. By Lemma 1.38, $P_{\ker(\lambda-T)}x = 0$ for all $\lambda \in \Omega_\xi$. □

Proposition 1.40. *Let $T \in \mathcal{L}(\mathcal{H})$ and let σ be a compact subset of $\rho_r(T) \cap \sigma_p(T)$; then there exists $x \in \mathcal{H}$ such that*

$$P_{\ker(\lambda-T)}x \neq 0 \quad \textit{for all } \lambda \in \sigma.$$

Proof. Let $\{\Omega_i\}_{i=1}^h$ be the components of $\rho_r(T)$ having non-empty intersection with σ and set $\sigma_i = \sigma \cap \Omega_i$. For each i $(i = 1, 2, \cdots, h)$, we choose

$$\mu_i \in \sigma_i \quad \text{and} \quad y_i \in \ker(\mu_i - T), \quad y_i \neq 0.$$

Claim. There exist non-zero complex numbers α_i $(i = 1, 2, \cdots, h)$ such that the vector $y = \sum_{i=1}^h \alpha_i y_i$ satisfies $P_{\ker(\mu_i - T)}y \neq 0$, $i = 1, 2, \cdots, h$.

Proof of the claim. Since σ is compact, σ intersects only finitely many components of $\rho_r(T)$, i.e., $h < \infty$. For simplicity, we prove the conclusion only in the case of $n = 2$. Since $\mu_1 \neq \mu_2$, y_1 and y_2 are linearly independent. Let $\mathcal{M}_1 = \ker(\mu_1 - T)$ and $\mathcal{M}_2 = \ker(\mu_2 - T)$; then $y_1 = P_{\mathcal{M}_2}y_1 + P_{\mathcal{M}_2^\perp}y_1$ and $y_2 = P_{\mathcal{M}_1}y_2 + P_{\mathcal{M}_1^\perp}y_2$. If $P_{\mathcal{M}_2}y_1 = P_{\mathcal{M}_1}y_2 = 0$, we can choose $\alpha_1 = \alpha_2 = 1$. If at least one of $P_{\mathcal{M}_2}y_1$ and $P_{\mathcal{M}_1}y_2$ is non-zero, then choose α_1, α_2 such that

$$\alpha_1 P_{\mathcal{M}_2}y_1 + \alpha_2 y_2 \neq 0 \quad \text{and} \quad \alpha_1 y_1 + \alpha_2 P_{\mathcal{M}_1}y_2 \neq 0.$$

Thus $y = \alpha_1 y_1 + \alpha_2 y_2$ satisfies $P_{\ker(\mu_i - T)}y \neq 0$ $(i = 1, 2)$.

Let $\Delta = \{\lambda \in \sigma : P_{\ker(\lambda-T)}y = 0\}$. If $\Delta = \varnothing$, we take $x = y$. If $\Delta \neq \varnothing$, we can apply Lemmas 1.38 and 1.39 in order to show that Δ is a finite set, $\Delta = \{\lambda_k\}_{k=1}^n$ and that there exist natural numbers $\{m_k\}_{k=1}^n$ such that

$$y \in \{\ker(\lambda_k - T)^{m_k}\}^\perp \text{ and } y \notin \ker\{(\lambda_k - T)^{m_k+1}\}^\perp, \quad k = 1, 2, \cdots, n.$$

By Lemma 1.37, $y \in \{\ker \prod_{k=1}^n (\lambda_k - T)^{m_k}\}^\perp$. On the other hand, since $\operatorname{ran}\{\prod_{k=1}^n (\lambda_k - T)^{m_k}\}^*$ is closed, we can find $x \in \mathcal{H}$ such that $y = \{\prod_{k=1}^n (\lambda_k - T)^{m_k}\}^* x$.

Assume that $P_{\ker(\mu-T)}x = 0$ for some $\mu \in \sigma$; then $x = (\mu - T)^* z$ for some $z \in \mathcal{H}$ and therefore

$$y = (\mu - T)^* \Big\{ \prod_{k=1}^{n} (\lambda_k - T)^{m_k} \Big\}^* x.$$

But this is impossible: if $\mu \neq \lambda_k$ for all $k = 1, 2, \cdots, n$, we have $P_{\ker(\mu-T)}y \neq 0$; if $\mu = \lambda_j$ for some j $(1 \leq j \leq n)$, $y \in \{\ker(\lambda_j - T)^{m_j+1}\}^\perp$. In either case, this contradicts our assumptions on y. □

Proposition 1.41. *Let Ω be a connected and bounded open subset of $\mathbb{C}$, n be a natural number and let $T \in \mathcal{L}(\mathcal{H})$ satisfying*

(a) $\Omega \subset \sigma(T)$;

(b) $\operatorname{ran}(\lambda - T) = \mathcal{H}$ *and* $\operatorname{nul}(\lambda - T) = n$ *for all* $\lambda \in \Omega$;

then the following are equivalent

(i) $\bigvee \{\ker(\lambda - T) : \lambda \in \Omega\} = \mathcal{H}$;

(ii) $\bigvee \{\ker(\lambda_0 - T)^n : n \geq 1\} = \mathcal{H}$ *for all* $\lambda_0 \in \Omega$;

(iii) $\bigvee \{\ker(\lambda_n - T) : n \geq 1\} = \mathcal{H}$ *for all sequences* $\{\lambda_n\}_{n=1}^\infty \subset \Omega$ *such that* $\lim_{n\to\infty} \lambda_n = \lambda_0 \in \Omega$;

(iv) $\bigvee \{\ker(\lambda_n - T)^k : n \geq 1,\, k \geq 1\} = \mathcal{H}$ *for all sequences* $\{\lambda_n\}_{n=1}^\infty \subset \Omega$ *such that* $\lim_{n\to\infty} \lambda_n = \lambda_0 \in \Omega$.

Proof. (i)⇒(ii). Let $\mathcal{M} = \bigvee \{\ker(\lambda_0 - T)^n : n \geq 1\}$ and $y \in \mathcal{H} \ominus \mathcal{M}$; then $P_{\ker(\lambda_0-T)^n}y = 0$ $(n = 1, 2, \cdots)$. Since $\lambda_0 \in \Omega \subset \rho_r(T)$, by Lemma 1.39, $P_{\ker(\lambda-T)}y = 0$ for all $\lambda \in \Omega$. But $\bigvee \{\ker(\lambda - T) : \lambda \in \Omega\} = \mathcal{H}$, therefore $y = 0$.

(ii)⇒(i). Let $\mathcal{M} = \bigvee \{\ker(\lambda - T) : \lambda \in \Omega\}$; then $\ker(\lambda_0 - T) \subset \mathcal{M}$ and $\lambda_0 \in \rho_r(T|_{\mathcal{M}})$. By Lemma 1.36, $\lambda_0 \in \rho_l(T_{\mathcal{M}^\perp})$. Thus,

$$\ker(\lambda_0 - T) = \ker(\lambda_0 - T|_{\mathcal{M}}) \subset \operatorname{ran}(\lambda_0 - T|_{\mathcal{M}}).$$

By induction, we can prove that $\ker(\lambda_0 - T)^n \subset \mathcal{M}$, therefore,

$$\mathcal{H} = \bigvee \{\ker(\lambda_0 - T)^k : k = 1, 2, \cdots\} \subset \mathcal{M}, \quad \text{i.e., } \mathcal{M} = \mathcal{H}.$$

(i)⇒(iii) can be proved by using Lemma 1.38.

(iii)⇒(i) is trivial.

From (i) and (ii) (or (i) and (iii)) we can get (iv). By an argument similar to the implication (ii)⇒(i), we can prove (iv)⇒(i). □

By Proposition 1.41, the condition (i) in Definition 1.31 of Cowen–Douglas operator $\mathcal{B}_n(\Omega)$ can be replaced by one of the conditions (ii), (iii) and (iv).

Corollary 1.42. *Let Ω_1, Ω_2 be two connected open subsets of $\mathbb{C}$ and $\Omega_1 \subset \Omega_2$; then $\mathcal{B}_n(\Omega_2) \subset \mathcal{B}_n(\Omega_1)$.*

Theorem 1.43. *Let $T \in \mathcal{B}_n(\Omega)$; then*

(i) $\sigma(T) = \sigma_W(T)$;

(ii) $\operatorname{ind}(\lambda - T) > 0$ *for all* $\lambda \in \rho_{s-F}(T) \supset \Omega$;

(iii) $\operatorname{ind}(\lambda - T) = \operatorname{nul}(\lambda - T) = n$ *for all* $\lambda \in \Omega$;

(iv) $\operatorname{nul}(\lambda - T)^* = 0$ *for all* $\lambda \in \mathbb{C}$.

Proof. Fix $\omega \in \Omega$, by Definition 1.31, $\operatorname{ind}(\omega - T) = n$ and $\operatorname{nul}(\omega - T)^* = \{0\}$. By Proposition 1.41, $\mathcal{H} = \bigvee \{\ker(\omega - T)^k : k = 1, 2, \cdots\}$. Therefore there exists an $ONB\ \{e_n\}_{n=1}^{\infty}$ of $\mathcal{H}$, with respect to which T admits a triangular matrix representation with diagonal $d(T) = \{\omega\}$. Assume that $\lambda \in \sigma_p(T^*)^*$; by Proposition 1.16, $\lambda = \omega$, contradicting $\operatorname{nul}(\omega - T)^* = \{0\}$. Thus $\sigma_p(T^*) = \varnothing$, $\operatorname{ind}(\lambda - T) > 0$ for all $\lambda \in \rho_{s-F}(T)$ and $\sigma(T) = \sigma_W(T)$. □

An operator T is *quasitriangular* if there exists an increasing sequence $\{P_n\}_{n=1}^{\infty}$ of finite rank orthogonal projections such that $P_n \to 1$ (strongly, as $n \to \infty$) and $\|(1 - P_n)TP_n\| \to 0$ ($n \to \infty$). It is well-known that an arbitrary small compact perturbation of a quasitriangular operator can be a triangular operator. The following theorem analyses the relation between quasitriangular operators and Cowen–Douglas operators.

Theorem 1.44. *Let $R \in \mathcal{L}(\mathcal{H})$ satisfy*

(i) *$\sigma(R)$ and $\sigma_W(R)$ are connected and contain a connected open set Ω;*

(ii) $\operatorname{ind}(\lambda - R) \geq 0$ *for all $\lambda \in \rho_{s-F}(R)$ (i.e., R is a quasitriangular operator);*

(iii) *$\rho_{s-F}(R) \supset \Omega$ and* $\operatorname{ind}(\lambda - R) = n$ *for all $\lambda \in \Omega$.*

Then for $\epsilon > 0$, there exists a compact operator K_ϵ, $\|K_\epsilon\| < \epsilon$, such that $R - K_\epsilon \in \mathcal{B}_n(\Omega)$.

In order to prove Theorem 1.44, we need one of D.A. Domingo's results (D.A. Herrero [2]). It will be stated here without proof.

Theorem 1.45. *Let $R \in \mathcal{L}(\mathcal{H})$ be quasitriangular such that $\sigma(R)$ and $\sigma_W(R)$ are connected, let $\epsilon > 0$ and $\omega \in \sigma_W(R)$; then there exists an $ONB\ \{e_n\}_{n=1}^{\infty}$ of $\mathcal{H}$ and $K_\epsilon \in \mathcal{K}(\mathcal{H})$, $\|K_\epsilon\| < \epsilon$, such that $R - K_\epsilon$ admits a triangular matrix representation with diagonal $d(R - K_\epsilon) = \{\omega\}$ with respect to $\{e_n\}_{n=1}^{\infty}$.*

Proof of Theorem 1.44. Take $\omega \in \Omega$. By Theorem 1.45, there exists $K_\epsilon \in \mathcal{K}(\mathcal{H})$ with $\|K_\epsilon\| < \epsilon$ such that $T := R - K_\epsilon$ admits a triangular representation $T = (t_{ij})_{i,j=1}^{\infty}$ with respect to some $ONB\,\{e_n\}_{n=1}^{\infty}$ of $\mathcal{H}$, and $d(T) = \{\omega\}$. We can choose K_ϵ properly so that $t_{i\,i+1} \neq 0$ $(i = 1, 2, \cdots)$. By Proposition 1.16 $\sigma_p(T^*)^* \subset \{\omega\}$. Since $t_{i\,i+1} \neq 0$ $(i = 1, 2, \cdots)$, computations indicate that $\ker(\omega - T)^* = \{0\}$. Thus, $\sigma_p(T^*) = \varnothing$ and $\operatorname{nul}(\lambda - T) = \operatorname{ind}(\lambda - T) = n$ and $\operatorname{ran}(\lambda - T) = \mathcal{H}$ for all $\lambda \in \Omega$. Since $\omega - T$ is triangular and $d(\omega - T) = \{0\}$, $\bigvee\{\ker(\omega - T)^k : k \geq 1\} = \mathcal{H}$. By Proposition 1.41, $T \in \mathcal{B}_n(\Omega)$. □

The famous Apostol, Foiaş and Voiculescu Theorem indicates that T is quasitriangular if and only if $\operatorname{ind}(\lambda - T) \geq 0$ for all $\lambda \in \rho_{s-F}(T)$. Thus, there exist a lot of quasitriangular operators. Theorem 1.44 implies that the Cowen–Douglas operators are also a richer class of operators.

Similar to $\mathcal{B}_n(\Omega)$ $(1 \leq n < \infty)$, we can define $B_\infty(\Omega)$.

Definition 1.46. Let Ω be a connected, bounded open subset of $\mathbb{C}$. Then $B_\infty(\Omega)$, the set of *Cowen–Douglas operators of infinite order*, is the class of operators B on $\mathcal{H}$ satisfying

(i) $\sigma(B) \supset \Omega$;

(ii) $\operatorname{ran}(\lambda - B) = \mathcal{H}$ for all $\lambda \in \Omega$;

(iii) $\operatorname{nul}(\lambda - B) = \infty$ for all $\lambda \in \Omega$;

(iv) $\bigvee\{\ker(\lambda - B) : \lambda \in \Omega\} = \mathcal{H}$.

The condition (iv) of Definition 1.46 can be replaced by each of the following (iv)$'$, (iv)$''$ and (iv)$'''$.

(iv)$'$ $\bigvee\{\ker(\lambda_0 - T)^n : n \geq 1\} = \mathcal{H}$ for all $\lambda_0 \in \Omega$.

(iv)$''$ $\bigvee\{\ker(\lambda_n - T) : n \geq 1\} = \mathcal{H}$ for all sequences $\{\lambda_n\}_{n=1}^{\infty} \subset \Omega$ such that $\lim_{n\to\infty} \lambda_n = \lambda_0 \in \Omega$.

(iv)$'''$ $\bigvee\{\ker(\lambda_n - T)^k : n \geq 1, k \geq 1\} = \mathcal{H}$ for all sequences $\{\lambda_n\}_{n=1}^{\infty} \subset \Omega$ such that $\lim_{n\to\infty} \lambda_n = \lambda_0 \in \Omega$.

Using arguments similar to Theorems 1.43 and 1.44 we can prove Theorems 1.47 and 1.48.

Theorem 1.47. *Let $T \in B_\infty(\Omega)$; then*

(i) $\sigma(T) = \sigma_W(T)$;

(ii) $\operatorname{ind}(\lambda - T) > 0$ *for all* $\lambda \in \rho_{s-F}(T) \supset \Omega$;

(iii) $\operatorname{ind}(\lambda - T) = \operatorname{nul}(\lambda - T) = \infty$ *for all* $\lambda \in \Omega$;

(iv) $\operatorname{nul}(\lambda - T)^* = 0$ *for all* $\lambda \in \mathbb{C}$.

Theorem 1.48. *If* $R \in \mathcal{L}(\mathcal{H})$ *satisfies*

(i) $\sigma(R)$ *and* $\sigma_W(R)$ *are connected and contain a connected open set* Ω;

(ii) $\operatorname{ind}(\lambda - R) \geq 0$ *for all* $\lambda \in \rho_{s-F}(R)$;

(iii) $\rho_{s-F}(R) \supset \Omega$ *and* $\operatorname{ind}(\lambda - R) = \infty$ *for all* $\lambda \in \Omega$;

then for $\epsilon > 0$, *there exists a compact operator* K_ϵ, $\|K_\epsilon\| < \epsilon$, *such that* $R - K_\epsilon \in B_\infty(\Omega)$.

The next theorem analyses the relation between $\mathcal{B}_n(\Omega)$ and $\mathcal{B}_1(\Omega)$.

Theorem 1.49. *Given* $B \in \mathcal{B}_n(\Omega)$ $(1 \leq n < \infty)$, *there exist* n $\mathcal{B}_1(\Omega)$ *operators* $B_1, B_2, \cdots, B_n$ *such that*

$$B = \begin{bmatrix} B_1 & & & * \\ & B_2 & & \\ & & \ddots & \\ & 0 & & B_n \end{bmatrix}$$

with respect to some decomposition $\mathcal{H} = \bigoplus_{i=1}^n \mathcal{H}_i$ *of the space* $\mathcal{H}$.

Proof. Since $\Omega \subset \rho_r(B)$, by Proposition 1.40 there exists $y \in \mathcal{H}$ such that $P_{\ker(\lambda - B)} y \neq 0$ for all $\lambda \in \sigma$, where σ is a perfect set in Ω. Set

$$\mathcal{M}_1 = \bigvee \{\ker(\lambda - B) \cap \{y\}^\perp : \lambda \in \sigma\};$$

then $\mathcal{M}_1$ is invariant under B and

$$B = \begin{bmatrix} B'_{n-1} & * \\ 0 & B_n \end{bmatrix} \begin{matrix} \mathcal{M}_1 \\ \mathcal{M}_1^\perp \end{matrix}$$

where $B'_{n-1} = B|_{\mathcal{M}_1}$ and $B_n = B_{\mathcal{M}_1^\perp}$. Since $\Omega \subset \rho_F(B)$, $\Omega \subset \rho(\pi(B))$. Thus $\Omega \subset \rho_{re}(B_n)$, $\Omega \subset \rho_{le}(B'_{n-1})$, and $\operatorname{ran}(\lambda - B_n)$ and $\operatorname{ran}(\lambda - B'_{n-1})$ are closed for $\lambda \in \Omega$. From the construction of $\mathcal{M}_1$ and $P_{\ker(\lambda - B)} y \neq 0$ $(\lambda \in \sigma)$, it is easily seen that $\operatorname{nul}(\lambda - B'_{n-1}) = n-1$ and $\operatorname{nul}(\lambda - B_n) = 1$. Thus $\Omega \subset \rho_{s-F}(B'_{n-1}) \cap \rho_{s-F}(B_n)$. By Theorem 1.43 (iv), $\sigma_p(B^*) = \varnothing$ and therefore $\sigma_p(B_n^*) = \varnothing$. This implies that $\operatorname{ind}(\lambda - B_n) = \operatorname{nul}(\lambda - B_n) = 1$ for $\lambda \in \Omega$. Since $\bigvee\{\ker(\lambda - B) : \lambda \in \Omega\} = \mathcal{H}$, $\bigvee\{\ker(\lambda - B_n) : \lambda \in \Omega\} = \mathcal{M}^\perp$. Thus $B_n \in \mathcal{B}_1(\Omega)$.

Computations show that $(\omega - B'_{n-1})\ker(\lambda - B)\cap\{y\}^\perp = \ker(\lambda - B)\cap\{y\}^\perp$ for all $\lambda \in \sigma$ and $\omega \in \Omega\backslash\sigma$. This implies that $\operatorname{ran}(\omega - B'_{n-1}) = \mathcal{M}_1$ for all $\omega \in \Omega\backslash\sigma$, i.e., $\Omega\backslash\sigma \subset \rho_r(B'_{n-1})$. Since $\partial\rho_r(B'_{n-1}) \subset \sigma_{lre}(B'_{n-1})$ and $\sigma_{lre}(B'_{n-1}) \cap \Omega = \varnothing$, $\Omega \subset \rho_r(B'_{n-1})$. Since $\bigvee\{\ker(\lambda - B)\cap\{y\}^\perp : \lambda \in \sigma\} = \mathcal{M}_1$, $\bigvee\{\ker(\lambda - B'_{n-1}) : \lambda \in \Omega\} = \mathcal{M}_1$. Thus $B'_{n-1} \in B_{n-1}(\Omega)$. Applying the same argument finitely many times, we get

$$B = \begin{bmatrix} B_1 & & & * \\ & B_2 & & \\ & & \ddots & \\ 0 & & & B_n \end{bmatrix},$$

where $B_i \in \mathcal{B}_1(\Omega)$ $(1 \le i \le n)$. $\square$

Let S^* be the backward shift given in Example 1.32. It is obvious that S^* is essentially normal. Let M be a normal operator with perfect spectrum $\sigma(M) \subset \partial\sigma(S^*)$ and $\epsilon > 0$; then by Corollary 1.30, there exist unitary operators U and $K \in \mathcal{K}(\mathcal{H})$ such that $\|K\| < \epsilon$ and

$$(S^* \oplus M + K) = U^* S U.$$

If the unitary equivalence is replaced by quasisimilarity, we have the following surprising result. (Interested reader can find the proof in J. Agler, E. Franks and D.A. Herrero [1].)

Theorem 1.50. *Let T_ϕ be an analytic Toeplitz operator on H^2, where $\phi \in H^\infty$ is a univalent analytic function, let M be a biquasitriangular operator, i.e.,* $\operatorname{ind}(\lambda - M) =$ *0 for all $\lambda \in \rho_{s-F}(M)$, such that $\sigma(M) = \sigma_e(M)$ and $\sigma(M) \cup \partial(\sigma(T^*_\phi))$ is connected, and let $\epsilon > 0$; then there exists $K \in \mathcal{K}(\mathcal{H})$, $\|K\| < \epsilon$, such that*

$$A = T^*_\phi \oplus M + K \text{ is quasisimilar to } T^*_\phi.$$

Two operators A, $B \in \mathcal{L}(\mathcal{H})$ are *quasisimilar* if there exist operators X, $Y \in \mathcal{L}(\mathcal{H})$ with trivial kernel and dense range such that

$$AX = XB \quad \text{and} \quad YA = BY.$$

Note that the S^* in Example 1.32 can be considered as T^*_z, thus if the T^*_ϕ in Theorem 1.50 is replaced by S^*, the conclusion is still valid.

To finish this section, we give a simpler and useful result.

Proposition 1.51. *Let $B \in \mathcal{B}_n(\Omega)$; then there exists a holomorphic $\mathcal{H}$-valued function $F(\lambda) = (e_1(\lambda), e_2(\lambda), \cdots, e_n(\lambda))$ on Ω such that $\{e_i(\lambda)\}_{i=1}^n$ is a basis of* $\ker(\lambda - B)$ $(\lambda \in \Omega)$.

Proof. Without loss of generality, we can assume that $0 \in \Omega$. Let P be the orthogonal projection from $\mathcal{H}$ onto $\ker B$. Since $\Omega \subset \rho_r(B)$, we can find $A \in \mathcal{L}(\mathcal{H})$ such that $AB = 1 - P$ and $BA = 1$. Set $\Delta = \{\lambda \in \mathbb{C} : |\lambda| < 1/\|A\|\}$. Thus if $\lambda \in \Delta$, $(1 - \lambda A)^{-1}$ exists and is analytic. Let $\{e_1, e_2, \cdots, e_n\}$ be an ONB of $\ker B$. Define

$$F(\lambda) = \{(1 - \lambda A)^{-1} e_1, (1 - \lambda A)^{-1} e_2, \cdots, (1 - \lambda A)^{-1} e_n\}.$$

Then $F(\lambda)$ is a holomorphic $\mathcal{H}$-valued function. It is easily seen that $\{e_i(\lambda)\}_{i=1}^n$ is a basis of $\ker(\lambda - B)$ $(\lambda \in \Delta)$, where $e_i(\lambda) = (1 - \lambda A)^{-1} e_i$ $(i = 1, 2, \cdots, n)$. By analytic continuation, $F(\lambda)$ can be extended to Ω. □

1.6 Similarity orbit theorem

In the theory of approximation of Hilbert space operators, one of the important problems is to characterize the norm-closure $\overline{R}$ of a subset R of $\mathcal{L}(\mathcal{H})$ invariant under similarity in "simple terms". Since the spectrum and its different parts are the most obvious similarity invariants of an operator, these "simple terms" will usually be expressed in terms of properties of the different subsets of the spectra of the operators in the closure of R.

Given $T \in \mathcal{L}(\mathcal{H})$, $S(T) := \{WTW^{-1} : W \text{ is invertible}\}$ is the *similarity orbit* of T. Let ρ be the faithful unital representation of $\mathcal{A}(\mathcal{H}) = \mathcal{L}(\mathcal{H})/\mathcal{K}(\mathcal{H})$ in $\mathcal{L}(\mathcal{H}_\rho)$ for the Hilbert space $\mathcal{H}_\rho$. If μ is an isolated point of $\sigma_e(T)$, then by the Riesz Theorem

$$\rho(\pi(T)) \sim (\mu + Q_\mu) \oplus R_\mu$$

where Q_μ is quasinilpotent and $\mu \notin \sigma(R_\mu)$.

For every $\lambda \in \mathbb{C}$ we define

$$K(\lambda, \pi(T)) = \begin{cases} 0, & \lambda \notin \sigma_e(T), \\ n, & \lambda \text{ is an isolated point of } \sigma_e(T) \\ & \text{and } Q_\lambda \text{ is nilpotent of order } n, \\ \infty, & \text{otherwise.} \end{cases}$$

A sufficient condition of the similarity orbit theorem is stated here without proof. The reader is referred to C.Apostol, L.A.Fialkow, D.A.Herrero and D.Voiculescu [1].

Theorem 1.52. *Given T, $A \in \mathcal{L}(\mathcal{H})$ satisfying*

(i) $\sigma_0(A) \subset \sigma_0(T)$, $\dim H(\lambda, A) = \dim H(\lambda, T)$ *for all* $\lambda \in \sigma_0(A)$;

(ii) *each component of* $\sigma_{lre}(A)$ *meets* $\sigma_e(T)$;

(iii) $\rho_{s-F}(A) \subset \rho_{s-F}(T)$, $\operatorname{ind}(\lambda - A) = \operatorname{ind}(\lambda - T)$, $\min\cdot\operatorname{ind}(\lambda - A)^k \geq \min\cdot\operatorname{ind}(\lambda - T)^k$ *for all* $\lambda \in \rho_{s-F}(A)$ *and* $k \geq 1$;

(iv) $K(\lambda, \pi(T)) = \infty$ *if* λ *is an isolated point of* $\sigma_e(T)$;

then $A \in S(T)^-$.

Example 1.53. *Let V be the example given in Example 3.22 below. Then* $S(V)^- = \{T \in \mathcal{L}(\mathcal{H}) : \sigma(T)$ *and* $\sigma_{lre}(T)$ *are connected,* $0 \in \sigma_{lre}(T)$ *and* $\operatorname{ind}(\lambda - T) = 0$ *for all* $\lambda \in \rho_{s-F}(T)\}$.

Proof. Suppose that $T \in S(V)^-$. Since $\sigma(V) = \{0\}$, by the upper semi-continuity of the spectrum, $\sigma(T)$ is connected and $0 \in \sigma(T)$. By the stability of the index of a semi-Fredholm operator under a small perturbation $\operatorname{ind}(\lambda - T) = 0$ for all $\lambda \in \rho_{s-F}(T) = \rho_F(T)$. In the Calkin algebra consider the upper semi-continuity of the spectrum; we get $\sigma_{lre}(T) = \sigma_e(T)$ is connected and $0 \in \sigma_{lre}(T)$.

Conversely, if $\sigma(T)$ and $\sigma_{lre}(T)$ are connected, $0 \in \sigma_{lre}(T)$ and $\operatorname{ind}(\lambda - T) = 0$ for all $\lambda \in \rho_{s-F}(T)$, then (i), (ii) and (iii) of Theorem 1.52 are satisfied. Since for each $k \in \mathbb{N}$, $V^k \notin \mathcal{K}(\mathcal{H})$, $K(0, \pi(V)) = \infty$. Thus $T \in S(V)^-$. □

1.7 Remarks

The operators τ_{AB} were first systematically studied by M. Rosenblum in [1]. Rosenblum's theorem has been proved there, except that the equality $\sigma(\tau_{AB}) = \sigma(A) - \sigma(B)$ is replaced by the weaker statement $\sigma(\tau_{AB}) \subset \sigma(A) - \sigma(B)$. The fact that this inclusion is actually an equality is due to D.C. Kleinecke (see G. Lumer and M. Rosenblum [1]). The reader is referred to Y.Q. Ji, C.L. Jiang and Z.Y. Wang [1], C. Davis and P. Rosenthal [1], H. Radjavi and P.Rosenthal [1], C. Rickart [1], S.K. Berberian [1] and L.A. Fialkow [1] for the other results in this chapter about the operator τ_{AB}. The research on the operator τ_{AB} is very rich. In this chapter we only chose part of it, most of which are chosen from D.A.Herrero [1].

1.2 and 1.3 are basically from Chapter 3 of D.A. Herrero [1]. The reader can find more details there.

Proposition 1.41 is due to C.L. Jiang [2]. Corollary 1.42 was first given by M.J. Cowen and R.G. Douglas [1]. Theorems 1.43 and 1.44 were contributed by

D.A. Herrero [2]. Theorem 1.50 appears in J. Agler, E. Franks and D.A. Herrero [1]. The results about $B_\infty(\Omega)$ are due to Y.Q. Ji, C.L. Jiang and Z.Y. Wang [1], [3]. Theorem 1.52 only gives a simplified version of the similarity orbit theorem, but it is enough for our purpose. The reader can find the general form of the theorem in C. Apostol, L.A. Fialkow, D.A. Herrero and D. Voiculescu [1]. The other results are from Chapter 3 of D.A. Herrero [1].

2. The irreducibility and strong irreducibility of operators

2.1 Irreducible operators

Let $\mathcal{A}$ be a C^*-algebra of bounded linear operators on a Hilbert space $\mathcal{H}$. If the commutant $\mathcal{A}'$ of $\mathcal{A}$ does not contain any non-trivial orthogonal projection, then $\mathcal{A}$ is *irreducible*. $T \in \mathcal{L}(\mathcal{H})$ is an *irreducible operator* if the commutant $\mathcal{A}'(T)$ of T is irreducible, or equivalently, there is no non-trivial reducing subspace of T (P.R. Halmos [1] [2]). Since irreducibility is an important unitary invariant, irreducible C^*-algebras and irreducible operators can be used as basic parts to describe some important properties of C^*-algebras and operators (D. Voiculescu [1]). Therefore, irreducibility plays an important role in operator theory and operator algebra.

Example 2.1. Let $T = \begin{bmatrix} \lambda_1 & 1 \\ 0 & \lambda_2 \end{bmatrix} \in \mathcal{L}(\mathbb{C}^2)$. If $\lambda_1 = \lambda_2$, T is a Jordan block. By the basic theory of linear algebra, T is irreducible if $\lambda_1 \neq \lambda_2$. Assume that $P = \begin{bmatrix} p_{11} & p_{12} \\ p_{21} & p_{22} \end{bmatrix}$ is an orthogonal projection commuting with T. Computation shows that $p_{21}\lambda_1 = \lambda_2 p_{21}$ and therefore $p_{21} = p_{12}^* = 0$ and $p_{11} = p_{22}$. Since $P^2 = P$, $p_{11} = p_{22} = 0$ or 1, i.e., P is trivial and T is irreducible.

Example 2.2. Let $\mathcal{H}$ be a Hilbert space and

$$T = \begin{bmatrix} \lambda_1 & 1 & D \\ 0 & \lambda_2 & 1 \\ 0 & 0 & \lambda_3 \end{bmatrix} \in \mathcal{L}(\mathcal{H}^{(3)}),$$

where $\{\lambda_i\}_{i=1}^3$ are pairwise distinct and D is irreducible.

Assume that

$$P = \begin{bmatrix} P_{11} & P_{12} & P_{13} \\ P_{21} & P_{22} & P_{23} \\ P_{31} & P_{32} & P_{33} \end{bmatrix} \in \mathcal{A}'(T)$$

is an orthogonal projection. By a similar argument to Example 2.2, we get $P = \bigoplus P_{ii}$, P_{ii} $(i = 1, 2, 3)$ is an orthogonal projection and $P_{11} = P_{22} = P_{33}$. Since $P_{11}D = DP_{33}$ and since D is irreducible, $P_{11} = P_{22} = P_{33} = 0$ or 1, i.e., T is irreducible.

Proposition 2.3. *Let $T \in \mathcal{L}(\mathcal{H})$; then the following are equivalent*

(i) *T is irreducible;*

(ii) *The von Neumann algebra $W(T)$ generated by T and 1 is $\mathcal{L}(\mathcal{H})$;*

(iii) *$\{T, T^*\}' = \mathbb{C}$, where $\{T, T^*\}'$ is the commutant of T and T^*.*

Proof. (i)⇒(ii). By the von Neumann double commutant theorem, it suffices to show that $[W(T)]' = \mathbb{C}$. Since T is irreducible, the only orthogonal projections contained in $[W(T)]'$ are 0 and 1. If $A \in [W(T)]'$ and $A \notin \mathbb{C}$, then $A = \alpha U$, where $\alpha \in \mathbb{C}$, U is a unitary operator. If $A = \alpha U$, $\sigma(U)$ contains more than one point. Let $E(\cdot)$ be the spectral projection of U; then there exists a subset σ of $\sigma(U)$ such that $E(\sigma) \neq 0, 1$. But $E(\sigma) \in [W(T)]'$, which contradicts the assumption that T is irreducible. Similarly, we can prove that $A^*A \notin \mathbb{C}$. Thus $[W(T)]' = \mathbb{C}$ and $W(T) = [W(T)]'' = \mathcal{L}(\mathcal{H})$.

(ii)⇒(iii) and (iii)⇒(i) are obvious. □

Whether every $\mathcal{C}^*$-algebra (operator) can be represented as an orthogonal direct sum of irreducible $\mathcal{C}^*$-algebras (operators, respectively) is one of the important topics of operator algebra and operator theory. Such a representation can be considered as a certain generalization of the Jordan theorem of linear algebra in separable Hilbert space operator theory. Since compact operators and $\mathcal{C}^*$-algebras generated by compact operators are the most direct generalization of matrices and matrix algebras in Hilbert space, they have an analogous Jordan irreducible representation theorem. Since the aim of this section is to analyse the irreducible representation of non-compact operators, we only state the following theorem without proof (W.B. Arveson [1]).

Theorem 2.4. *Every $\mathcal{C}^*$-subalgebra $\mathcal{A}$ of $\mathcal{K}(\mathcal{H})$ is unitarily equivalent to an orthogonal direct sum of at most countably many irreducible $\mathcal{C}^*$-algebras $\mathcal{A}_n$: $\mathcal{A} \simeq \bigoplus \mathcal{A}_n$. Furthermore, each $\mathcal{A}_n$ is the ideal of all compact operators on some Hilbert space $\mathcal{H}_n$.*

Theorem 2.5. *Let $T \in \mathcal{K}(\mathcal{H})$; then T is unitarily equivalent to an orthogonal direct sum of at most countably many irreducible operators.*

Proof. Let $\mathcal{A} \subset \mathcal{K}(\mathcal{H})$ be the $\mathcal{C}^*$-algebra generated by T and T^*; then $\mathcal{A} \simeq \bigoplus_{i=1}^{l} \mathcal{A}_i$ ($1 \leq l \leq \infty$), where $\mathcal{A}_i$ is an irreducible $\mathcal{C}^*$-algebra. Let the acting space of $\mathcal{A}_i$ be $\mathcal{H}_i$; then $T \simeq \bigoplus_{i=1}^{l} T_i$, where $T_i \in \mathcal{A}_i$. Since the algebra generated by T_i and T_i^* is $\mathcal{A}_i$, T_i is irreducible. □

Theorem 2.6. *Let $T \in \mathcal{L}(\mathcal{H})$ be essentially normal; then there is a decomposition of the space: $\mathcal{H} = \bigoplus_{i=0}^{l} \mathcal{H}_i$ ($1 \leq l \leq \infty$) such that*

(i) *each $\mathcal{H}_i$ reduces T, $T = \bigoplus_{i=0}^{l} T_i$;*

(ii) *$T_0 = T|_{\mathcal{H}_0}$ is normal;*

(iii) *For $n \geq 1$, $\mathcal{H}_n$ is infinite dimensional and $T_n = T|_{\mathcal{H}_n}$ is an irreducible essentially normal operator; furthermore, the decomposition is unique in the sense of unitary equivalence.*

Proof. Denote $\mathcal{A} = \mathcal{C}^*(T)$ and $\mathcal{F} = \mathcal{A} \cap \mathcal{K}(\mathcal{H})$; then $\mathcal{F}$ is an ideal of $\mathcal{A}$ and

$$\bigvee(\mathcal{F}\mathcal{H}) := \bigvee\{\operatorname{ran} A : A \in \mathcal{F}\}$$

is invariant under every operator in $\mathcal{A}$. Since $\mathcal{A}$ is a $\mathcal{C}^*$-algebra, $\mathcal{H}_0 = (\mathcal{F}\mathcal{H})^{\perp}$ reduces $\mathcal{A}$. Set $T_0 = T|_{\mathcal{H}_0}$. Since $(T^*T - TT^*)|_{\mathcal{H}_0} = T_0^*T_0 - T_0T_0^*$,

$$(T^*T - TT^*)\mathcal{H}_0 \subset \mathcal{H}_0.$$

But $T^*T - TT^* \in \mathcal{F}$ and therefore

$$(T^*T - TT^*)\mathcal{H}_0 \subset \bigvee(\mathcal{F}\mathcal{H}) = \mathcal{H}_0^{\perp}.$$

Thus, $(T^*T - TT^*)\mathcal{H}_0 = \{0\}$ and T_0 is normal.

Since $\mathcal{F}|_{\mathcal{H}_0^{\perp}} := \{A|_{\mathcal{H}_0^{\perp}} : A \in \mathcal{F}\}$ is a $\mathcal{C}^*$-algebra which consists of compact operators, by Theorem 2.4, $\mathcal{F}|_{\mathcal{H}_0^{\perp}} = U_1(\bigoplus \mathcal{K}(\mathcal{G}_n))U_1^* = \bigoplus \mathcal{F}_n$, where $\mathcal{K}(\mathcal{G}_n)$ is the ideal of compact operators on the Hilbert space $\mathcal{G}_n$, U_1 is a unitary operator and $\mathcal{F}_n = U_1\mathcal{K}(\mathcal{G}_n)U_1^*$. Set $U = \begin{bmatrix} 1 & 0 \\ 0 & U_1 \end{bmatrix} \begin{matrix} \mathcal{H}_0 \\ \mathcal{H}_0^{\perp} \end{matrix}$. Since $\mathcal{F}$ is an ideal and $(\mathcal{K}(\mathcal{G}_n)\mathcal{G}_n)^- = \mathcal{G}_n$, for each $A \in \mathcal{A}$,

$$AU\mathcal{G}_n = AU\mathcal{K}(\mathcal{G}_n)U^*U\mathcal{G}_n = A\mathcal{F}_nU\mathcal{G}_n \subset A\mathcal{F}U\mathcal{G}_n \subset \mathcal{F}U\mathcal{G}_n \subset U\mathcal{G}_n.$$

Thus $\mathcal{H}_n = U\mathcal{G}_n$ reduces $\mathcal{A}$ ($n = 1, 2, \cdots$).

Set $T_n = T|_{\mathcal{H}_n}$ ($n = 1, 2, \cdots$). Since the mapping $A \mapsto A|_{\mathcal{H}_n}$ is a *-isomorphism from $\mathcal{A}$ to $\mathcal{L}(\mathcal{H}_n)$, it has closed range. Therefore $\mathcal{A}_{\mathcal{H}_n}$ is a $\mathcal{C}^*$-subalgebra of $\mathcal{L}(\mathcal{H}_n)$.

For each polynomial $p(x, y)$, $p(T, T^*)|_{\mathcal{H}_n} = p(T_n, T_n^*)$ implies that $\mathcal{A}|\mathcal{H}_n = \mathcal{C}^*(T_n)$ and T_n is essentially normal. Since $\mathcal{C}^*(T_n) \supset \mathcal{F}_n = U\mathcal{K}(\mathcal{G}_n)U^*$, T_n is irreducible. $(n = 1, 2, \cdots)$.

By Theorem 2.5, the decomposition is unique. □

$T \in \mathcal{L}(\mathcal{H})$ is *m-multicyclic* if there are m non-zero vectors $x_1, x_2, \cdots, x_m$ such that

$$\bigvee\{T^k x_i : k \geq 0, i = 1, 2, \cdots, m\} = \mathcal{H}$$

and there are no $m - 1$ vectors satisfying this property. The famous Berger–Shaw theorem asserts that every hyponormal m-multicyclic operator is essentially normal. Thus we have

Corollary 2.7. *Every hyponormal m-multicyclic operator can be uniquely written in the sense of unitary equivalence as an orthogonal direct sum of a normal operator and irreducible operators.*

An operator $T \in \mathcal{L}(\mathcal{H})$ is called *completely non-normal* if it has no non-trivial reducing subspace $\mathcal{M}$ such that $T|_{\mathcal{M}}$ is normal.

Corollary 2.8. *Every completely non-normal operator can be uniquely written in the sense of unitary equivalence as an orthogonal direct sum of irreducible operators.*

It should be pointed out here that not every operator is a direct sum of irreducible operators. For instance, let M be a normal operator with empty point spectrum, i.e., $\sigma_p(M) = \varnothing$, and let $\mathcal{M}$ be a reducing subspace of M. Then $M|_{\mathcal{M}}$ is normal and $\sigma_p(M|_{\mathcal{M}}) = \varnothing$, therefore $M|_{\mathcal{M}}$ is still reducible.

On the other hand, the following two theorems indicate that $\mathcal{L}(\mathcal{H})$ contains a large number of irreducible operators.

Theorem 2.9. *The set of all irreducible operators is dense in $\mathcal{L}(\mathcal{H})$. Furthermore, given T in $\mathcal{L}(\mathcal{H})$, $p > 1$ and $0 < \epsilon < 1$, there exists K in $\mathcal{C}^p(\mathcal{H})$, $|K|_p < \epsilon$, such that $T + K$ is irreducible.*

Proof. Let $T = A + iB$ (Cartesian decomposition). According to the Weyl–von Neumann–Kuroda theorem (see, e.g., Theorem 2.3 of T. Kato [1]), there exists $K_1 \in \mathcal{C}^p(\mathcal{H})$, $|K_1| < \epsilon/3$, such that $A + K_1$ is a diagonal hermitian operator with respect to some $ONB\ \{e_n\}_{n=1}^\infty$ of $\mathcal{H}$. Let $\{\lambda_n\}_{n=1}^\infty$ be the sequence of corresponding eigenvalues, i.e., $(A+K_1)e_n = \lambda_n e_n$, $n = 1, 2, \cdots$. We can "slightly" modify the λ_n's and get a $K_2 \in \mathcal{C}^1(\mathcal{H})$, $|K_2|_1 < \epsilon/3$, such that K_2 is a hermitian diagonal operator (with respect to the same ONB), $(A + K_1 + K_2)e_n = \mu_n e_n$, $n = 1, 2, \cdots$, and the μ_n's are pairwise distinct.

Computation indicates that the commutant

$$\mathcal{A}'(A + K_1 + K_2) = \left\{C \in \mathcal{L}(\mathcal{H}) : C = \sum_{n=1}^{\infty} \gamma_n e_n \otimes e_n\right\}$$

is the set of all diagonal normal operators with respect to the basis. In particular, if P is an orthogonal projection commuting with $A + K_1 + K_2$, $P = \sum_{n=1}^{\infty} \delta_n e_n \otimes e_n$, where $\delta_n = 0$ or 1.

Let $B = (b_{jk})_{j,k=1}^{\infty}$ be the matrix representation of B with respect to the ONB $\{e_n\}_{n=1}^{\infty}$. Choose a sequence $\{l_k\}_{k=1}^{\infty}$ of positive numbers satisfying

(a) $2\sum_{k=1}^{\infty} l_k < \epsilon/3$;

(b) $b_{1k} + l_k \neq 0$, $k = 1, 2, \cdots$; $b_{k2} + l_k \neq 0$, $k = 2, 3, \cdots$.

Set

$$K_3 = \begin{bmatrix} l_1 & l_2 & l_3 & \cdots \\ l_2 & & & \\ l_3 & & 0 & \\ \vdots & & & \end{bmatrix} \begin{matrix} e_1 \\ e_2 \\ e_3 \\ \vdots \end{matrix}$$

Then K_3 is hermitian with $|K_3|_p < |K_3|_1 < \epsilon/3$.

Note that the Cartesian decomposition of $T + K_1 + K_2 + iK_3$ is $A + K_1 + K_2 + i(B + K_3)$. Let P be an orthogonal projection commuting with $T + K_1 + K_2 + iK_3$; then P commutes with $A + K_1 + K_2$ and $B + K_3$. Thus $P = \sum_{n=1}^{\infty} \delta_n e_n \otimes e_n$ with $\delta_n = 0$ or 1. Comparing the first columns and first rows of both sides of $P(B+K_3) = (B+K_3)P$, we get $\delta_n = \delta_1$ $(n = 2, 3, \cdots)$. Therefore, $T+K_1+K_2+iK_3$ is irreducible and $K = K_1 + K_2 + iK_3$ satisfies the requirements of the theorem. □

We will state the following theorem without proof. The reader can find the proof in P.R. Halmos [2] or D.A. Herrero [1].

Theorem 2.10. *The set of all irreducible operators is an F_σ dense subset of $\mathcal{L}(\mathcal{H})$.*

2.2 The similarity of irreducible operators

Given an $n \times n$ matrix $T \in \mathcal{L}(\mathbb{C}^n)$, the famous Jordan theorem asserts that T is similar to an orthogonal direct sum of finitely many Jordan blocks. This theorem has important applications in many mathematical fields, especially in the theory of ordinary differential equations. Let $\mathcal{H}$ be a complex, separable, infinite dimensional Hilbert space. It is reasonable to expect a Jordan decomposition theorem in $\mathcal{L}(\mathcal{H})$. A natural question is what the analogue of the Jordan blocks are in $\mathcal{L}(\mathcal{H})$. From the results of the last section, it seems reasonable to let irreducible operators play this role. Since the Jordan representations of matrices which are similar to each other are the same, we need to consider the similar transformations of irreducible operators. The irreducible operator $T = \begin{bmatrix} \lambda_1 & 1 \\ 0 & \lambda_2 \end{bmatrix}$ given in Example 2.1 is similar to $\lambda_1 \oplus \lambda_2$ when $\lambda_1 \neq \lambda_2$ by Corollary 1.7. In the infinite dimensional case, we have the following

Example 2.11. Let $\{e_n\}_{n=1}^{\infty}$ be an ONB of the Hilbert space $\mathcal{H}$. Set

$$\phi_i = a_{i1}e_1 + a_{i2}e_2 + \cdots + a_{ii}e_i, \qquad i = 1, 2, \cdots,$$

where $\{a_{ik} : k = 1, 2, \cdots, i;\ i = 1, 2, \cdots\}$ are chosen so that each $a_{ik} > 0$ and $\delta := \sum\limits_{i=1}^{\infty} \|\phi_i - e_i\|^2 < 1$. Thus $(\phi_i, e_j) \neq 0$, $(i \neq j)$ and

$$\bigvee\{\phi_i : 1 \leq i \leq n\} = \bigvee\{e_i : 1 \leq i \leq n\} \qquad n = 1, 2, \cdots.$$

Therefore, $\bigvee\{\phi_i : 1 \leq i < \infty\} = \mathcal{H}$.

Define $X : \mathcal{H} \to \mathcal{H}$ by

$$Xx = \sum_{i=1}^{\infty} \xi_i \phi_i \qquad \text{where } x = \sum_{i=1}^{\infty} \xi_i e_i.$$

Note that

$$\begin{aligned} \|Xx\| = \left\| \sum_{i=1}^{\infty} \xi_i \phi_i \right\| &\leq \left\| \sum_{i=1}^{\infty} \xi_i e_i \right\| + \left\| \sum_{i=1}^{\infty} \xi_i (\phi_i - e_i) \right\| \\ &\leq \|x\| + \Big(\sum_{i=1}^{\infty} |\xi_i|^2 \Big)^{1/2} \Big(\sum_{i=1}^{\infty} \|\phi_i - e_i\|^2 \Big)^{1/2} \\ &\leq (1 + \delta^{1/2}) \|x\|, \end{aligned}$$

and

$$\begin{aligned} \|Xx\| &\geq \left\| \sum_{i=1}^{\infty} \xi_i e_i \right\| - \left\| \sum_{i=1}^{\infty} \xi_i (\phi_i - e_i) \right\| \\ &\geq \|x\| - \Big(\sum_{i=1}^{\infty} |\xi_i|^2 \Big)^{1/2} \Big(\sum_{i=1}^{\infty} \|\phi_i - e_i\|^2 \Big)^{1/2} \\ &= (1 - \delta^{1/2}) \|x\|. \end{aligned}$$

Thus $X \in \mathcal{L}(\mathcal{H})$ is injective and has closed range. Since $\phi_n = Xe_n \in \operatorname{ran} X$ for all n, $\mathcal{H} = \bigvee\{\phi_n : n \geq 1\} \subset \operatorname{ran} X$, i.e., X is surjective and therefore is invertible.

Choose a sequence $\{\lambda_j\}_{j=1}^{\infty}$ of real numbers such that $\lambda_j \neq \lambda_k$ $(j \neq k)$ and $\sum_{j=1}^{\infty}|\lambda_j| < \infty$.

Define A, $N \in \mathcal{L}(\mathcal{H})$ by

$$Ax = \sum_{j=1}^{\infty} \lambda_j(X^{-1}x,\, e_j)\phi_j \quad \text{and} \quad Nx = \sum_{j=1}^{\infty} \lambda_j(x,\, e_j)e_j \quad \text{for } x \in \mathcal{H}.$$

Then it is easily seen that N is a diagonal self-adjoint operator and $XNX^{-1} = A$.

If a non-trivial subspace $\mathcal{M}$ reduces A, then

$$\sigma_p(A) = \sigma_p(A|_{\mathcal{M}}) \cup \sigma_p(A|_{\mathcal{M}^\perp}).$$

It is obvious that $\sigma_p(A) = \{\lambda_i\}_{i=1}^{\infty}$, $\operatorname{nul}(\lambda_i - A) = 1$ and the eigenvector corresponding to λ_i is ϕ_i, $(i = 1, 2, \cdots)$. Thus, $\sigma_p(A|_{\mathcal{M}})$ cannot contain the whole of $\{\lambda_i\}_{i=1}^{\infty}$, otherwise $\{\phi_i\}_{i=1}^{\infty} \subset \mathcal{M}$ and $\mathcal{M} \supset \bigvee\{\phi_i : i \geq 1\} = \mathcal{H}$. Hence there are λ_i and λ_j $(i \neq j)$ such that $\lambda_i \in \sigma_p(A|_{\mathcal{M}})$, $\lambda_j \in \sigma_p(A|_{\mathcal{M}^\perp})$ and $\phi_i \in \mathcal{M}$, $\phi_j \in \mathcal{M}^\perp$. This contradicts $(\phi_i,\, \phi_j) \neq 0$. Therefore the reducible operator N is similar to the irreducible operator A.

Theorem 2.12. *Let $N \in \mathcal{L}(\mathcal{H})$ be a normal operator. Then N is similar to an irreducible operator if and only if*

(i) *$\lambda - N$ is not finite rank for all $\lambda \in \mathcal{C}$;*

(ii) *N does not satisfy any quadratic equation $ax^2 + bx + c = 0$, where $|a| + |b| + |c| \neq 0$.*

In order to prove Theorem 2.12, we need several lemmas.

Lemma 2.13. *Assume that each operator in the similarity orbit $S(A)$ of $A \in \mathcal{L}(\mathcal{H})$ is reducible. Then $S(\mathcal{A}''(A))$ consists of reducible operators, where $S(\mathcal{A}''(A))$ denotes the similarity orbit of the double commutant $\mathcal{A}''(A)$ of A.*

Proof. Let $B \in \mathcal{A}''(A)$ and X be invertible. Since XAX^{-1} is reducible, XAX^{-1} commutes with some non-trivial orthogonal projection P, i.e., $P \in \mathcal{A}'(XAX^{-1})$. Thus $X^{-1}PX \in \mathcal{A}'(A)$ and therefore $BX^{-1}PX = X^{-1}PXB$ or $XBX^{-1}P = PXBX^{-1}$, i.e., XBX^{-1} is reducible. □

Lemma 2.14. *Let $A_i \in \mathcal{L}(\mathcal{H})$ $(i = 1,2,3)$ and $A = A_1 \oplus A_2 \oplus A_3$ such that* $\ker \tau_{A_i A_j} = \{0\}$ $(i \neq j)$; *then A is similar to an irreducible operator.*

Proof. By Lemma 2.13, it is sufficient to prove that there is some operator in $\mathcal{A}''(A)$ that is similar to an irreducible operator. Since $\ker \tau_{A_i A_j} = \{0\}$ $(i \neq j)$, calculations show that

$$\mathcal{A}'(A) = \{B_1 \oplus B_2 \oplus B_3 : B_i \in \mathcal{L}(\mathcal{H}),\ i = 1,2,3\}.$$

Therefore, it is sufficient to prove that $B = \alpha_1 \oplus \alpha_2 \oplus \alpha_3$ with pairwise distinct $\alpha_i \in \mathbb{C}$ is similar to an irreducible operator.

Define

$$T = \begin{bmatrix} \alpha_1 & 1 & D \\ 0 & \alpha_2 & 1 \\ 0 & 0 & \alpha_3 \end{bmatrix},$$

where D is an irreducible operator. By Example 2.2, T is irreducible. By Corollary 1.7, $T \sim \alpha_1 \oplus \alpha_2 \oplus \alpha_3$. □

Lemma 2.15. *Let $A_1 \in \mathcal{L}(\mathbb{C}^m)$ $(m < \infty)$ and $A_2, A_3 \in \mathcal{L}(\mathcal{H})$ such that* $\ker \tau_{A_i A_j} = \{0\}$, $(i \neq j)$; *then $A = A_1 \oplus A_2 \oplus A_3$ is similar to an irreducible operator.*

Proof. By Lemma 2.13, it is sufficient to prove that $B = \alpha_1 \oplus \alpha_2 \oplus \alpha_3$ is similar to an irreducible operator, where $\{\alpha_i\}_{i=1}^3$ are pairwise distinct. Without loss of generality, we assume that $\mathcal{H} = L^2(0,\,1)$.

Define

$$T = \begin{bmatrix} \alpha_1 & F & 0 \\ 0 & \alpha_2 & M \\ 0 & 0 & \alpha_3 \end{bmatrix},$$

where M is "multiplication by the independent variable t", i.e., $(Mf)(t) = t\,f(t)$, and F is the surjective mapping from $L^2(0,\,1)$ to $\mathbb{C}^m$ given by

$$Ff = \left(\int_0^1 t\,f(t)\,dt,\ \int_0^1 t^2\,f(t)\,dt,\ \cdots,\ \int_0^1 t^m\,f(t)\,dt \right), \quad \text{for } f \in L^2(0,\,1).$$

It is easily seen that $T \sim B$.

Assume that $P \in \mathcal{A}'(T)$ is an orthogonal projection. Since $\alpha_i \neq \alpha_j$ $(i \neq j)$, calculations indicate that $P = P_1 \oplus P_2 \oplus P_3$, where P_i is an orthogonal projection satisfying $P_2 M = M P_3$ and $P_1 F = F P_2$. Thus

$$P_2 M^2 = M P_3 M = M(MP_3)^* = M(P_2 M)^* = M^2 P_2.$$

Therefore, $P_2M^{2k} = M^{2k}P_2$ for all $k \geq 1$. Let $p_n(t)$ be a sequence of polynomials in t such that $p_n(t) \to \sqrt{t}$ uniformly in [0, 1]. Since M is a positive operator,

$$P_2M = \lim_{n\to\infty} P_2p_n(M^2) = \lim_{n\to\infty} p_n(M^2)P_2 = MP_2.$$

Hence $P_2 = P_3$. The spectral theory of self-adjoint operators asserts that there exists a Borel subset E of [0, 1] such that $P_2f = \chi_E f$ for all $f \in \mathcal{H}$, where χ_E is the characteristic function of E. If the Lebesgue measure $\mu(E)$ of E is 1, then $P_2 = P_3 = 1$. Since $P_1F = FP_2$ and F is surjective, $P_1 = 1$, i.e., $P = 1$. If $\mu(E) < 1$, let $\lambda_1, \lambda_2, \cdots, \lambda_m$ be non-zero, pairwise distinct Lebesgue points in $F = [0,1]\backslash E$ ($\lambda \in F$ is a Lebesgue point if $\lim_{\epsilon\to 0} \mu(F \cap [\lambda - \epsilon, \lambda + \epsilon])/2\epsilon = 1$). Since the matrix

$$\begin{bmatrix} \lambda_1 & \lambda_2 & \cdots & \lambda_m \\ \lambda_1^2 & \lambda_2^2 & \cdots & \lambda_m^2 \\ \vdots & \vdots & \ddots & \vdots \\ \lambda_1^m & \lambda_2^m & \cdots & \lambda_m^m \end{bmatrix}$$

is invertible, there is a positive number δ such that the matrix $(a_{ij})_{i,j=1}^m$ is invertible provided $|a_{ij} - \lambda_j^i| < \delta$ $(i, j = 1, 2, \cdots, m)$. Choose a sufficiently small $\epsilon > 0$ so that

$$\left| \frac{1}{2\epsilon} \int_{[\lambda_j-\epsilon,\lambda_j+\epsilon]\cap F} t^i \, dt - \lambda_j^i \right| < \delta, \quad i, j = 1, 2, \cdots, m.$$

Set $f_j = \frac{1}{2\epsilon}\chi_{[\lambda_j-\epsilon,\lambda_j+\epsilon]\cap F}$. Since $E \cap F = \varnothing$,

$$P_1Ff_j = FP_2f_j = F\chi_Ef_j = 0, \qquad (j = 1, 2, \cdots, m).$$

On the other hand, if $\{e_k\}_{k=1}^m$ is the ONB of $\mathbb{C}^m$, we have

$$\begin{aligned} 0 = P_1Ff_j &= P_1\left(\int_0^1 tf_j \, dt, \int_0^1 t^2 f_j \, dt, \cdots, \int_0^1 t^m f_j \, dt\right) \\ &= P_1\left[\sum_{k=1}^m \left(\frac{1}{2\epsilon}\int_{[\lambda_j-\epsilon,\lambda_j+\epsilon]\cap F} t^k \, dt\right) e_k\right] \\ &= \sum_{k=1}^m \left(\frac{1}{2\epsilon}\int_{[\lambda_j-\epsilon,\lambda_j+\epsilon]\cap F} t^k \, dt\right) P_1e_k. \end{aligned}$$

Denote $a_{ij} = \frac{1}{2\epsilon}\int_{[\lambda_j-\epsilon,\lambda_j+\epsilon]\cap F} t^i \, dt$; then the matrix $(a_{ij})_{i,j=1}^m$ is invertible. Thus $P_1e_k = 0$ $(k = 1, 2, \cdots, m)$, i.e., $P_1 = 0$. Since $FP_2 = P_1F$, $FP_2 = 0$. Note that the first component of FP_2e is $\int_0^1 t\chi_E \, dt$ where $e \in \mathcal{H}_2$, $e(t) \equiv 1$, therefore $\mu(E) = 0$, i.e., $P_2 = P_3 = 0$. Hence $P = 0$ and T is irreducible. □

Lemma 2.16. *Let $A \in \mathcal{L}(\mathcal{H})$ be a finite rank operator; then A is reducible.*

Proof. Since A, A^* are finite rank, $\dim \mathcal{M} < \infty$, where $\mathcal{M} = \bigvee\{\operatorname{ran} A, \operatorname{ran} A^*\}$. It is apparent that $\mathcal{M}$ deduces A. □

Lemma 2.17. *Assume that $Q \in \mathcal{L}(\mathcal{H})$ satisfies a quadratic equation; then Q is reducible.*

Proof. We prove the lemma only when $Q^2 = 0$. The reader can see P.R. Halmos [1] for the general case.

Since $Q^2 = 0$, there is a decomposition $\mathcal{H} = \mathcal{M}_1 \oplus \mathcal{M}_2 \oplus \mathcal{M}_3$ such that $\mathcal{M}_1 \oplus \mathcal{M}_2 = \ker Q$ and

$$Q = 0 \oplus \begin{bmatrix} 0 & A \\ 0 & 0 \end{bmatrix},$$

where $A \in \mathcal{L}(\mathcal{M}_3, \mathcal{M}_2)$ is injective and has dense range. Thus it suffices to assume that $Q = \begin{bmatrix} 0 & A \\ 0 & 0 \end{bmatrix}$. If at least one of $\dim \mathcal{M}_2$, $\dim \mathcal{M}_3$ is finite, Q is of finite rank. By Lemma 2.16, Q is reducible. Thus, without loss of generality, we can assume that $Q = \begin{bmatrix} 0 & A \\ 0 & 0 \end{bmatrix}$ is defined on $\mathcal{H} \oplus \mathcal{H}$ and $\dim \mathcal{H} = \infty$. Let $A = U_1 B$ be the polar decomposition of A, where B is positive and U_1 is a unitary operator. Then

$$\begin{bmatrix} 0 & B \\ 0 & 0 \end{bmatrix} = \begin{bmatrix} U_1^* & 0 \\ 0 & 1 \end{bmatrix} \begin{bmatrix} 0 & A \\ 0 & 0 \end{bmatrix} \begin{bmatrix} U_1 & 0 \\ 0 & 1 \end{bmatrix}.$$

Since unitary equivalence preserves reducibility, we can assume that A is a positive operator. By the spectral theory of self-adjoint operators, A is reducible, i.e., there is a non-trivial orthogonal projection P_1 such that P_1 commutes with A. Therefore the orthogonal projection $P = P_1 \oplus P_1$ commutes with Q and Q is reducible. □

Now we are in a position to prove Theorem 2.12.

Proof of Theorem 2.12. By Lemmas 2.16, 2.17 we can get the necessary condition.

If (i) and (ii) of Theorem 2.12 are satisfied, then there are three possibilities:

(a) $\sigma(N)$ consists of infinitely many points;

(b) $\sigma_e(N)$ contains at least three points;

(c) $\sigma(N)$ is a finite set, $\sigma_e(N)$ consists of two points and $\sigma(N) \backslash \sigma_e(N) \neq \varnothing$.

In cases (a) or (b), there exist pairwise disjoint Borel sets σ_1, σ_2 and σ_3 such that $\sigma(N) = \sigma_1 \cup \sigma_2 \cup \sigma_3$ and $\dim E(\sigma_i)\mathcal{H} = \infty$ $(i = 1, 2, 3)$, where $E(\cdot)$ is the spectral measure of N. Set $N_j = N|_{E(\sigma_j)\mathcal{H}}$; then we have $\ker \tau_{N_k N_j} = \{0\}$ $(k \neq j)$. By Lemma 2.14, N is similar to an irreducible operator.

In case (c), assume that $\sigma_e(N) = \{\lambda_1, \lambda_2\}$. Thus $\dim E(\{\lambda_j\})\mathcal{H} = \infty$ $(i = 1, 2)$ and $0 < \dim E(\sigma(E)\backslash\sigma_e(N)) < \infty$. By Lemma 2.15, N is similar to an irreducible operator. □

It is very difficult to get a necessary and sufficient condition for an operator to be similar to an irreducible operator. Here we have two sufficient conditions.

Proposition 2.18. *Given $T \in \mathcal{L}(\mathcal{H})$ such that $\mathcal{A}'(T)$ is abelian, then T is similar to an irreducible operator.*

Proof. (i) Suppose that there exists an orthogonal projection P in $\mathcal{A}'(T)$ such that $\dim P\mathcal{H} = \infty$ and $A = T|_{P\mathcal{H}}$ is irreducible. Fix a $\lambda \notin \sigma(A)$, then $\lambda \oplus A \in \mathcal{L}((P\mathcal{H})^\perp \oplus P\mathcal{H})$ is in $\mathcal{A}'(T)$. Since $\mathcal{A}'(T)$ is abelian, $\mathcal{A}'(T) = \mathcal{A}''(T)$. Thus $\lambda \oplus A \in \mathcal{A}''(T)$. Choose $D \in \mathcal{L}(P\mathcal{H}, (P\mathcal{H})^\perp)$ such that D is surjective. By Corollary 1.7, $\lambda \oplus A$ is similar to

$$B = \begin{bmatrix} \lambda & D \\ 0 & A \end{bmatrix} \begin{matrix} (P\mathcal{H})^\perp \\ P\mathcal{H}. \end{matrix}$$

By Lemma 2.13, it suffices to prove that B is irreducible. Let $P = \begin{bmatrix} P_{11} & P_{12} \\ P_{21} & P_{22} \end{bmatrix}$ be an orthogonal projection in $\mathcal{A}'(B)$. Calculation indicates that $\lambda P_{21} = AP_{21}$. Since $\lambda \notin \sigma(A)$, $P_{21} = P_{12} = 0$. Since A is irreducible, $P_{22} = 0$ or 1. $DP_{22} = P_{11}D$ and D is surjective imply that $P_{22} = P_{11} = 0$ or 1. Therefore B is irreducible.

(ii) If there does not exist any infinite rank orthogonal projection P in $\mathcal{A}'(T)$ such that $T|_{P\mathcal{H}}$ is irreducible, T is reducible. Therefore there exists an infinite rank orthogonal projection P_1 such that $P_1\mathcal{H}$ reduces T and $T|_{P_1\mathcal{H}}$ is reducible. Similarly, there is an orthogonal projection P_2 such that $P_2\mathcal{H} \subset P_1\mathcal{H}$, $\dim P_2\mathcal{H} = \infty$ and $P_2\mathcal{H}$ reduces $T|_{P_1\mathcal{H}}$. By induction, we get a sequence $\{E_1, E_2, \cdots\}$ of orthogonal projections such that each $E_i\mathcal{H}$ reduces T, E_i, E_j are orthogonal $(i \neq j)$, and $E = \sum_{i=1}^\infty E_i \in \mathcal{A}'(T)$.

Set $Q_1 = \sum_{k=1}^\infty E_{3k}$, $Q_2 = \sum_{k=1}^\infty E_{3k+1}$ and $Q_3 = (1 \ominus E) \oplus [E \ominus (Q_1 \oplus Q_2)]$; then $\dim \operatorname{ran} Q_i = \infty$, $Q_i\mathcal{H} = \mathcal{H}_i$ reduces T $(i = 1, 2, 3)$ and $T = T_1 \oplus T_2 \oplus T_3$, where $T_i = T|_{\mathcal{H}_i}$. Thus

$$F = \begin{bmatrix} \lambda_1 & & 0 \\ & \lambda_2 & \\ 0 & & \lambda_3 \end{bmatrix} \begin{matrix} \mathcal{H}_1 \\ \mathcal{H}_2 \\ \mathcal{H}_3 \end{matrix} \in \mathcal{A}'(T) = \mathcal{A}''(T),$$

where $\{\lambda_i\}_{i=1}^3$ are distinct numbers. By Lemma 2.14, F is similar to an irreducible operator. By Lemma 2.13, T is similar to an irreducible operator. □

Proposition 2.19. *Each Cowen–Douglas operator is similar to an irreducible operator.*

In order to prove Proposition 2.19, we need two lemmas.

Lemma 2.20. *Given $B \in \mathcal{B}_n(\Omega)$ and $\lambda \in \Omega$, there is a decomposition $\mathcal{H} = \bigoplus \mathcal{H}_k$ of the space such that*

$$B = \begin{bmatrix} \lambda & \overline{B}_{12} & \overline{B}_{13} & \cdots \\ & \lambda & \overline{B}_{23} & \cdots \\ & & \lambda & \cdots \\ & 0 & & \ddots \end{bmatrix} \tag{2.1}$$

with respect to the decomposition, where each $\mathcal{H}_k$ has dimension n and each $\overline{B}_{k\,k+1}$ is invertible. Furthermore, there is a unitary operator U such that

$$UBU^* = \begin{bmatrix} \lambda & B_{12} & B_{13} & \cdots \\ & \lambda & B_{23} & \cdots \\ & & \lambda & \cdots \\ & 0 & & \ddots \end{bmatrix} \begin{matrix} \mathcal{H}_1 \\ \mathcal{H}_2 \\ \mathcal{H}_3 \\ \vdots \end{matrix},$$

where each $B_{k\,k+1}$ is a positive invertible operator.

Proof. Without loss of generality, we assume that $\lambda = 0 \in \Omega$. Set

$$\mathcal{H}_1 = \ker B, \quad \mathcal{H}_2 = \ker B^2 \ominus \ker B, \quad \cdots, \quad \mathcal{H}_k = \ker B^k \ominus \ker B^{k-1}, \quad \cdots.$$

Then $\dim \mathcal{H}_k = n$, $\mathcal{H} = \bigoplus_{k=1}^\infty \mathcal{H}_k$ and B admits the representation

$$B = \begin{bmatrix} 0 & \overline{B}_{12} & & * \\ & 0 & \overline{B}_{23} & \\ & & 0 & \ddots \\ & 0 & & \ddots \end{bmatrix}.$$

Let $k_0 = \min\{k : \ker \overline{B}_{k\,k+1} \neq \{0\},\ k \geq 1\}$. Then there is a vector $x_{k_0+1} \neq 0$ such that $\overline{B}_{k_0 k_0+1} x_{k_0+1} = 0$. Note that $\mathcal{M} = \bigoplus_{k=1}^{k_0} \mathcal{H}_k = \ker B^{k_0}$, and $\dim \mathcal{M} = k_0 n$. But $y = (0, 0, \cdots, x_{k_0+1}, 0, \cdots) \in \ker B^{k_0}$ and $y \notin \mathcal{M}$. The contradiction implies that

$\overline{B}_{k\,k+1}$ is invertible for $k \geq 1$. Thus $\overline{B}_{12} = U_1 B'_{12}$, where U_1 is unitary and B'_{12} is positive and invertible. Similarly,

$$U_1\overline{B}_{23} = U_2 B'_{23}, \quad U_2\overline{B}_{34} = U_3 B'_{34}, \quad \cdots, \quad U_k\overline{B}_{k+1\,k+2} = U_{k+1}B'_{k+1\,k+2}, \quad \cdots,$$

where U_k is unitary, and $B'_{k\,k+1}$ is positive and invertible ($k \geq 1$).

Define $U = \bigoplus_{k=0}^{\infty} U_k$, where $U_0 = 1$. Then U is unitary and

$$UBU^* = \begin{bmatrix} 0 & B_{12} & & & * \\ & 0 & B_{23} & & \\ & & 0 & B_{34} & \\ & & & 0 & \ddots \\ & 0 & & & \ddots \end{bmatrix},$$

where $B_{k\,k+1} = U_k B'_{k\,k+1} U_k^*$ ($k \geq 1$) is positive and invertible. □

For simplicity, in what follows we say that each $\mathcal{B}_n(\Omega)$ operator B has the matrix representation (2.1).

Lemma 2.21. *Given $B \in \mathcal{B}_n(\Omega)$, if B_{1k_0} in the representation (2.1) is irreducible for some $k_0 \geq 2$, then B is irreducible.*

Proof. Let $P \in \mathcal{A}'(B)$ be an orthogonal projection that admits the matrix representation $P = (P_{ij})_{i,j=1}^{\infty}$ with respect to the decomposition of the space. For all x in $\mathcal{H}_1 = \ker(B-\lambda)$, since $0 = P(B-\lambda)x = (B-\lambda)Px$, $Px \in \mathcal{H}_1$. Thus $P_{k1} = 0\ (k \geq 2)$. Similarly, for all x in $\mathcal{H}_1 \oplus \mathcal{H}_2 = \ker(B-\lambda)^2$, since $0 = P(B-\lambda)^2 x = (B-\lambda)^2 Px$, $P_{k2} = 0$, $(k \geq 3)$, $\cdots$, $P_{kj} = 0$, $(k > j)$. $P = P^*$ implies that $P_{ij} = 0\ (i \neq j)$ and $P = \bigoplus_{i=1}^{\infty} P_{ii}$. Since $P_{kk}B_{k\,k+1} = B_{k\,k+1}P_{k+1\,k+1}$ and since $B_{k\,k+1}$ is positive and invertible, $P_{kk} = P_{k+1\,k+1}$ ($k = 1, 2, \cdots$). But we have $B_{1k_0}P_{k_0k_0} = P_{11}B_{1k_0}$. Thus B_{1k_0} is irreducible implies that $P_{kk} = 0$ or 1 ($k = 1, 2, \cdots$), i.e. $P = 0$ or 1 and therefore B is irreducible. □

Proof of Proposition 2.19. Without loss of generality, we can assume that $0 \in \Omega$. By Lemma 2.20,

$$B = \begin{bmatrix} 0 & B_{12} & & & * \\ & 0 & B_{23} & & \\ & & 0 & B_{34} & \\ & & & 0 & \ddots \\ & 0 & & & \ddots \end{bmatrix},$$

where $B_{k\,k+1}$ is invertible. Set $B_1 = (B_{12}, B_{13}, \cdots) \in \mathcal{L}(\mathcal{H}_1^\perp, \mathcal{H}_1)$ and

$$B_2 = \begin{bmatrix} 0 & B_{23} & & * \\ & 0 & B_{34} & \\ & & 0 & \\ & 0 & & \ddots \end{bmatrix} \in \mathcal{L}(\mathcal{H}_1, \mathcal{H}_1^\perp),$$

where $\mathcal{H}_k = \ker B^k \ominus \ker B^{k-1}$. Let J_n be $n \times n$ Jordan nilpotent. Then $J_n \in \mathcal{L}(\mathbb{C}^n)$ is irreducible. Set $E = B_{12}^{-1}(B_{13} - J_n)$ and

$$X_1 = \begin{bmatrix} 1 & E & 0 & \dots \\ & 1 & 0 & \dots \\ & & 1 & \\ & 0 & & \ddots \end{bmatrix} \in \mathcal{L}(\mathcal{H}_1^\perp, \mathcal{H}_1^\perp)$$

and $X = \begin{bmatrix} 1 & 0 \\ 0 & X_1 \end{bmatrix} \in \mathcal{L}(\mathcal{H})$. Then X is invertible and

$$\begin{aligned} XBX^{-1} &= \begin{bmatrix} 1 & 0 \\ 0 & X_1 \end{bmatrix} \begin{bmatrix} 0 & B_1 \\ 0 & B_2 \end{bmatrix} \begin{bmatrix} 1 & 0 \\ 0 & X_1^{-1} \end{bmatrix} \\ &= \begin{bmatrix} 0 & B_1 X_1^{-1} \\ 0 & X_1 B_2 X_1^{-1} \end{bmatrix} = \begin{bmatrix} 0 & B_{12} & J_n & \\ & 0 & B_{23} & \\ & & 0 & \ddots \\ & 0 & & \ddots \end{bmatrix}. \end{aligned}$$

Thus XBX^{-1} is irreducible by Lemma 2.21. □

Corollary 2.22. *Let T_ϕ be an analytic Toeplitz operator on H^2 with symbol $\phi \in H^\infty$ such that $\rho_F(T_\phi) \cap \sigma(T_\phi) \neq \varnothing$; then T_ϕ is similar to an irreducible operator.*

Proof. Since $\rho_F(T_\phi) \cap \sigma(T_\phi) \neq \varnothing$, we can find a connected open subset Ω of $\rho_F(T_\phi)$ such that $-\infty < \operatorname{ind}(T_\phi - \lambda) = -n < 0$ for $\lambda \in \Omega$ (R.G. Douglas [1]). Without loss of generality, we assume that $0 \in \Omega$. By the Factorization Theorem there is a finite Blaschke product $B(z)$ and an invertible function $f(z)$ such that $\phi(z) = B(z)f(z)$, where

$$B(z) = \prod_{i=1}^{n} \left[(z - \lambda_i)/(1 - \lambda_i^* z)\right], \quad |\lambda_i| < 1 \ \ (i = 1, 2, \cdots, n).$$

Therefore, T_f is invertible and T_B is a partial isometry. Note that T_{z^n} is also a partial isometry and $\operatorname{codim}\operatorname{ran} T_B = \operatorname{codim}\operatorname{ran} T_{z^n} = n$. Thus there exists a unitary operator U such that $UT_BU^* = T_{z^n}$. Set

$$A := UT_\phi U^* = UT_BU^*UT_fU^* = T_{z^n}UT_fU^*.$$

Then A satisfies $\operatorname{nul} A^* = n$ and $\bigvee\{\ker A^k : k \geq 1\} = H^2$. Thus $A^* \in \mathcal{B}_n(\Omega)$. By Proposition 2.19, A^*, and therefore T_ϕ, is similar to an irreducible operator. □

2.3 Strongly irreducible operators and examples

We know now that irreducibility is only unitary-invariant and not similarity-invariant. In the theory of $\mathcal{C}^*$-algebras or self-adjoint algebras, it is one of the core topics to discuss unitary-invariants and find representation theorems on unitary-invariants. But the Jordan Decomposition Theorem asserts that linear transformations can be classified up to similarity by their Jordan representations. Therefore, it seems that we should pay more attention to the study of similarity-invariants in the theory of non-self-adjoint algebras and the irreducible operators cannot play a role in $\mathcal{L}(\mathcal{H})$ similar to that which the Jordan blocks do in matrix theory. Naturally, people raise the concept of strongly irreducible operators.

Definition 2.23. An operator $T \in \mathcal{L}(\mathcal{H})$ is called *strongly irreducible* if XTX^{-1} is irreducible for all invertible operator X; otherwise T is *strongly reducible.*

From the definition, strongly irreducible operators are irreducible and strong irreducibility is invariant under similarity.

Let T be strongly reducible, i.e., there exists an invertible operator X and two non-trivial subspaces $\mathcal{H}_1$ and $\mathcal{H}_2$ such that $\mathcal{H} = \mathcal{H}_1 \oplus \mathcal{H}_2$ and $XTX^{-1}\mathcal{H}_i \subset \mathcal{H}_i$ $(i = 1, 2)$. Set $\mathcal{M} = X^{-1}\mathcal{H}_1$, $\mathcal{N} = X^{-1}\mathcal{H}_2$. Then $\mathcal{M}$ and $\mathcal{N}$ are subspaces of $\mathcal{H}$, $\mathcal{M} \cap \mathcal{N} = \{0\}$, $\mathcal{M} \dot{+} \mathcal{N} = \mathcal{H}$ and $T\mathcal{M} \subset \mathcal{M}$ and $T\mathcal{N} \subset \mathcal{N}$. Conversely, assume that T has two non-trivial invariant subspaces $\mathcal{M}$ and $\mathcal{N}$, such that $\mathcal{M} \cap \mathcal{N} = \{0\}$ and $\mathcal{M} \dot{+} \mathcal{N} = \mathcal{H}$. Set $\mathcal{K} = \mathcal{M} \oplus \mathcal{N}$: if $x, y \in \mathcal{M}$ (or $\mathcal{N}$), $(x, y)_\mathcal{K} = (x, y)_\mathcal{H}$, if $x \in \mathcal{M}$ and $y \in \mathcal{N}$, $(x, y)_\mathcal{K} = 0$. Then $\mathcal{K}$ is a Hilbert space. Define $X : \mathcal{H} \to \mathcal{K}$ by $X(x+y) = x \oplus y$, where $x \in \mathcal{M}$ and $y \in \mathcal{N}$. It is obvious that X is invertible and $\mathcal{M}$ and $\mathcal{N}$ are reducing subspaces of XTX^{-1}, i.e., T is strongly reducible. Therefore, Definition 2.23 is equivalent to the following definition.

Definition 2.24. An operator $T \in \mathcal{L}(\mathcal{H})$ is called *strongly irreducible* if there does not exist two non-trivial invariant subspaces $\mathcal{M}$ and $\mathcal{N}$ of T such that $\mathcal{M} \cap \mathcal{N} = \{0\}$ and $\mathcal{M} \dot{+} \mathcal{N} = \mathcal{H}$; otherwise, T is *strongly reducible.*

Proposition 2.25. *Let $T \in \mathcal{L}(\mathcal{H})$. Then the following are equivalent:*

(i) *T is strongly irreducible;*

(ii) *There does not exist a non-trivial idempotent P such that $PT = TP$;*

(iii) *T^* is strongly irreducible.*

Proof. (i)$\Rightarrow$(ii). If there is a non-trivial idempotent P such that $PT = TP$, set $\mathcal{M} = \operatorname{ran} P$ and $\mathcal{N} = \operatorname{ran}(1-P)$. Then $\mathcal{M}$ and $\mathcal{N}$ are non-trivial invariant subspaces of T, $\mathcal{M} \cap \mathcal{N} = \{0\}$ and $\mathcal{M} \dot{+} \mathcal{N} = \mathcal{H}$. This contradicts the assumption that T is strongly irreducible.

(ii)$\Rightarrow$(i). If T is strongly reducible, T has two non-trivial invariant subspaces $\mathcal{M}$ and $\mathcal{N}$ such that $\mathcal{M} \cap \mathcal{N} = \{0\}$ and $\mathcal{M} \dot{+} \mathcal{N} = \mathcal{H}$. Define P as the projection of $\mathcal{H}$ onto $\mathcal{M}$ along $\mathcal{N}$. Then $P^2 = P$, $P \neq 1$ or 0 and $PT = TP$. This contradicts (ii).

The equivalence between (ii) and (iii) is obvious. $\square$

Example 2.26. Each unicellular operator is strongly irreducible.

An operator $T \in \mathcal{L}(\mathcal{H})$ is a *unicellular operator* if the lattice $\operatorname{Lat} T$ of invariant subspaces of T is a totally ordered set.

Example 2.27. Let $\{e_n\}_{n=1}^{\infty}$ be an ONB of the Hilbert space $\mathcal{H}$ and let $\{\omega_n\}_{n=1}^{\infty}$ be a sequence of non-zero bounded complex numbers. The unilateral weighted shift $T \in \mathcal{L}(\mathcal{H})$ with weight $\{\omega_n\}_{n=1}^{\infty}$ is defined by $Te_n = \omega_n e_{n+1}$ $(n = 1, 2, \cdots)$. Computation shows that $\ker T^* = \bigvee\{e_1\}$. If $X \in \mathcal{L}(\mathcal{H})$ and $XT^* = T^*X$, then $T^*Xe_1 = XT^*e_1 = 0$, i.e., $Xe_1 = x_{11}e_1$ for some complex number x_{11}. Similarly, $T^{*2}Xe_2 = XT^{*2}e_2 = 0$ and $Xe_2 = x_{12}e_1 + x_{22}e_2$ for $x_{12}, x_{22} \in \mathbb{C}$. In general, $Xe_n = \sum_{i=1}^{n} x_{in}e_i$, i.e., X admits an upper triangular matrix representation with respect to the $ONB\,\{e_n\}$. Therefore, if P is an idempotent commuting with T, P admits a lower triangular matrix representation $P = (p_{ij})_{i,j=1}^{\infty}$, i.e., $p_{ij} = 0$ $(i < j)$. Since $P^2 = P$, $p_{ii} = 0$ or 1 $(i = 1, 2, \cdots)$. Assume that $p_{11} = 0$ (if $p_{11} = 1$, consider $1 - P$). Since $\omega_1 p_{22} = 0$ and $\omega_1 \neq 0$, $p_{22} = 0$. Since $P^2 = P$, $p_{21} = 0$. In general $p_{nn} = 0$ and $p_{ni} = 0$ $(i \leq n,\ n = 1, 2, \cdots)$, i.e., $P = 0$. Therefore, T is strongly irreducible.

Proposition 2.28. *Each operator in $\mathcal{B}_1(\Omega)$ is strongly irreducible.*

Proof. Let $B \in \mathcal{B}_1(\Omega)$ and without loss of generality, we can assume that $0 \in \Omega$. By

Lemma 2.20, B admits a matrix representation

$$B = \begin{bmatrix} 0 & b_{12} & & * \\ & 0 & b_{23} & \\ & & 0 & \ddots \\ & 0 & & \ddots \end{bmatrix}$$

with respect to some ONB $\{e_n\}_{n=1}^{\infty}$ of $\mathcal{H}$ and $|b_{k\,k+1}| > 0$ for all $k \geq 1$. Let P be an idempotent commuting with B. Since $0 = PBe_1 = BPe_1$, $Pe_1 \in \ker B$. Since $\operatorname{nul} B = 1$, $Pe_1 = p_{11}e_1$ for some $p_{11} \in \mathbb{C}$. Similarly, $0 = PB^2e_2 = B^2Pe_2$ and $\operatorname{nul} B^2 = 2$ imply that $Pe_2 = p_{12}e_1 + p_{22}e_2$. By induction, $Pe_n = \sum_{i=1}^{n} p_{in}e_n$, $(n = 1, 2, \cdots)$, i.e., P admits an upper triangular matrix representation with respect to the ONB $\{e_n\}$. By the same argument as Example 2.27, we can prove that B is strongly irreducible. □

Corollary 2.29. *Let $\phi \in H^\infty$ be a univalent analytic function; then T_ϕ is strongly irreducible.*

Proof. Assume that ϕ maps the unit disc D onto the bounded connected open domain Ω. Thus $\Omega \subset \rho_F(T_\phi) \cap \sigma(T_\phi)$ and $\operatorname{ind}(\lambda - T_\phi) = -1$ for all $\lambda \in \Omega$. Let $\phi(z) = \sum_{n=0}^{\infty} a_n z^n$. Note that $\{z^n : n \geq 0\}$ is an ONB of H^2, therefore T_ϕ admits a matrix representation

$$T_\phi = \begin{bmatrix} a_0 & & & 0 \\ a_1 & a_0 & & \\ a_2 & a_1 & a_0 & \\ \vdots & \ddots & \ddots & \ddots \end{bmatrix}$$

with respect to $\{z^n\}$. Note that T_ϕ^* has an upper triangular matrix representation, $\Omega^* \subset \rho_F(T_\phi^*)$ and $\operatorname{ind}(\lambda - T_\phi)^* = 1$ for all $\lambda \in \Omega$. Since $\sigma_p(T_\phi) = \varnothing$, $\operatorname{ran}(\lambda - T_\phi)^* = H^2$ for all $\lambda \in \Omega$. Since ϕ is a univalent analytic function, $a_1 \neq 0$. Therefore

$$\bigvee \{\ker(\lambda - T_\phi)^{*k} : k \geq 1, \lambda \in \Omega\} = H^2,$$

and $T_\phi^* \in \mathcal{B}_1(\Omega^*)$. By Propositions 2.28 and 2.25, T_ϕ is strongly irreducible. □

Let $\{e_n\}_{n=-\infty}^{\infty}$ be a two sided ONB of $\mathcal{H}$, and $\{\omega_n\}_{n=-\infty}^{\infty}$ be a sequence of non-zero bounded complex numbers. Then the bilateral weighted shift T with weight

$\{\omega_n\}$ is defined by $Te_n = \omega_n e_{n+1}$ $(n = 0, \pm 1, \cdots)$. Thus T admits the following matrix representation

$$T = \begin{bmatrix} \ddots & & & & & & \\ & \ddots & 0 & & & 0 & \\ & & \omega_{-2} & 0 & & & \\ & & & \omega_{-1} & 0 & & \\ & & & & \omega_0 & 0 & \\ & & 0 & & & \omega_1 & 0 \\ & & & & & & \ddots & \ddots \end{bmatrix} \begin{matrix} \vdots \\ e_{-2} \\ e_{-1} \\ e_0 \\ e_1 \\ e_2 \\ \vdots \end{matrix}$$

$$= \begin{bmatrix} B_1 & 0 \\ \omega_0 e_1 \otimes e_0 & B_2 \end{bmatrix} \simeq \begin{bmatrix} B_2 & \omega_0 e_1 \otimes e_0 \\ 0 & B_1 \end{bmatrix}$$

where

$$B_1 = \begin{bmatrix} 0 & \omega_{-1} & & 0 \\ & 0 & \omega_{-2} & \\ & & 0 & \ddots \\ & 0 & & \ddots \end{bmatrix} \begin{matrix} e_0 \\ e_{-1} \\ e_{-2} \\ \vdots \end{matrix}, \quad B_2 = \begin{bmatrix} 0 & & & 0 \\ \omega_1 & 0 & & \\ & \omega_2 & 0 & \\ & 0 & \ddots & \ddots \end{bmatrix} \begin{matrix} e_1 \\ e_2 \\ e_3 \\ \vdots \end{matrix}$$

Computations indicate that the spectral radii of B_1 and B_2 are

$$r(B_1) = \lim_k \sup_n \left(\prod_{i=0}^{k-1} |\omega_{-(n+i)}| \right)^{1/k},$$

$$r(B_2) = \lim_k \sup_n \left(\prod_{i=0}^{k-1} |\omega_{n+i}| \right)^{1/k}.$$

Set $r_1(B_1) = \lim_k \inf_n \left(\prod_{i=0}^{k-1} |\omega_{-(n+i)}| \right)^{1/k}$, $r_1(B_2) = \lim_k \inf_n \left(\prod_{i=0}^{k-1} |\omega_{n+i}| \right)^{1/k}$. Denote

$$\Delta(B_i) = \{z \in \mathbb{C} : |z| < r(B_i)\},$$
$$\Delta_1(B_i) = \{z \in \mathbb{C} : |z| < r_1(B_i)\} \qquad (i = 1, 2).$$

It is not difficult to see that

(a) If $r_1(B_1) > 0$, then $\Delta_1(B_1) \subset \rho_F(B_1) \cap \sigma(B_1)$, $\text{nul}\,(\lambda - B_1) = \text{ind}\,(\lambda - B_1) = 1$ for all $\lambda \in \Delta_1(B_1)$ and $B_1 \in \mathcal{B}_1(\Delta_1(B_1))$.

(b) If $r_1(B_2) > 0$, then $\Delta_1(B_2) \subset \rho_F(B_2) \cap \sigma(B_2)$, $\text{nul}\,(\lambda - B_2)^* = -\text{ind}\,(\lambda - B_2) = 1$ for all $\lambda \in \Delta_1(B_2)$ and $B_2^* \in \mathcal{B}_1(\Delta_1(B_2)^*)$.

(c) $P_{\ker(\lambda - B_1)} e_0 \neq 0$ for all $\lambda \in \Delta_1(B_1)$.

Proposition 2.30. *Let T be the bilateral weighted shift with weight $\{\omega_n\}_{n=-\infty}^{\infty}$. If $r_1(B_1) > r(B_2)$, then $T \in \mathcal{B}_1(\Delta_1(B_1) \backslash \sigma(B_2))$; if $r_1(B_2) > r(B_1)$, then $T^* \in \mathcal{B}_1(\Delta_1(B_2) \backslash \sigma(B_1))$.*

Proof. Assume that $r_1(B_1) > r(B_2)$ and denote $\Omega = \Delta_1(B_1) \backslash \sigma(B_2)$. Then

$$\Omega \subset \rho_F(B_2 \oplus B_1) \quad \text{and} \quad \text{ind}\,(\lambda - B_2 \oplus B_1) = \text{nul}\,(\lambda - B_2 \oplus B_1) = 1$$

for all $\lambda \in \Omega$. Since $T = B_2 \oplus B_1 + \omega_0 e_1 \otimes e_2$ and $\omega_0 e_1 \otimes e_2$ is compact, $\Omega \subset \rho_F(T)$ and $\text{ind}\,(\lambda - T) = 1$ for all $\lambda \in \Omega$. If $\lambda \in \Omega$ and $x = x_2 + x_1 \in \ker(\lambda - T)^*$, where $x_1 \in \mathcal{H}_1 := \bigvee\{e_n : n \leq 0\}$ and $x_2 \in \mathcal{H}_2 := \bigvee\{e_n : n \geq 1\}$, since $(\lambda - B_2)^*$ is invertible, $x_2 = 0$. $(\lambda - B_1) \in \mathcal{B}_1(\Delta_1(B_1))$ implies that $x_1 = 0$. Therefore $\text{ind}\,(\lambda - T) = \text{nul}\,(\lambda - T) = 1$ for all $\lambda \in \Omega$. Now it suffices to prove

$$\mathcal{H} = \bigvee\{\ker(\lambda - T) : \lambda \in \Omega\}$$

in order to show $T \in \mathcal{B}_1(\Omega)$. By Proposition 1.51, there is a non-zero $\mathcal{H}_1$-valued analytic function $f(\lambda)$ on $\Delta_1(B_1)$ such that $(\lambda - B_1)f(\lambda) = 0$, $\lambda \in \Delta_1(B_1)$.

Set $g(\lambda) = \omega_0(f(\lambda), e_0)(\lambda - B_2)^{-1}e_1$, $\lambda \in \Omega$; then $g(\lambda)$ is a vector valued analytic function on Ω and $(\lambda - T)(g(\lambda) \oplus f(\lambda)) = 0$, for all $\lambda \in \Omega$. Thus it suffices to prove that

$$\mathcal{M} = \bigvee\{g(\lambda) \oplus f(\lambda) : \lambda \in \Omega\} = \mathcal{H}.$$

Assume that $y \in \mathcal{H} \ominus \mathcal{M}$, and $y = \phi + \psi$, where $\phi \in \mathcal{H}_2$ and $\psi \in \mathcal{H}_1$. We have

$$0 = \big(g(\lambda) \oplus f(\lambda), y\big) = \big(g(\lambda), \phi\big) + \big(f(\lambda), \psi\big)$$

or

$$\begin{aligned}\big(f(\lambda), \psi\big) &= -\big(g(\lambda), \phi\big) \\ &= -\omega_0\big(f(\lambda), e_0\big)\big((\lambda - B_2)^{-1}e_1, \phi\big).\end{aligned}$$

Since $P_{\ker(\lambda - B_1)} e_0 \neq 0$ for all $\lambda \in \Delta_1(B_1)$, $(f(\lambda), e_0) \neq 0$ for all $\lambda \in \Omega$. Thus,

$$(f(\lambda), \psi)/(f(\lambda), e_0) = -\omega_0((\lambda - B_2)^{-1} e_1, \phi) \text{ for all } \lambda \in \Omega.$$

Note that $(f(\lambda), \psi)/(f(\lambda), e_0)$ is analytic on $\Delta_1(B_1)$ and $((\lambda - B_2)^{-1} e_1, \phi)$ is analytic out of $\sigma(B_2)$. Hence both of them are analytic on the whole plane $\mathbb{C}$. Since $((\lambda - B_2)^{-1} e_1, \phi) \to 0$ $(\lambda \to \infty)$, by Liouville's theorem they identically equal zero. Since $B_1 \in \mathcal{B}_1(\Delta_1(B_1)) \subset \mathcal{B}_1(\Omega)$, $\bigvee\{f(\lambda) : \lambda \in \Omega\} = \mathcal{H}_1$, therefore $\psi = 0$. When $|\lambda| > r(B_2)$,

$$\begin{aligned} 0 &= ((\lambda - B_2)^{-1} e_1, \phi) \\ &= \left(\sum_{n=0}^{\infty} \lambda^{-(n+1)} B_2^n e_1, \phi \right) \\ &= \sum_{n=0}^{\infty} (B_2^n e_1, \phi) \lambda^{-(n+1)}, \end{aligned}$$

thus $(B_2^n e_1, \phi) = 0$, $n = 0, 1, 2, \cdots$. Since e_1 is a cyclic vector of B_2, $\phi = 0$. Therefore $y = 0$ and $\mathcal{M} = \mathcal{H}$.

The second part of the proposition can be proved similarly. □

We give the following proposition without proof. The interested reader can refer to R.L. Kelley [1].

Proposition 2.31. *Every bilateral weighted shift T with an annulus spectrum is strongly irreducible.*

Proposition 2.32. *Given A, $B \in \mathcal{L}(\mathcal{H})$ such that $A \in \mathcal{B}_1(\Omega)$, B has a cyclic vector x and $[\sigma(B)]^\wedge$, the polynomially convex hull of $\sigma(B)$, is contained in Ω; there exists a vector $e_0 \in \mathcal{H}$ such that*

$$T = \begin{bmatrix} B & x \otimes e_0 \\ 0 & A \end{bmatrix} \begin{matrix} \mathcal{H} \\ \mathcal{H} \end{matrix} \in \mathcal{B}_1(\Omega \backslash [\sigma(B)]^\wedge).$$

Proof. Choose a connected open set Ω_0 such that

$$[\sigma(B)]^\wedge) \subset \Omega_0 \subset \overline{\Omega}_0 \subset \Omega.$$

By Proposition 1.40, there exists a vector $e_0 \in \mathcal{H}$ such that $P_{\ker(\lambda - A)} e_0 \neq 0$ for all $\lambda \in \overline{\Omega}_0$. Then by the same arguments as Proposition 2.30, we can prove the proposition. □

More bilateral weighted shifts will be analysed in Chapter 3.

Now, let $\Gamma = \partial\Omega$, where Ω is an analytic Cauchy domain and let $L^2(\Gamma)$ be the Hilbert space of complex functions on Γ which are square integrable with respect to $(1/2\pi)$-times the arc-length measure on Γ; $M_\lambda(\Gamma)$ will stand for the "multiplication by λ" operator acting on $L^2(\Gamma)$. The subspace $H^2(\Gamma)$ spanned by the rational functions with poles outside $\overline{\Omega}$ is invariant under $M_\lambda(\Gamma)$. Thus $M_\lambda(\Gamma)$ admits the following block matrix representation

$$M_\lambda(\Gamma) = \begin{bmatrix} M_+(\Gamma) & Z \\ 0 & M_-(\Gamma) \end{bmatrix} \begin{matrix} H^2(\Gamma) \\ L^2(\Gamma) \ominus H^2(\Gamma). \end{matrix}$$

If $\Omega = D$, the unit disc, the Beurling Theorem asserts that every invariant subspace H_1 of the unilateral shift $S\,\big(= M_+(\partial D)\big)$ on $H^2\,\big(= H^2(\partial D)\big)$ is of the form $H_1 = \phi H^2$ for some inner function ϕ. Thus the intersection $H_1 \cap H_2$ of two arbitrary non-trivial invariant subspaces H_1 and H_2 of S is non-trivial. This implies that S has some irreducibility stronger than strong irreducibility. In fact, we have

Proposition 2.33. *Let Ω be an analytic Cauchy domain, H_1 and H_2 be two non-trivial invariant subspaces of $M_+(\Gamma)$ on $H^2(\Gamma)$; then $H_1 \cap H_2 \neq \{0\}$.*

Proof. Fix a $\lambda \in \Omega$. Since $(M_+(\Gamma)-\lambda)H^2(\Gamma)$ is closed with codimension 1, $(M_+(\Gamma)-\lambda)\mathcal{H}_1$ is closed. It is obvious that $(M_+(\Gamma)-\lambda)\mathcal{H}_1$ is an invariant subspace of $M_+(\Gamma)$ contained in $\mathcal{H}_1$. Thus $(M_+(\Gamma)-\lambda)\mathcal{H}_1 \neq \mathcal{H}_1$. In fact, if $(M_+(\Gamma)-\lambda)\mathcal{H}_1 = \mathcal{H}_1$, then for each $f \in \mathcal{H}_1$ and each $n \in \mathbb{N}$, $f(z) = (z-\lambda)^n g_n(z)$ for $g_n \in \mathcal{H}_1$. This implies that $f^{(n)}(\lambda) = 0$ $(n = 0, 1, 2, \cdots)$ and $f = 0$, a contradiction.

Pick a non-zero function $\phi \in \mathcal{H}_1 \ominus [(M_+(\Gamma)-\lambda)\mathcal{H}_1]$. For each $h \in R(\Gamma)$,

$$0 = \big((M_+(\Gamma)-\lambda)\,h(M_+(\Gamma))\phi,\, \phi\big) = \int_\Gamma (z-\lambda)\,h(z)|\phi(z)|^2\,d\mu,$$

where $R(\Gamma)$ is the set of rational functions with poles outside $\overline{\Omega}$. Thus, as an element of $L^2(\Gamma)$, $|\phi|^2$ is orthogonal to $\{g \in \operatorname{Re} R(\Gamma) : g(\lambda) = 0\}$ $(\subset \operatorname{Re} L^2(\Gamma))$. It is known (S.D. Fisher [1]) that the orthogonal complement of $\operatorname{Re} R(\Gamma)$ in $\operatorname{Re} L^2(\Gamma)$ consists of linear combinations of finitely many C^∞ functions, say $Q_1, \cdots, Q_m$. It is clear that the constant functions are orthogonal to $\{g \in \operatorname{Re} R(\Gamma) : g(\lambda) = 0\}$. Therefore $|\phi|^2$ is a linear combination of $Q_1, \cdots, Q_m$ and 1 and $\sup_{z\in\Gamma} |\phi(z)| < \infty$. For each $h \in R(\Gamma)$, we have

$$\phi h = h(M_+(\Gamma))\phi \in \mathcal{H}_1.$$

Since $R(\Gamma)$ is dense in $H^2(\Gamma)$ and since ϕ is bounded, $\phi H^2(\Gamma) \subset \mathcal{H}_1$.

Similarly, we can obtain a non-zero bounded function $\psi \in \mathcal{H}_2$ such that $\psi H^2(\Gamma) \subset \mathcal{H}_2$. Therefore, $H_1 \cap H_2 \supset \phi\psi H^2(\Gamma) \neq \{0\}$. □

2.4 The basic properties of strongly irreducible operators

Proposition 2.34. *Let $T \in \mathcal{L}(\mathcal{H})$ be strongly irreducible. Then*

(i) *$\sigma(T)$ is connected;*

(ii) *$\lambda - T$ is not of finite rank for all $\lambda \in \mathbb{C}$;*

(iii) *T is not an algebraic operator, i.e., there is no non-zero polynomial $p(z)$ such that $p(T) = 0$;*

(iv) *$\rho_{s-F}^{s}(T) = \varnothing$.*

Proof. By the Riesz decomposition theorem we can prove (i). By Lemma 2.16 we get (ii).

(iii) Assume that $p(T) = 0$ for some non-zero polynomial $p(z)$. By the spectral mapping theorem $p(\sigma(T)) = \{0\}$. Thus $\sigma(T)$ consists of finitely many numbers. Since $\sigma(T)$ is connected, $\sigma(T) = \{\lambda_0\}$. Without loss of generality, we can assume that $\lambda_0 = 0$ and $p(z) = z^n q(z)$, where $q(z)$ is a polynomial such that $q(0) \neq 0$. Since $q(T)$ is invertible, $T^n = 0$. Let $k = \min\{m : T^m = 0\}$ and pick $x \in \mathcal{H} \ominus \ker T^{k-1}$. Set $\mathcal{M} = \bigvee\{T^j x : j \geq 0\}$. Then $\dim \mathcal{M} = k$ and $\mathcal{M}$ is invariant under T. It is easily seen that $T|_{\mathcal{M}} \sim q_k$, where q_k is $k \times k$ Jordan nilpotent.

Set

$$\mathcal{M}_j = \mathcal{M} \cap \ker T^j = \bigvee\{T^{k-1}x,\ T^{k-2}x, \cdots, T^{k-j}x\}, \quad (j = 1, 2, \cdots, k).$$

We will prove by induction that there exists a sequence $\{\mathcal{N}_j\}_{j=1}^{k}$ of invariant subspaces of T such that

(a) $\mathcal{N}_j \subset \ker T^j$;

(b) $\mathcal{N}_j \cap \mathcal{M}_j = \{0\}$;

(c) $\mathcal{N}_j \dot{+} \mathcal{M}_j = \ker T^j$, $(j = 1, 2, \cdots, k)$.

If $j = 1$, set $\mathcal{N}_1 = \ker T \ominus \mathcal{M}_1$.

Assume that $\mathcal{N}_j$, $j < k$, is defined and satisfies (a),(b) and (c). Define

$$\mathcal{N}_{j+1} = \left[T^{-1}(\mathcal{N}_j) \ominus (\mathcal{N}_j + \mathcal{M}_1)\right] \oplus \mathcal{N}_j,$$

where $T^{-1}(\mathcal{N}_j) = \{z \in \mathcal{H} : Tz \in \mathcal{N}\}$. It is obvious that $T^{-1}(\mathcal{N}_j)$ and therefore $\mathcal{N}_{j+1}$ is contained in $\ker T^{j+1}$. Computation shows that $T^{-1}(\mathcal{N}_j) = \mathcal{N}_{j+1} + \mathcal{M}_1$, and $\mathcal{N}_{j+1}$, $\mathcal{M}_{j+1}$ are invariant subspaces of T and $T\mathcal{N}_{j+1} \subset \mathcal{N}_j$, $T\mathcal{M}_{j+1} \subset \mathcal{M}_j$.

Assume that $y \in \mathcal{N}_{j+1} \cap \mathcal{M}_{j+1}$. Thus

$$y = \alpha_1 T^{k-1}x + \alpha_2 T^{k-2}x + \cdots + \alpha_{j+1}T^{k-j-1}x$$

for some complex numbers $\{\alpha_i\}_{i=1}^{j+1}$. Since

$$Ty = \alpha_2 T^{k-1}x + \cdots + \alpha_{j+1}T^{k-j}x \in \mathcal{M}_j \cap \mathcal{N}_j,$$

$Ty = 0$ by the inductive assumption. Since $\{T^{k-i}x\}_{i=1}^{j}$ are linearly independent, $\alpha_i = 0$, $(i = 2, 3, \cdots, j+1)$, i.e., $y = \alpha_1 T^{k-1}x \in \mathcal{M}_1$. On the other hand,

$$y \in \mathcal{N}_{j+1} = \left[T^{-1}(\mathcal{N}_j) \ominus (\mathcal{N}_j + \mathcal{M}_1)\right] \oplus \mathcal{N}_j,$$

thus $y \in \mathcal{N}_j$ and therefore $y \in \mathcal{N}_j \cap \mathcal{M}_j = \{0\}$. Hence $\mathcal{N}_{j+1} \cap \mathcal{M}_{j+1} = \{0\}$.

Let $z \in \ker T^{j+1}$. Then $Tz = y_j + z_j$ by the inductive assumption, where $y_j \in \mathcal{M}_j$, $z_j \in \mathcal{N}_j$. Since $T|_{\mathcal{M}} \sim q_k$, there exists $y_{j+1} \in \mathcal{M}_{j+1}$ such that $y_j = Ty_{j+1}$. Therefore $z - y_{j+1} \in T^{-1}(\mathcal{N}_j)$ or $z \in \mathcal{M}_{j+1} + \mathcal{N}_{j+1} + \mathcal{M}_1 = \mathcal{M}_{j+1} + \mathcal{N}_{j+1}$. This implies that $\mathcal{N}_{j+1} \dot{+} \mathcal{M}_{j+1} = \ker T^{j+1}$.

When $j = k$, we get $\mathcal{H} = \mathcal{N}_k \dot{+} \mathcal{M}_k$ and $\mathcal{N}_k$, $\mathcal{M}_k$ are invariant under T. This contradicts the strong irreducibility of T.

(iv) If $\lambda \in \rho_{s-F}^{s}(T)$, by Theorem 1.19,

$$T = \begin{bmatrix} T_r & * & * \\ 0 & T_0 & * \\ 0 & 0 & T_l \end{bmatrix}$$

with respect to $\mathcal{H} = \mathcal{H}_r(T) \oplus \mathcal{H}_0(T) \oplus \mathcal{H}_l(T)$, and $\lambda \in \sigma_0(T_0)$. The Riesz decomposition theorem asserts that $T_0 = \begin{bmatrix} A & T_{23} \\ 0 & B \end{bmatrix}$, where A is on a finite dimensional subspace, $\sigma(A) = \{\lambda\}$ and $\lambda \notin \sigma(B)$. Thus

$$T = \begin{bmatrix} T_r & T_{12} & T_{13} & T_{14} \\ 0 & A & T_{23} & T_{24} \\ 0 & 0 & B & T_{34} \\ 0 & 0 & 0 & T_l \end{bmatrix} = \begin{bmatrix} T_r & T_{12} & B_{13} \\ 0 & A & B_{23} \\ 0 & 0 & C_l \end{bmatrix},$$

where $C_l = \begin{bmatrix} B & T_{34} \\ 0 & T_l \end{bmatrix}$, $B_{13} = (T_{13}, T_{14})$ and $B_{23} = (T_{23}, T_{24})$. By Theorem 1.19, $\lambda \in \rho_r(T_r) \cap \rho_l(T_l)$. Thus

$$\sigma_r(T_r) \cap \sigma_l(A) = \sigma_r(A) \cap \sigma_l(C_l) = \varnothing.$$

By Theorem 1.11 there exist operators X, Y such that $T_rX - XA = T_{12}$ and $AY - YC_l = B_{23}$, whence we have

$$T \sim \begin{bmatrix} 1 & 0 & 0 \\ 0 & 1 & Y \\ 0 & 0 & 1 \end{bmatrix} \begin{bmatrix} 1 & X & 0 \\ 0 & 1 & 0 \\ 0 & 0 & 1 \end{bmatrix} \begin{bmatrix} T_r & T_{12} & B_{13} \\ 0 & A & B_{23} \\ 0 & 0 & C_l \end{bmatrix} \left\{ \begin{bmatrix} 1 & 0 & 0 \\ 0 & 1 & Y \\ 0 & 0 & 1 \end{bmatrix} \begin{bmatrix} 1 & X & 0 \\ 0 & 1 & 0 \\ 0 & 0 & 1 \end{bmatrix} \right\}^{-1}$$

$$= \begin{bmatrix} T_r & 0 & * \\ 0 & A & 0 \\ 0 & 0 & C_l \end{bmatrix} \simeq A \oplus \begin{bmatrix} T_r & * \\ 0 & C_l \end{bmatrix},$$

which contradicts the assumption. □

It is obvious that similarity preserves strong irreducibility. But the next example indicates that quasisimilarity does not preserve it.

Example 2.35. Let S be the unilateral shift on $\mathcal{H}$. Let

$$A = \begin{bmatrix} 1+S^* & 1 \\ 0 & -(1+S^*) \end{bmatrix} \quad \text{and} \quad B = \begin{bmatrix} 1+S^* & 0 \\ 0 & -(1+S^*) \end{bmatrix}$$

on $\mathcal{H} \oplus \mathcal{H}$. Set

$$X = \begin{bmatrix} 2(1+S^*) & 1 \\ 0 & 1 \end{bmatrix} \quad \text{and} \quad Y = \begin{bmatrix} 1 & 1 \\ 0 & -2(1+S^*) \end{bmatrix}.$$

Then we have $XA = BX$ and $AY = YB$. Since $\ker(1+S^*) = \{0\}$ and $\operatorname{ran}(1+S^*)^- = \mathcal{H}$, it is easily seen that X and Y have trivial kernel and dense range, i.e., A is quasisimilar to B. If $P = \begin{bmatrix} P_{11} & P_{12} \\ P_{21} & P_{22} \end{bmatrix}$ is an idempotent commuting with A, then we have $P_{21}(1+S^*) = -(1+S^*)P_{21}$ or $P_{21}S^* = -(2+S^*)P_{21}$. Note that $\sigma_p(S^*) \cap \sigma_p(2+S^*) = \varnothing$ and $S^* \in \mathcal{B}_1(D)$, so by Proposition 1.14, $P_{21} = 0$. Thus P_{11} and P_{22} are idempotents commuting with S^*. Since S^* is strongly irreducible by Proposition 2.28, P_{11} and P_{22} are 0 or 1. Assume that $P_{11} = 0$ (otherwise consider $1-P$), since $PA = AP$, $P_{12}(1+S^*) + (1+S^*)P_{12} = -P_{22}$. If $P_{22} = 1$, we have $P_{12}S^* + S^*P_{12} + 2P_{12} = -1$.

Let $\{e_n\}_{n=1}^{\infty}$ be the ONB of $\mathcal{H}$ such that $Se_n = e_{n+1}, (n = 1, 2, \cdots)$; then calculation indicates that

$$P_{12}e_n = -\frac{1}{2}\sum_{k=1}^{n}(-1)^{k-1}e_{n+1-k}, \qquad (n = 1, 2, \cdots).$$

Thus $\|P_{12}e_n\| = \sqrt{n}/2 \to \infty$ $(n \to \infty)$. This contradiction implies that $P_{22} = 0$. Therefore, A is strongly irreducible.

Although quasisimilarity cannot preserve strong irreducibility, it still preserves the strong irreducibility of some special classes of operators.

Theorem 2.36. *Let $B \in \mathcal{B}_1(\Omega)$ and let the operator T be quasisimilar to B; then T is strongly irreducible.*

Proof. Assume that there exist two non-trivial invariant subspaces $\mathcal{M}$ and $\mathcal{N}$ of T such that $\mathcal{M} \cap \mathcal{N} = \{0\}$ and $\mathcal{M} \dot{+} \mathcal{N} = \mathcal{H}$. Since T is quasisimilar to B, $\sigma_p(T) = \sigma_p(B) \supset \Omega$ and $\text{nul}\,(\lambda - T) = 1$ for all $\lambda \in \Omega$. Thus the eigenvector x of $\lambda - T$ belongs to $\mathcal{M}$ or $\mathcal{N}$ and Ω is the disjoint union of $\sigma_p(T_1) \cap \Omega$ and $\sigma_p(T_2) \cap \Omega$, where $T_1 = T|_{\mathcal{M}}$ and $T_2 = T|_{\mathcal{N}}$. Since $\mathcal{H} = \bigvee\{\ker(\lambda - B) : \lambda \in \Omega\}$, $\mathcal{H} = \bigvee\{\ker(\lambda - T) : \lambda \in \Omega\}$. Therefore $\sigma_p(T_1) \cap \Omega$ and $\sigma_p(T_2) \cap \Omega$ are non-empty. By the connectedness of Ω, we can find a sequence $\{\lambda_n\}_{n=1}^{\infty}$ in $\sigma_p(T_1) \cap \Omega$ such that $\lim_{n\to\infty} \lambda_n = \lambda_0 \in \sigma_p(T_2) \cap \Omega$. By Corollary 1.18, there exists a sequence $\{y_n\}_{n=0}^{\infty}$ of non-zero vectors such that $y_n \in \ker(\lambda_n - B)$, $\lim_{n\to\infty} y_n = y_0$ and $y_0 \in \ker(\lambda_0 - B)$. Let X be affine (has trivial kernel and dense range) such that $XB = TX$. Then $Xy_n \in \ker(\lambda_n - T) \subset \mathcal{M}$ and Xy_n converges to a non-zero vector $Xy_0 \in \ker(\lambda_0 - T) \subset \mathcal{N}$, i.e., $Xy_0 \in \mathcal{M} \cap \mathcal{N}$, a contradiction. □

Let S be the unilateral shift and ϕ be a bounded univalent analytic function on D. Then it is easily seen that $\phi(S) = T_\phi$ is the analytic Toeplitz operator with symbol ϕ. In Section 2.3, we have proved that T_ϕ is strongly irreducible. The next theorem strengthens the result.

Theorem 2.37. *Let $T \in \mathcal{L}(\mathcal{H})$ be strongly irreducible and let $\phi(z)$ be a univalent analytic function on $\Omega = \{z : |z| \leq \|T\|\}$; then $\phi(T)$ is strongly irreducible.*

Proof. Without loss of generality, we assume that $\|T\| = 1$. Since ϕ is a univalent analytic function on Ω, there exist δ, $\delta > 0$, such that ϕ is a univalent analytic function on $D_\delta = \{z \in \mathbb{C} : |z| \leq 1 + \delta\}$ and $\|f(T)\| \leq \gamma \max\{|f(z)| : z \in D_\delta\}$ for all functions f analytic on D_δ. If $A = \phi(T)$ is strongly reducible, there is a nontrivial idempotent $P \in \mathcal{L}(\mathcal{H})$ such that $PA = AP$. Note that $T = \phi^{-1}(A)$, where ϕ^{-1} is a univalent analytic function from $\phi(D_\delta)$ onto D_δ such that $\phi^{-1} \circ \phi(z) = z$. Since $\phi(D_\delta)$ is simply connected, there is a sequence $\{p_n(z)\}_{n=1}^{\infty}$ of polynomials such that $p_n(z) \to \phi^{-1}(z)$ uniformly on $\phi(D_\delta)$ as $n \to \infty$. Thus $\|p_n(A) - \phi^{-1}(A)\| \to 0$ $(n \to \infty)$. Since $Pp_n(A) = p_n(A)P$, $PT = TP$, which contradicts the assumption.

$\square$

2.5 Open problems

Problem 2.1. *What kinds of strongly irreducible operators preserve strong irreducibility under quasisimilarity?*

Problem 2.2. *If T does not satisfy any quadratic equation and $\lambda - T$ is not of finite rank for all $\lambda \in \mathbb{C}$, is T similar to an irreducible operator?*

Problem 2.3. *Let T be a strongly irreducible operator with $\|T\| \leq 1$ and let $\phi \in H^\infty$. Is $\phi(T)$ strongly irreducible?*

Problem 2.4. *Let $B \in \mathcal{B}_n(\Omega)$ such that $\sigma(B) = \overline{\Omega}$, let M be a normal operator such that $\sigma(M) = \sigma_e(M)$, $\sigma(M) \cup \partial\Omega$ is connected and let $\epsilon > 0$. Does there exist a compact operator K with $\|K\| < \epsilon$ such that $B \oplus M$ is quasisimilar to $B + K$?*

2.6 Remarks

The concept of irreducible operator was given by P.R. Halmos [1]; related work appeared in H.Radjavi and P.Rosenthal [1]. P.R.Halmos [2], D.Voiculescu [1], C.K.Fong and C.L. Jiang [1], [2] and F. Gilfeather [1]. Example 2.2 is due to F. Gilfeather [1]. Proposition 2.3 is Proposition 5.3 of C.C.Cowen [1]. Theorem 2.5 appears in V.B.Arveson [1]. Theorem 2.6 was proved by H.Behnke [1] and then generalized by F.Gilfeather [2] and N.Suzuki [1]. Theorem 2.9 is due to D.A. Herrero [1] and later, C.K. Fong and C.L. Jiang proved that given $T \in \mathcal{L}(\mathcal{H})$ if $\mathrm{Re}\,(T)$ or $\mathrm{Im}\,(T)$ is finitely cyclic then the conclusion $K \in C^p(\mathcal{H})$ in Theorem 2.9 can be replaced by $K \in C^1(\mathcal{H})$. Example 2.11 was given by Z.J. Jiang [1]. Theorem 2.12 was proved by C.K. Fong and C.L. Jiang [1]. Proposition 2.19 is due to C.L. Jiang and P.Y. Wu [1].

The concept of strongly irreducible operator was given by F. Gilfeather [1] in 1972. Proposition 2.28 is due to M.J. Cowen and R.G. Douglas [1]. R.L. Kelly first proved that if T is a bilateral weighted shift with an annulus spectrum, then T is strongly irreducible (R.L. Kelly [1]). Proposition 2.30 was proved by C.L. Jiang and Z.Y. Wang. Proposition 2.33, Example 2.35 and Theorem 2.36 are due to C.K. Fong and C.L. Jiang (C.K. Fong and C.L. Jiang [3], C.L. Jiang [2]). Theorem 2.37 was given by Z.J. Jiang and S.L. Sun [1].

3. The strongly irreducible operator in some classes of operators

Recall that each $k \times k$ Jordan block

$$J = \begin{bmatrix} \lambda & & & 0 \\ 1 & \lambda & & \\ & \ddots & \ddots & \\ 0 & & 1 & \lambda \end{bmatrix}$$

has the following properties:

(i) J is strongly irreducible on $\mathbb{C}^k$;

(ii) $\mathcal{A}'(J)$ is strictly cyclic;

(iii) $\mathcal{A}'(J)$ consists of all $k \times k$ matrices B that have the form

$$B = \begin{bmatrix} \alpha_1 & & & 0 \\ \alpha_2 & \alpha_1 & & \\ \vdots & \ddots & \ddots & \\ \alpha_k & & \alpha_2 & \alpha_1 \end{bmatrix},$$

where $\alpha_i \in \mathbb{C}$, $i = 1, 2, \cdots, k$.

(iv) $\mathcal{A}^a(J)$, the algebra generated by rational functions of J with poles outside $\sigma(J) = \{\lambda\}$, is an algebra of strict multiplicity k.

A weakly closed operator algebra $\mathcal{A}$ on a Hilbert space $\mathcal{H}$ is an *algebra of strict multiplicity* k if there exist k vectors $e_1, \cdots, e_k$ in $\mathcal{H}$ such that

$$\sum_{i=1}^{k} \mathcal{A} e_i = \mathcal{H},$$

and there are no $k-1$ vectors satisfying this property. $\mathcal{A}$ is *strictly cyclic* if $\mathcal{A}$ is an algebra of strict multiplicity 1. An operator $T \in \mathcal{L}(\mathcal{H})$ is a strictly cyclic operator if the weakly closed operator algebra $\mathcal{A}(T)$ generated by T and 1 is a strictly cyclic algebra.

If strongly irreducible operators can be considered as the natural generalization of the Jordan blocks in infinite dimensional Hilbert space, we need to analyse their structure and properties. A natural question is raised. Do there exist strongly irreducible operators which have the properties (i)–(iv) the $k \times k$ Jordan blocks have; or precisely, given a connected, bounded open subset Ω in the complex plane and a natural number n, does there exist a strongly irreducible operator $A(\Omega)$ such that $\rho_F(A(\Omega)) \supset \Omega$, $\operatorname{ind}(\lambda - A(\Omega)) = -n$ for $\lambda \in \Omega$, $\mathcal{A}'(A(\Omega))$ is strictly cyclic and $\mathcal{A}^a(A(\Omega))$ is an algebra of strict multiplicity n?

An operator $T \in \mathcal{L}(\mathcal{H})$ is said to have the *nice property* if either $\mathcal{A}'(T)$ or $\mathcal{A}'(T^*)$ is a commutative strictly cyclic operator algebra.

In this chapter we will use the theory of Sobolev spaces to prove the existence of the strongly irreducible operator with the nice property. Furthermore, we will use Cowen–Douglas operators to construct some strongly irreducible operators and describe the norm closure of strongly irreducible operators.

In what follows, $(SI)(\mathcal{H})$ denotes the set of strongly irreducible operators on a Hilbert space $\mathcal{H}$. $T \in (SI)(\mathcal{H})$, or briefly $T \in (SI)$, means that T is a strongly irreducible operator.

3.1 The multiplication operators on Sobolev spaces

Let Ω be an analytic Cauchy domain in the complex plane and let $W^{22}(\Omega)$ denote the Sobolev space

$$W^{22}(\Omega) = \left\{ f \in L^2(\Omega, dm) : \begin{array}{l} \text{the distributional partial derivatives of first} \\ \text{and second order of } f \text{ belong to } L^2(\Omega, dm) \end{array} \right\},$$

where dm denotes the planar Lebesgue measure.

Set

$$W(\Omega) = \{M_f : f \in W^{22}(\Omega)\},$$

where M_f is "multiplication by f" on $W^{22}(\Omega)$, i.e., $M_f \in \mathcal{L}(W^{22}(\Omega))$ is defined by $M_f g = fg$ for each $g \in W^{22}(\Omega)$. It is easily seen that $W(\Omega)$ is a strictly cyclic operator algebra with strictly cyclic vector e ($e(s, t) \equiv 1$).

Proposition 3.1. *Let M_λ be multiplication by the independent variable λ $(= s + it)$ on $W^{22}(\Omega)$; then*

(i) $\sigma(M_\lambda) = \sigma_{lre}(M_\lambda) = \overline{\Omega}$, $\mathrm{nul}(\lambda_0 - M_\lambda) = 0$ *and* $\mathrm{nul}(\lambda_0 - M_\lambda)^* = 1$ *for all* $\lambda_0 \in \overline{\Omega}$;

(ii) $\mathcal{A}'(M_\lambda) = W(\Omega)$, *i.e.,* $\mathcal{A}'(M_\lambda)$ *is strictly cyclic;*

(iii) $M_\lambda \in (SI)$.

Proof. (i) It is easily seen that $\sigma(M_\lambda) = \overline{\Omega}$, $\mathrm{nul}(\lambda_0 - M_\lambda) = 0$ and $\mathrm{nul}(\lambda_0 - M_\lambda)^* = 1$ for all $\lambda_0 \in \overline{\Omega}$. Since $W(\Omega)$ is strictly cyclic, $W(\Omega)^*$, the dual space of the Banach space $W(\Omega)$, has the form

$$W(\Omega)^* = \{g^* : g^*(T_f) = (T_f e, g) = (f, g), \ f, g \in W^{22}(\Omega)\}. \quad \text{A. Lambert [1]}$$

In particular, if $k^*_{\lambda_0}$ ($\lambda_0 \in \overline{\Omega}$) is the multiplicative functional on $W(\Omega)$ such that $k^*_{\lambda_0}(M_f) = f(\lambda_0)$, $f \in W^{22}(\Omega)$, then there exists a vector $k_{\lambda_0} \in W^{22}(\Omega)$ such that $k^*_{\lambda_0}(M_f) = f(\lambda_0) = (f, k_{\lambda_0})$. Since $(g, (\lambda_0 - M_\lambda)^* k_{\lambda_0}) = ((\lambda_0 - M_\lambda)M_g, k_{\lambda_0}) = 0$, for all $g \in W^{22}(\Omega)$, $k_{\lambda_0} \in \ker(\lambda_0 - M_\lambda)^*$.

Set $f(s, t) = [(s - s_0)^2 + (t - t_0)^2]^{2/3} = |\lambda - \lambda_0|^{4/3}$, where $\lambda_0 = s_0 + it_0$, $\lambda = s + it \in \Omega$. Calculations indicate that $f \in W^{22}(\Omega)$ and $(\lambda - \lambda_0)^{-1} f \notin W^{22}(\Omega)$, i.e., $f \notin \mathrm{ran}(\lambda_0 - M_\lambda)$. Since $(f\ k_{\lambda_0}) = f(\lambda_0) = 0$, $f \in \left[\ker(\lambda_0 - M_\lambda)^*\right]^\perp = \mathrm{ran}(\lambda_0 - M_\lambda)^-$, i.e., $\mathrm{ran}(\lambda_0 - M_\lambda)$ is not closed and $\lambda_0 \in \sigma_{lre}(M_\lambda)$. Therefore $\sigma_{lre}(M_\lambda) = \overline{\Omega}$.

(ii) Given $T \in \mathcal{A}'(M_\lambda)$, we have $M^*_\lambda T^* k_\mu = T^* M^*_\lambda k_\mu = \mu^* T^* k_\mu$ for all $\mu \in \overline{\Omega}$. Since $\mathrm{nul}(\mu - M_\lambda)^* = 1$, $T^* k_\mu = h(\mu)^* k_\mu$ for some $h(\mu) \in \mathbb{C}$. Therefore, for each $\lambda \in \Omega$ and each $f \in W^{22}(\Omega)$,

$$(Tf)(\lambda) = (Tf, k_\lambda) = (f, T^* k_\lambda) = \big(f, h(\lambda)^* k_\lambda\big) = h(\lambda) f(\lambda).$$

This implies that $h(\lambda) = (Te)(\lambda) \in W^{22}(\Omega)$ and $T = M_h \in W(\Omega)$.

(iii) Assume that P is an idempotent commuting with M_λ. Then by (ii) $P = M_g$ for some g in $W^{22}(\Omega)$. Since $M_g^2 = M_{g^2} = M_g$, $g^2 = g$. Thus $g(\lambda) = 0$ or 1 for all $\lambda \in \Omega$. From the connectedness of Ω, we can conclude that $g(\lambda) = 0$ or 1, i.e., $P = 0$ or 1. Therefore $M_\lambda \in (SI)$. □

Let $\mathcal{A}^a(M_\lambda)$ be the algebra generated by the rational functions of M_λ with poles outside $\overline{\Omega}$ and denote $R(\Omega) = \mathcal{A}^a(M_\lambda)e$. Then it is obvious that $R(\Omega)$ is the subspace of $W^{22}(\Omega)$ generated by rational functions with poles outside $\overline{\Omega}$ and $R(\Omega)$ is invariant under M_λ. Denote $M_\lambda(\Omega) = M_\lambda|_{R(\Omega)}$. Then $M_\lambda(\Omega)$ is a rationally strictly cyclic operator on the Hilbert space $R(\Omega)$ with strictly cyclic vector e. An operator T is *rationally strictly cyclic* if $\mathcal{A}^a(T)$ is a strictly cyclic algebra.

Proposition 3.2. (i) $\sigma(M_\lambda(\Omega)) = \overline{\Omega}$, $\sigma_e(M_\lambda(\Omega)) = \partial\Omega$, $\text{nul}\,(\lambda_0 - M_\lambda(\Omega)) = 0$ *and* $\text{ind}\,(\lambda_0 - M_\lambda(\Omega)) = -1$ *for all* $\lambda_0 \in \Omega$;

(ii) $\mathcal{A}'(M_\lambda(\Omega)) = \mathcal{A}^a(M_\lambda(\Omega)) = \{M_f(\Omega) : f \in R(\Omega)\}$, *where* $M_f(\Omega) = M_f|_{R(\Omega)}$, *i.e.,* $\mathcal{A}'(M_\lambda(\Omega))$ *is strictly cyclic and* $M_\lambda(\Omega)$ *is rationally strictly cyclic;*

(iii) $M_\lambda(\Omega)$ *is a strongly irreducible operator.*

Proof. (i) If $\lambda_0 \notin \overline{\Omega}$, then $(\lambda_0 - \lambda)^{-1} \in R(\Omega)$, $M_{(\lambda_0-\lambda)^{-1}} = (\lambda_0 - M_\lambda)^{-1}$ and $M_{(\lambda_0-\lambda)^{-1}}(\Omega) = (\lambda_0 - M_\lambda(\Omega))^{-1}$, i.e. $\sigma(M_\lambda(\Omega)) \subset \overline{\Omega}$. If $\lambda_0 \in \overline{\Omega}$, since $\ker(\lambda_0 - M_\lambda) = \{0\}$, $\ker(\lambda_0 - M_\lambda(\Omega)) = \{0\}$. There exists a rational function $r(\lambda)$ with poles outside $\overline{\Omega}$ such that $r(\lambda_0) \neq 0$, i.e. $(r, k_{\lambda_0}) \neq 0$. This implies that $P_{R(\Omega)} k_{\lambda_0} \neq 0$. Thus $\ker(\lambda_0 - M_\lambda(\Omega))^* \neq \{0\}$ and $\sigma(M_\lambda(\Omega)) = \overline{\Omega}$.

Let $\lambda_0 \in \Omega$, and assume that $[M_\lambda(\Omega) - \lambda_0] f_n = F_n \xrightarrow{W^{22}(\Omega)} F$, where $f_n, F \in R(\Omega)$. By Sobolev' imbedding theorem, F_n and F are analytic in Ω and continuous on $\partial\Omega$. Since $F(\lambda_0) = 0$, there exist a number $\delta > 0$ and a function h analytic in $\Gamma = \{\lambda \in \mathbb{C} : |\lambda - \lambda_0| < \delta\}$ such that $\overline{\Gamma} = \{\lambda \in \mathbb{C} : |\lambda - \lambda_0| \leq \delta\} \subset \Omega$ and $F(\lambda) = (\lambda - \lambda_0)\, h(\lambda)$ for $\lambda \in \Gamma$. Denote $\Sigma = \Omega \backslash (1/3\overline{\Gamma}) = \Omega \backslash \{\lambda \in \mathbb{C} : |\lambda - \lambda_0| \leq \delta/3\}$ and define

$$f(\lambda) = \begin{cases} h(\lambda), & \lambda \in \Gamma, \\ F(\lambda)/(\lambda - \lambda_0), & \lambda \in \Sigma. \end{cases}$$

Then $f(\lambda)$ is an analytic extension of h satisfying $F(\lambda) = (\lambda - \lambda_0)\, f(\lambda)$ for all $\lambda \in \Omega$. Note that

$$\begin{aligned} \int_\Sigma |f_n - f|^2 dm &\leq \frac{9}{\delta^2} \int_\Sigma |(\lambda - \lambda_0)(f_n(\lambda) - f(\lambda))|^2 dm \\ &\leq \frac{9}{\delta^2} \|F_n - F\|_{W^{22}(\Omega)} \to 0, \qquad (n \to \infty) \end{aligned}$$

and

$$\begin{aligned} \frac{\partial}{\partial s} F_n(\lambda) &= f_n(\lambda) + (\lambda - \lambda_0)\, f_n'(\lambda), \\ \frac{\partial}{\partial s} F(\lambda) &= f(\lambda) + (\lambda - \lambda_0)\, f'(\lambda). \end{aligned}$$

From $\frac{\partial}{\partial s} F_n \xrightarrow{L^2(\Sigma)} \frac{\partial}{\partial s} F$ and $f_n \xrightarrow{L^2(\Sigma)} f$, we have

$$(\lambda - \lambda_0)\, f_n'(\lambda) \xrightarrow{L^2(\Sigma)} (\lambda - \lambda_0) f'.$$

Thus

$$\int_\Sigma |f_n' - f'|^2\, dm \leq \frac{9}{\delta^2} \int_\Sigma |(\lambda - \lambda_0)(f_n' - f')|^2\, dm \to 0 \qquad (n \to \infty).$$

Similarly, from $\frac{\partial^2}{\partial s^2}F_n \xrightarrow{L^2(\Sigma)} \frac{\partial^2}{\partial s^2}F$ and $f_n' \xrightarrow{L^2(\Sigma)} f'$, we have

$$(\lambda-\lambda_0)f_n''(\lambda) \xrightarrow{L^2(\Sigma)} (\lambda-\lambda_0)f'',$$

therefore $f_n''(\lambda) \xrightarrow{L^2(\Sigma)} f''$. Finally we get $f_n \xrightarrow{W^{22}(\Sigma)} f$. From Sobolev's imbedding theorem, f_n converge to f uniformly on $r=\partial\Gamma$. Thus, for $\lambda \in 2/3\Gamma$ and $0 \le k \le 2$,

$$|f_n^{(k)}(\lambda) - f^{(k)}(\lambda)| \le \frac{1}{2\pi}\int_r \frac{|f_n(\xi)-f(\xi)|}{|\xi-\lambda|^{k+1}}\,d|\xi| \le \frac{3^{k+1}}{\delta^k}\max_{\xi\in r}|f_n(\xi)-f(\xi)|.$$

Therefore, $f_n \xrightarrow{W^{22}(\frac{2}{3}\Gamma)} f$. Note that $f_n \xrightarrow{W^{22}(\Sigma)} f$ and $f_n \xrightarrow{W^{22}(\frac{2}{3}\Gamma)} f$ imply that $f_n \xrightarrow{W^{22}(\Omega)} f$ and $f \in R(\Omega)$. Thus the range of $M_\lambda(\Omega)-\lambda_0$ is closed. The remainder of (i) is obvious.

By the same arguments as Proposition 3.1, we can prove (ii) and (iii). $\square$

Corollary 3.3. *Let M_λ and $M_\lambda(\Omega)$ be given as in Propositions* 3.1 *and* 3.2; *then*

(i) $\bigvee\{\ker(\lambda_n - M_\lambda)^* : \lambda_n \in \Delta\} = W^{22}(\Omega)$, *where* $\Delta = \{\lambda_n\}_{n=1}^\infty \subset \Omega$ *and* $\overline{\Delta} = \overline{\Omega}$;

(ii) $M_\lambda(\Omega)^* \in \mathcal{B}_1(\Omega^*)$.

Proof. (i) Denote $\mathcal{M} = \bigvee\{\ker(\lambda_n - M_\lambda)^* : \lambda_n \in \Delta\}$ and pick $g \in \mathcal{M}^\perp$. Let k_{λ_i} $(i=1,2,\cdots)$ be the vector corresponding to the multiplicative functional: $f \to f(\lambda_i) = (f\ k_{\lambda_i})$ on $W^{22}(\Omega)$. Then $g(\lambda_i)=0$ $(i=1,2,\cdots)$, since $k_{\lambda_i} \in \ker(\lambda_i - M_\lambda)^*$ and $g \in \mathcal{M}^\perp$. Since Δ is dense in Ω, $g=0$.

(ii) It suffices to prove that $\bigvee\{\ker(\mu - M_\lambda(\Omega))^* : \mu \in \Omega\} = R(\Omega)$. Just like the k_μ's in the case of $W(\Omega)$, for each $\mu \in \Omega$, there is a vector $h_\mu \in R(\Omega)$ corresponding to the multiplicative functional h_μ^* on the strictly cyclic algebra $\mathcal{A}^a(M_\lambda(\Omega)) = \mathcal{A}'(M_\lambda(\Omega))$, such that $h_\mu^*(M_f(\Omega)) = (M_f(\Omega)e,\ h_\mu) = (f,\ h_\mu) = f(\mu)$ for all $f \in R(\Omega)$ and $h_\mu \in \ker(\mu - M_\lambda(\Omega))^*$ (in fact, $h_\mu = P_{R(\Omega)}k_\mu$). By the same argument as (i) we can prove $\bigvee\{\ker(\mu - M_\lambda(\Omega))^* : \mu \in \Omega\} = R(\Omega)$. $\square$

Proposition 3.2 and Corollary 3.3 assert that for each analytic Cauchy domain Ω we have constructed a strongly irreducible operator with the nice property in $\mathcal{B}_1(\Omega)$ such that its spectrum is $\overline{\Omega}$. For stronger results, we need the following lemma.

Lemma 3.4. *Let Ω be an analytic Cauchy domain. Then*

$$[\operatorname{ran}\tau_{M_\lambda(\Omega),M_\lambda(\Omega)}] \cap [\ker\tau_{M_\lambda(\Omega),M_\lambda(\Omega)}] = \{0\},$$

i.e., if there exists $X \in \mathcal{L}(R(\Omega))$ *satisfying* $XM_\lambda(\Omega) = M_\lambda(\Omega)X$ *and* $M_\lambda(\Omega)Y - YM_\lambda(\Omega) = X$ *for some* $Y \in \mathcal{L}(R(\Omega))$, *then* $X = 0$.

Proof. Without loss of generality, we can assume that $\Omega \subset D = \{\lambda \in \mathbb{C} : |\lambda| < 1\}$. Otherwise, consider $M_\lambda(\Omega')$, where $\Omega' = \{(\lambda - \lambda_0)/a : \lambda \in \Omega\}$, λ_0 is a fixed point in Ω and $a =$ diameter of Ω.

By Proposition 3.2, $X = M_g(\Omega)$ for some g in $R(\Omega)$. Thus $M_\lambda(\Omega)Yf - YM_\lambda(\Omega)f = gf$ for all $f \in R(\Omega)$. $f = e$ implies that

$$\lambda Y(e) - Y(\lambda) = g,$$

or

$$Y(\lambda) = \lambda h - g, \quad \text{where } h = Y(e) \in R(\Omega).$$

$f = \lambda$ implies that

$$\lambda Y(\lambda) - Y(\lambda^2) = \lambda g,$$

or

$$Y(\lambda^2) = \lambda(\lambda h - g) - \lambda g = \lambda^2 h - 2\lambda g, \ \cdots .$$

In general, $f = \lambda^{n-1}$ implies that

$$\lambda Y(\lambda^{n-1}) - Y(\lambda^n) = \lambda^{n-1} g,$$

or

$$Y(\lambda^n) = \lambda^n h - n\lambda^{n-1} g \quad (n = 1, 2, \cdots).$$

Since $|\lambda|^{2n} \le |\lambda|^{2(n-1)}$ ($\lambda \in \Omega$ and $n = 1, 2, \cdots$),

$$\frac{n^2 \int_\Omega |\lambda|^{2(n-1)} dm}{\int_\Omega |\lambda|^{2n} dm} \to \infty \quad (n \to \infty).$$

Let a_n and b_n denote $n^2 \int_\Omega |\lambda|^{2(n-1)} dm$ and $\int_\Omega |\lambda|^{2n} dm$ respectively ($n = 1, 2, \cdots$). Then for each positive number M there exists N such that $a_n/b_n > M$ if $n > N$. Therefore, when $n > N + 2$,

$$\begin{aligned}
\frac{\|n\lambda^{n-1}\|^2_{W^{22}(\Omega)}}{\|\lambda^n\|^2_{W^{22}(\Omega)}} &= \frac{\int_\Omega [n^2|\lambda|^{2(n-1)} + 2n^2(n-1)^2|\lambda|^{2(n-2)}]\, dm}{\int_\Omega [|\lambda|^{2n} + 2n^2|\lambda|^{2(n-1)} + 3n^2(n-1)^2|\lambda|^{2(n-2)}]\, dm} \\
&\quad + \frac{\int_\Omega [3n^2(n-1)^2(n-2)^2|\lambda|^{2(n-3)}]\, dm}{\int_\Omega [|\lambda|^{2n} + 2n^2|\lambda|^{2(n-1)} + 3n^2(n-1)|\lambda|^{2(n-2)}]\, dm} \\
&= \frac{a_n + 2n^2 a_{n-1} + 3n^2(n-1)^2 a_{n-2}}{b_n + 2n^2 b_{n-1} + 3n^2(n-1)^2 b_{n-2}} \\
&> \frac{Mb_n + 2n^2 M b_{n-1} + 3n^2(n-1)^2 M b_{n-2}}{b_n + 2n^2 b_{n-1} + 3n^2(n-1)^2 b_{n-2}} = M,
\end{aligned}$$

i.e.,

$$\frac{\|n\lambda^{n-1}\|^2_{W^{22}(\Omega)}}{\|\lambda^n\|^2_{W^{22}(\Omega)}} \to \infty \quad (n \to \infty).$$

This implies that if $g \neq 0$, Y is unbounded. □

For each natural number n $(1 \leq n < \infty)$, we define

$$M_n(\Omega) = \begin{bmatrix} M_\lambda(\Omega) & & & 0 \\ 1 & M_\lambda(\Omega) & & \\ & \ddots & \ddots & \\ 0 & & 1 & M_\lambda(\Omega) \end{bmatrix}$$

with respect to the orthogonal direct sum $R(\Omega)^{(n)}$ of n copies of $R(\Omega)$.

Theorem 3.5. *Let $M_n(\Omega)$ be defined as above. Then for each $A \in \mathcal{A}'(M_n(\Omega))$*

(i)

$$A = \begin{bmatrix} M_{f_1(\Omega)} & & & 0 \\ M_{f_2(\Omega)} & M_{f_1(\Omega)} & & \\ \vdots & \ddots & \ddots & \\ M_{f_n(\Omega)} & \cdots & M_{f_2(\Omega)} & M_{f_1(\Omega)} \end{bmatrix},$$

where $f_i \in R(\Omega)$ $(i = 1, 2, \cdots, n)$. Furthermore, $\mathcal{A}'(M_n(\Omega))$ is strictly cyclic with the cyclic vector $e \oplus e \oplus \cdots \oplus e$;

(ii) $M_n(\Omega) \in (SI)$;

(iii) $\sigma(M_n(\Omega)) = \overline{\Omega}$, $\sigma_e(M_n(\Omega)) = \partial\Omega$, $\mathrm{nul}(\lambda_0 - M_n(\Omega)) = 0$ *and* $\mathrm{ind}(\lambda_0 - M_n(\Omega)) = -n$ *for all* $\lambda_0 \in \Omega$;

(iv) *$\mathcal{A}^a(M_n(\Omega))$ is an operator algebra of strict multiplicity n;*

(v) $M_n(\Omega)^* \in B_n(\Omega^*)$.

Proof. (i) Assume that

$$A = \begin{bmatrix} A_{11} & \dots & A_{1n} \\ \dots & & \dots \\ A_{n1} & \dots & A_{nn} \end{bmatrix} \in \mathcal{A}'(M_n(\Omega)),$$

i.e.,

$$\begin{bmatrix} A_{11} & \dots & A_{1n} \\ \dots & \dots & \\ A_{n1} & \dots & A_{nn} \end{bmatrix} \begin{bmatrix} M_\lambda(\Omega) & & & 0 \\ 1 & M_\lambda(\Omega) & & \\ & \ddots & \ddots & \\ 0 & & 1 & M_\lambda(\Omega) \end{bmatrix} =$$

$$\begin{bmatrix} M_\lambda(\Omega) & & & 0 \\ 1 & M_\lambda(\Omega) & & \\ & \ddots & \ddots & \\ 0 & & 1 & M_\lambda(\Omega) \end{bmatrix} \begin{bmatrix} A_{11} & \dots & A_{1n} \\ \dots & \dots & \\ A_{n1} & \dots & A_{nn} \end{bmatrix}.$$

At the $(1, n)$, $(1, n-1)$ entries we have

$$A_{1n} M_\lambda(\Omega) = M_\lambda(\Omega) A_{1n}$$

and

$$A_{1\,n-1} M_\lambda(\Omega) + A_{1n} = M_\lambda(\Omega) A_{1\,n-1}.$$

By Lemma 3.4, $A_{1n} = 0$ and $A_{1\,n-1} \in \mathcal{A}'(M_\lambda(\Omega))$. Similarly, we can conclude that $A_{ij} = 0$, $(1 \le i < j \le n)$. Consider the (i, i) entries of both sides; we have

$$A_{ii} \in \mathcal{A}'(M_\lambda(\Omega)), \quad (1 \le i \le n).$$

Compare the $(i+1, i)$ entries $(1 \le i \le n-1)$; we get

$$A_{i+1\,i} M_\lambda(\Omega) + A_{i+1\,i+1} = A_{ii} + M_\lambda(\Omega) A_{i+1\,i}.$$

By Lemma 3.4, $A_{i+1\,i+1} = A_{ii}$ and $A_{i+1\,i} \in \mathcal{A}'(M_\lambda(\Omega))$. At the $(i+2, i)$ entries $(1 \le i \le n-2)$,

$$A_{i+2\,i} M_\lambda(\Omega) + A_{i+2\,i+1} = A_{i+1\,i} + M_\lambda(\Omega) A_{i+2\,i}.$$

Thus $A_{i+2\,i+1} = A_{i+1\,i}$ and $A_{i+2\,i} \in \mathcal{A}'(M_\lambda(\Omega))$. Using this argument repeatedly and by Proposition 3.2, we get the general form of A. It is easily seen that $\mathcal{A}'(M_n(\Omega))$ is a strictly cyclic algebra with cyclic vector $e \oplus \cdots \oplus e$.

(ii) Let P be an idempotent commuting with $M_n(\Omega)$. By (i) P admits the form indicated in (i). Since $P^2 = P$, $f_1^2 = f_1$. Since Ω is connected, $f_1 \equiv 1$ or $f_1 \equiv 0$. In either case, $f_2 = f_3 = \cdots = f_n = 0$, i.e., $P = 1$ or 0 and $M_n(\Omega) \in (SI)$.

(iii) Let $\lambda_0 \in \Omega$. Note that $\ker(\lambda_0 - M_\lambda(\Omega)) = 0$, $\lambda_0 \in \rho_{s-F}(M_\lambda(\Omega))$ and $\operatorname{ran}(\lambda_0 - M_\lambda(\Omega))^* = R(\Omega)$. Computations show that $\operatorname{ran}(\lambda_0 - M_n(\Omega))^* = R(\Omega)^{(n)}$, i.e., $\lambda_0 \in \rho_{s-F}(M_n(\Omega))$ and $\operatorname{nul}(\lambda_0 - M_n(\Omega)) = 0$. It is not difficult to see that $\operatorname{nul}(\lambda_0 - M_n(\Omega))^* = n$.

(iv) Computation shows that

$$\mathcal{A}^a(M_n(\Omega))\, e_1 + \mathcal{A}^a(M_n(\Omega))\, e_2 + \cdots + \mathcal{A}^a(M_n(\Omega))\, e_n = R(\Omega)^{(n)},$$

where e_i is the vector in $R(\Omega)^{(n)}$ whose i-th coordinate is e and other coordinates are 0. Thus the strict multiplicity of $\mathcal{A}^a(M_n(\Omega))$ is less than or equal to n.

On the other hand, $\mathcal{R} := \operatorname{ran}(\lambda_0 - M_n(\Omega)) = \left[\ker(\lambda_0 - M_n(\Omega))^*\right]^\perp$ is invariant under $\mathcal{A}'(M_n(\Omega))$ and its co-dimension is n, where λ_0 is a point in Ω. A fortiori, it is invariant under $\mathcal{A}^a(M_n(\Omega))$. Moreover, if $\mathcal{M}$ is invariant under $\mathcal{A}(M_n(\Omega))$ and $\mathcal{M} \supset \mathcal{R}$, then since $\mathcal{M}^\perp \subset \ker(\lambda_0 - M_n(\Omega))^*$, $\mathcal{M}$ is also invariant under $\mathcal{A}^a(M_n(\Omega))$.

Thus for arbitrary vectors $y_1, y_2, \cdots, y_k \in \mathcal{R}(\Omega)^{(n)}$, since

$$M_n(\Omega)y_j = (M_n(\Omega) - \lambda_0)y_j + \lambda_0 y_j \in \bigvee\{\mathcal{R}, y_j\},$$

$\left[\mathcal{A}^a(M_n(\Omega))y_j\right]^- \subset \bigvee\{\mathcal{R}, y_j\}$, for each $j = 1, 2, \cdots, k$. Therefore,

$$\begin{aligned}\dim \mathcal{R}(\Omega)^{(n)} \ominus \left[\sum_{j=1}^{k} \mathcal{A}^a(M_n(\Omega))y_j\right]^- &\geq \dim\left[\mathcal{R}(\Omega)^{(n)} \ominus \bigvee\{\mathcal{R};\, y_1, y_2, \cdots, y_k\}\right] \\ &\geq n - k.\end{aligned}$$

This implies that the strict multiplicity of $\mathcal{A}^a(M_n(\Omega))$ is greater than or equal to n.

(v) is obvious. □

It is still an unsolved problem to construct a strongly irreducible operator A with the nice property such that $\sigma(A) = \overline{\Omega}$ and $A \in \mathcal{B}_\infty(\Omega)$, where Ω is an analytic Cauchy domain. But we have the following result. The reader is referred to D.A. Herrero [4] for the proof.

Proposition 3.6. *Given an analytic Cauchy domain Ω and $\eta > 0$ such that $\mathbb{C}\backslash\overline{\Omega}$ and $\mathbb{C}\backslash \bigwedge(\Omega,\eta)$ have the same number of components, where $\bigwedge(\Omega,\eta) := \bigcup\{\overline{\Omega} + r\eta : 0 \leq r \leq 1\}$, then there exists a strongly irreducible operator A such that $\sigma(A) = \bigwedge(\Omega,\eta)$, $\sigma_{lre}(A) = \bigcup\{\partial\Omega + r\eta : 0 \leq r \leq 1\}$, $\ker(\lambda - A)^* = \{0\}$ and $\operatorname{ind}(\lambda - A) = \infty$ for all $\lambda \in \Omega\backslash\sigma_{lre}(A)$.*

We are going to construct a strongly irreducible operator with the nice property with arbitrary spectrum. Let Γ be a perfect subset of $\mathbb{C}$ and assume that $\Gamma \subset (a,b)^2 = G$. Set

$$m_0(\Gamma) = \{f \in W_0^{22}(G) : f|_\Gamma \equiv 0\},$$

where $W_0^{22}(G) = \{f \in W^{22}(G) : f|_{\partial G} \equiv 0\}$. It is clear that $W_0^{22}(G)$ is a subspace of $W^{22}(G)$. For each $f \in W^{22}(G)$, denote $M_f^\circ =$ "multiplication by f" on $W_0^{22}(G)$. Then M_f° admits the representation

$$M_f^\circ = \begin{bmatrix} M_f^\circ|_{m_0(\Gamma)} & * \\ 0 & M_f^\circ(\Gamma) \end{bmatrix}$$

with respect to the decomposition $W_0^{22}(G) = m_0(\Gamma) \oplus m_0(\Gamma)^\perp$. Denote $W_0(\Gamma) = \{M_f^\circ(\Gamma) : f \in W^{22}(G)\}$. Then it is easily seen that $W_0(\Gamma)$ is a strictly cyclic operator algebra with cyclic vector $e_0(\Gamma) =$ orthogonal projection of e onto $W_0^{22}(G) \ominus m_0(\Gamma)$.

Theorem 3.7. (i) $\sigma(M_\lambda^\circ(\Gamma)) = \sigma_{lre}(M_\lambda^\circ(\Gamma)) = \Gamma$, $\operatorname{nul}(\mu - M_\lambda^\circ(\Gamma)) = 0$ *and* $\operatorname{nul}(\mu - M_\lambda^\circ(\Gamma))^* = 1$ *for all* $\mu \in \Gamma$;

(ii) $\mathcal{A}'(M_\lambda^\circ(\Gamma)) = W_0(\Gamma)$, *i.e.,* $\mathcal{A}'(M_\lambda^\circ(\Gamma))$ *is strictly cyclic;*

(iii) *If* Γ *is connected, then* $M_\lambda^\circ(\Gamma)$ *is a* (SI) *operator with the nice property;*

(iv) *If* $\{\lambda_i\}_{i=1}^\infty$ *is a dense subset of* Γ, *then* $\bigvee\{\ker(\lambda_i - M_\lambda^\circ(\Gamma))^* : i \geq 1\} = W_0^{22}(G) \ominus m_0(\Gamma)$.

Proof. (i) Let $\mu \in \Gamma$ and k_μ be in $W_0^{22}(G) \ominus m_0(\Gamma)$ corresponding to the multiplicative functional on $W_0(\Gamma)$: $M_f^\circ(\Gamma) \to (M_f^\circ(\Gamma)e_0(\Gamma), k_\mu) = (f, k_\mu) = f(\mu)$ for $f \in W^{22}(G)$. Then $k_\mu \in \ker(\mu - M_\lambda^\circ(\Gamma))^*$ and $\Gamma \subset \sigma_p(M_\lambda^\circ(\Gamma)^*)^* \subset \sigma(M_\lambda^\circ(\Gamma))$.

On the other hand, if $\mu \notin \Gamma$, by the Extension theorem we can find $f \in W_0^{22}(G) \ominus m_0(\Gamma)$ such that $f|_\Gamma \equiv 1/(\mu - \lambda)$, i.e., $\mu - M_\lambda^\circ(\Gamma)$ is invertible and $(\mu - M_\lambda^\circ(\Gamma))^{-1} = M_f^\circ(\Gamma)$. Therefore $\sigma(M_\lambda^\circ(\Gamma)) = \Gamma$.

If Γ°, the interior of Γ, is empty, then $\sigma(M_\lambda^\circ(\Gamma)) = \sigma_{lre}(M_\lambda^\circ(\Gamma))$. If $\mu \in \Gamma^\circ$, then there is an open disc $C = \{\lambda \in \mathbb{C} : |\lambda - \mu| < \delta\} \subset \Gamma^\circ$. By Proposition 3.1(i), $\operatorname{ran}(\mu - M_\lambda)$ is not closed in $W^{22}(C)$. Here, the operator "multiplication by λ" on $W^{22}(C)$ is still denoted by M_λ. Thus, there are $\{f_n\}_{n=1}^\infty$ in $W^{22}(C)$ such that

$$F_n(\lambda) = (\mu - \lambda) f_n(\lambda) \xrightarrow{W^{22}(C)} F(\lambda),$$

but $F(\lambda)/(\mu - \lambda) \notin W^{22}(C)$. By the Extension theorem, there exist $M > 0$ and $\widehat{F}_n$, $\widehat{F}$ in $W_0^{22}(G) \ominus m_0(\Gamma)$ such that $\|\widehat{F}_n - \widehat{F}\|_{W^{22}(G)} \leq M\|F_n - F\|_{W^{22}(C)}$ and $\widehat{F}_n|_C = F_n$,

$\widehat{F}|_C = F$, and $\widehat{F}_n$ can be written as $[\mu - M_\lambda^\circ(\Gamma)]\,\widehat{f}_n(\lambda)$, where $f_n \in W_0^{22}(G) \ominus m_0(\Gamma)$ and

$$\widehat{f}_n(\lambda) = \begin{cases} f_n(\lambda), & \lambda \in C, \\ \widehat{F}_n(\lambda)/(\mu - \lambda), & \lambda \in \Gamma \backslash C. \end{cases}$$

Thus $[\mu - M_\lambda^\circ(\Gamma)]\widehat{f}_n \xrightarrow{W^{22}(G)} \widehat{F}$. If $\operatorname{ran}(\mu - M_\lambda^\circ(\Gamma))$ is closed, $\widehat{F} = [\mu - M_\lambda^\circ(\Gamma)]\widehat{f}$ for some $\widehat{f} \in W_0^{22}(G) \ominus m_0(\Gamma)$. Thus $(\mu - \lambda)\, f(\lambda) = F(\lambda)$ for $\lambda \in C$, i.e., $F(\lambda) = (\mu - M_\lambda)f'$ for $f' = f|_C \in W^{22}(C)$. The contradiction implies that $\operatorname{ran}(\mu - M_\lambda^\circ(\Gamma))$ is not closed. Therefore $\sigma_{lre}(M_\lambda^\circ(\Gamma)) = \Gamma$.

The remainder of the proposition can be proved by the same arguments as Proposition 3.1. $\square$

Define a mapping $\Delta:\ W_0^{22}(G) \mapsto L^2(G, dm)$ by

$$\Delta f = \left(\frac{\partial^2}{\partial s^2} + \frac{\partial^2}{\partial t^2}\right) f, \quad f \in W_0^{22}(G).$$

Since $W_0^{22}(G)$ contains no non–zero harmonic functions, Δ is injective. Furthermore, for each $g \in L^2(G, dm)$, assume that

$$g(s,t) = \sum_{m,n=1}^{\infty} c_{mn} \sin \frac{\pi}{l} m(s-a) \sin \frac{\pi}{l} n(t-a),$$

where $l = b - a$, the c_{mn}'s are the Fourier coefficients of g and $\sum |c_{mn}|^2 < \infty$.

Set

$$u_k(s,t) = \left(\frac{l}{\pi}\right)^2 \sum_{m+n \le k} \frac{-c_{mn}}{m^2 + n^2} \sin \frac{\pi}{l} m(s-a) \sin \frac{\pi}{l} n(t-a).$$

Since $\{D^\alpha u_k\}_{k=1}^\infty$ ($|\alpha| \le 2$) is a Cauchy sequence in $L^2(G, dm)$, there is a function f in $W_0^{22}(G)$ such that

$$f(s,t) = -\left(\frac{l}{\pi}\right)^2 \sum \frac{c_{mn}}{m^2 + n^2} \sin \frac{\pi}{l} m(s-a) \sin \frac{\pi}{l} n(t-a),$$

and $\Delta f = g$. Thus Δ is surjective. It is easy to see that the $W^{22}(G)$-norm, restricted to $W_0^{22}(G)$, is actually equivalent to the norm

$$\|f\|_0 = \left[\int_G |\Delta f|^2 dm\right]^{1/2}.$$

Thus Δ is an isometric isomorphism from the Hilbert space $W_0^{22}(G)$ onto the Hilbert space $L^2(G, dm)$.

Theorem 3.8. (i) *$M^\circ_\lambda \simeq m_\lambda + K$, where m_λ is the normal operator "multiplication by λ" on $L^2(G, dm)$ and K is compact in the Schatten class C^3.*
(ii) *$M^\circ_\lambda(\Gamma) \simeq$ Normal + Compact and $[M^\circ_\lambda(\Gamma)^*, M^\circ_\lambda(\Gamma)] \in C^3$.*

Proof. (i) Consider the ONB

$$\left\{e_{mn}(s,t) = \frac{2}{l}\sin\frac{\pi}{l}m(s-a)\sin\frac{\pi}{l}n(t-a)\right\}_{m,n=1}^{\infty}$$

of $L^2(G, dm)$ and the corresponding ONB

$$\left\{\Delta^{-1}e_{mn} = -\left(\frac{l}{\pi}\right)^2\frac{l}{m^2+n^2}\,e_{mn}\right\}_{m,n=1}^{\infty}$$

of $W_0^{22}(G)$. Note that

$$\begin{aligned}
\Delta M^\circ_s \Delta^{-1} e_{mn} &= -\left(\frac{l}{\pi}\right)^2 (m^2+n^2)^{-1}\Delta M^\circ_s e_{mn}\\
&= -\left(\frac{l}{\pi}\right)^2 (m^2+n^2)^{-1}\Delta\left(\frac{2}{l}\,s\sin\frac{\pi}{l}m(s-a)\sin\frac{\pi}{l}n(t-a)\right)\\
&= s\,e_{mn} - \frac{l}{\pi}2m(m^2+n^2)^{-1}f_{mn}, \quad m,n = 1,2,\cdots,
\end{aligned}$$

where

$$\left\{f_{mn} = \frac{2}{l}\cos\frac{\pi}{l}m(s-a)\sin\frac{\pi}{l}n(t-a)\right\}_{m,n=1}^{\infty}$$

is another ONB of $L^2(G, dm)$.

Thus

$$M^\circ_s \simeq m_s + \sum\frac{-2lm}{\pi(m^2+n^2)}\,f_{mn}\otimes e_{mn} = m_s + K_s,$$

where m_s is the self-adjoint operator "multiplication by s" on $L^2(G, dm)$ and K_s is in C^3.

A similar argument shows that

$$M^\circ_t \simeq m_t + \sum\frac{-2ln}{\pi(m^2+n^2)}\,g_{mn}\otimes e_{mn} = m_t + K_t,$$

where m_t is the self-adjoint operator "multiplication by t" on $L^2(G, dm)$, $\left\{g_{mn} = \frac{2}{l}\sin\frac{\pi}{l}m(s-a)\cos\frac{\pi}{l}n(t-a)\right\}_{m,n=1}^{\infty}$ is an ONB of $L^2(G, dm)$ and K_t is in C^3.

Thus $M^\circ_\lambda \simeq m_\lambda + K$ and K is in C^3.

(ii) Since M°_s and M°_t have compact imaginary parts, $M^\circ_s(\Gamma)$ and $M^\circ_t(\Gamma)$ also have compact imaginary parts. A simply computation shows that $[M^\circ_\lambda(\Gamma)^*, M^\circ_\lambda(\Gamma)] \in C^3$. By the Brown–Douglas–Fillmore theorem $M^\circ_\lambda(\Gamma) =$ Normal + Compact. □

Proposition 3.9. *Given $\lambda \in \mathbb{C}$, there exists a strongly irreducible operator $T \in \mathcal{L}(\mathcal{H})$ with the nice property such that*

$$\sigma(T) = \sigma_p(T^*)^* = \{\lambda\} \quad \text{and} \quad \bigvee\{\ker(\lambda - T)^{*n} : n \geq 1\} = \mathcal{H}.$$

Proof. Let $\{e_n\}_{n=1}^{\infty}$ be an ONB of $\mathcal{H}$ and $\{\omega_n\}_{n=1}^{\infty}$ be a sequence of positive numbers such that $\omega_1 > \omega_2 > \cdots$ and $\lim_{n\to\infty} \omega_n = 0$. Let A be the unilateral weighted shift with weight $\{\omega_n\}$. Then computations show that $A \in (SI)(\mathcal{H})$. Thus $T = \lambda - A$ satisfies the requirements of the proposition. □

3.2 The strongly irreducible operators in Cowen–Douglas operators

In 3.1 we have proved that for a natural number n and an analytic Cauchy domain Ω, there exists a (SI) operator $M_n(\Omega) \in \mathcal{B}_n(\Omega)$ with nice property. In this section Ω always denotes a bounded, connected, open subset of $\mathbb{C}$. We will construct (SI) operators in $\mathcal{B}_n(\Omega)$, $(1 \leq n \leq \infty)$.

Lemma 3.10. *Given $A \in \mathcal{B}_1(\Omega)$ and $\lambda_0 \in \Omega$, there exists an ONB $\{e_n\}_{n=1}^{\infty}$ of $\mathcal{H}$ and $r > 0$ such that A admits the matrix representation*

$$A = \begin{bmatrix} \lambda_0 & a_{12} & a_{13} & \cdots \\ & \lambda_0 & a_{23} & \cdots \\ & & \lambda_0 & \cdots \\ & 0 & & \ddots \end{bmatrix} \tag{3.1}$$

with respect to $\{e_n\}_{n=1}^{\infty}$ and $|a_{k\,k+1}| > r > 0$ for all $k = 1, 2, \cdots$.

Proof. Assume that $e_1 = \ker(A - \lambda_0)$ and $\|e_1\| = 1$. Let B be the right inverse of $A - \lambda_0$. Since $\ker B^k = \{0\}$ $(k = 1, 2, \cdots)$ and $e_1 \notin \operatorname{ran} B$, $\{e_1, Be_1, B^2e_1, \cdots\}$ is linearly independent. Since $B^{k-1}e_1 \in \ker(A - \lambda_0)^k$ and $\operatorname{nul}(A - \lambda_0)^k = k$, $\ker(A - \lambda_0)^k = \bigvee\{e_1, Be_1, B^2e_1, \cdots, B^{k-1}e_1\}$ $(k = 1, 2, \cdots)$. Since $\bigvee\{\ker(A - \lambda_0)^k : k = 1, 2, \cdots\} = \mathcal{H}$, $\bigvee\{B^ke_1 : k = 0, 1, 2, \cdots\} = \mathcal{H}$. Let $\{e_k\}_{k=1}^{\infty}$ be the Gram–Schmidt orthonormalization of $\{B^ke_1\}_{k=0}^{\infty}$. Then A has an upper triangular matrix representation (3.1) with respect to the ONB $\{e_n\}_{n=1}^{\infty}$. Note that $A - \lambda_0$ is bounded from below on $\ker(A - \lambda_0)^{\perp}$, thus there exists $r > 0$ such that $\|(A - \lambda_0)y\| \geq r\|y\|$ for all $y \in \left[\bigvee\{e_1\}\right]^{\perp}$. Set $x_k = a_{k\,k+1}e_k$ and $x_k^1 = -\sum_{i=1}^{k-1} a_{i\,k+1}e_i$ $(k = 1, 2, \cdots)$. Since

$A-\lambda_0$ is onto, there is a vector $y_k \in \bigvee\{e_i : i = 2, 3, \cdots, k\}$ such that $x_k^1 = (A-\lambda_0)y_k$. Thus $x_k = (A - \lambda_0)(e_{k+1} + y_k)$ and

$$\begin{aligned}|a_{k\,k+1}| = \|x_k\| &= \|(A - \lambda_0)(e_{k+1} + y_k)\| \\ &\geq r\|e_{k+1} + y_k\| \\ &= r\sqrt{\|e_{k+1}\|^2 + \|y_k\|^2} \geq r \qquad (k = 1, 2, \cdots).\end{aligned}$$ □

Lemma 3.11. *Given $\epsilon > 0$, there exists a sequence $\{\alpha_{ik} : i = 1, 2, \cdots; k = 1, 2, \cdots\}$ of positive numbers satisfying*

(i) $\overline{\lim}_{n\to\infty} \prod_{k=1}^{n} \frac{\alpha_{ik}}{\alpha_{jk}} = \infty, \ \underline{\lim}_{n\to\infty} \prod_{k=1}^{n} \frac{\alpha_{ik}}{\alpha_{jk}} = 0 \quad (i \neq j);$

(ii) $\sup\{|\beta_{ik}| : i, k = 1, 2, \cdots\} < \epsilon$, $\lim_{i\to\infty} \sup_k\{|\beta_{ik}|\} = \lim_{k\to\infty} \sup_i\{|\beta_{ik}|\} = 0$, *where* $\beta_{ik} = \alpha_{ik} - 1$.

Proof. Observe that

$$\begin{aligned}&\lim_{k\to\infty} \left(\frac{k+1}{k}\right)^{\alpha} = \lim_{k\to\infty} \left(\frac{k}{k+1}\right)^{\alpha} = 1 \\ &\lim_{m\to\infty} \prod_{k=n}^{m} \left(\frac{k+1}{k}\right)^{\alpha} = \infty \qquad\qquad (0 < \alpha < 1). \\ &\lim_{m\to\infty} \prod_{k=n}^{m} \left(\frac{k}{k+1}\right)^{\alpha} = 0.\end{aligned} \tag{3.2}$$

There exists a natural number n_1 such that $M_1 = \prod_{k=1}^{n_1} a_{1k}/a_{2k} > 1$, where $a_{1k} = [(k+1)/k]^{1/2}$, $a_{2k} = 1$ $(k = 1, 2, \cdots, n_1)$. By (3.2), we can find $n_2 > n_1$ such that $M_2 = (\prod_{k=n_1+1}^{n_2} a_{1k})/(\prod_{k=n_1+1}^{n_2} a_{2k}) < 1/(2M_1)$, where $a_{1k} = 1$, $a_{2k} = [(k+1)/k]^{1/2}$ $(k = n_1 + 1, \cdots, n_2)$. Similarly, we can find a sequence $\{n_j\}_{j=1}^{\infty}$ of natural numbers such that

$$\prod_{k=1}^{n_j} \frac{a_{1k}}{a_{2k}} \begin{cases} > j, & j \text{ is odd,} \\ < 1/j, & j \text{ is even.} \end{cases}$$

Define

$$a_{3k} = \begin{cases} \left(\frac{k+1}{k}\right)^{1/2+1/4}, & n_{2j} + 1 \leq k \leq n_{2j+1}, \\ \left(\frac{k}{k+1}\right)^{1/2+1/4}, & n_{2j+1} + 1 \leq k \leq n_{2(j+1)}, \end{cases} \qquad j = 0, 1, 2, \cdots.$$

In general

$$a_{ik} = \begin{cases} \left(\frac{k+1}{k}\right)^{1/2+1/4+\cdots+1/2^{i-1}}, & n_{2j}+1 \le k \le n_{2j+1}, \\ \left(\frac{k}{k+1}\right)^{1/2+1/4+\cdots+1/2^{i-1}}, & n_{2j+1}+1 \le k \le n_{2(j+1)}, \end{cases} \quad j = 0, 1, 2, \cdots,$$

where $n_0 = 0$.

Calculations show that there exists a natural number N such that $|a_{ik} - 1| < \epsilon$ when $i \ge 3$ and $k \ge N$. Thus $\alpha_{ik} := a_{i+2,k+N}$ $(i, k \ge 1)$ satisfy the requirements of the lemma. □

Lemma 3.12. *Given* $B \in \mathcal{B}_1(\Omega)$ *and* $\epsilon > 0$, *there exists a sequence* $\{B_i\}_{i=1}^{\infty}$ *in* $\mathcal{B}_1(\Omega)$ *such that*

(i) $B_i = B + K_i$, *where* $K_i \in \mathcal{K}(\mathcal{H})$, $\sup_i\{\|K_i\|\} < \epsilon$ *and* $\lim_{i\to\infty} \|K_i\| = 0$;

(ii) $\ker \tau_{B_i B_j} = \{0\}$ $(i \ne j)$.

Proof. Without loss of generality, we can assume that $0 \in \Omega$. By Lemmas 2.20 and 3.10 there is an ONB $\{e_n\}_{n=1}^{\infty}$ of $\mathcal{H}$ such that B admits the matrix representation

$$B \simeq \begin{bmatrix} 0 & b_1 & & * \\ & 0 & b_2 & \\ & & 0 & \ddots \\ & 0 & & \ddots \end{bmatrix}$$

with respect to $\{e_k\}$ with $b_k > 0$ $(k \ge 1)$ and $\inf_k\{b_k\} \ge r > 0$.

For each $i \ge 1$, let $K_i \in \mathcal{L}(\mathcal{H})$ such that $K_i e_1 = 0$ and $K_i e_k = \beta_{ik} b_k e_{k-1}$ $(k > 1)$, where $\beta_{ik} = \alpha_{ik} - 1$ is given in Lemma 3.11. Thus $K_i \in \mathcal{K}(\mathcal{H})$, $\sup\{\|K_i\|\} < \epsilon$ and $\lim_{i\to\infty} \|K_i\| = 0$.

Set $B_i = B + K_i$. It is easily seen that $B_i \in \mathcal{B}_1(\Omega)$ $(i \ge 1)$. Let $X \in \mathcal{L}(\mathcal{H})$ and $B_i X = X B_j$ $(i \ne j)$. Thus $B_i^n X e_n = X B_j^n e_n = 0$ $(n \ge 1)$, $X e_n \in \ker B_i^n = \bigvee\{e_k : k = 1, 2, \cdots, n\}$, and

$$X = \begin{bmatrix} x_{11} & x_{12} & x_{13} & \cdots \\ & x_{22} & x_{23} & \cdots \\ & & x_{33} & \\ & 0 & & \ddots \end{bmatrix}$$

with respect to $\{e_k\}_{k=1}^{\infty}$.

A simple calculation indicates that

$$\alpha_{in} b_n x_{n+1\, n+1} = x_{nn} \alpha_{jn} b_n$$

and

$$x_{n+1\, n+1} = \prod_{k=1}^{n} \frac{\alpha_{jk}}{\alpha_{ik}} x_{11} \quad (n \geq 1).$$

If $x_{11} \neq 0$, by Lemma 3.11 $\overline{\lim}_{n\to\infty} |x_{nn}| = \infty$. Thus $x_{nn} = 0$ $(n = 1, 2, \cdots)$.

Similarly, $\alpha_{in} b_n x_{n+1\, n+2} = x_{n\, n+1} \alpha_{j\, n+1} b_{n+1}$ and

$$x_{n+1\, n+2} = \frac{\alpha_{i\, n+1} b_{n+1}}{\alpha_{j1} b_1} \prod_{k=1}^{n+1} \frac{\alpha_{jk}}{\alpha_{ik}} x_{12} \quad (n \geq 1).$$

Therefore $x_{n\, n+1} = 0$ $(n = 1, 2, \cdots)$.

By the same argument, $x_{ij} = 0$ $(i, j = 1, 2, \cdots)$ and $X = 0$. Thus $\ker \tau_{B_i B_j} = \{0\}$ $(i \neq j)$. □

Observe that since $\sigma_r(B_1) \cap \sigma_l(B_j) \neq \varnothing$, by Theorem 1.11 there exists $C_j \in \mathcal{K}(\mathcal{H})$ with $\|C_j\| < \epsilon/2^{j+1}$ such that $C_j \notin \operatorname{ran} \tau_{B_1 B_j}$ $(j = 1, 2, \cdots)$.

Proposition 3.13. *Given $B \in \mathcal{B}_1(\Omega)$, let $\{B_k\}_{k=1}^{\infty}$ be given as in Lemma 3.12 and*

$$T = \begin{bmatrix} B_1 & C_2 & C_3 & \cdots & C_{n-1} \\ & B_3 & & 0 & \\ & & B_5 & & \\ & 0 & & \ddots & \\ & & & & B_{2n-1} \end{bmatrix} \in \mathcal{L}(\mathcal{H}^{(n-1)}) \quad (1 < n \leq \infty),$$

where $C_j \in \mathcal{K}(\mathcal{H})$ such that $\|C_j\| < \epsilon/2^{j+1}$ and $C_j \notin \operatorname{ran} \tau_{B_1 B_{2j-1}}$ $(j = 1, 2, \cdots, n-1)$. Then $T \in \mathcal{B}_n(\Omega)$ and $T \in (SI)$, $\sigma(T) = \sigma(B)$ and $\ker \tau_{T, B_{2k}} = \ker \tau_{B_{2k}, T} = \{0\}$ $(k = 1, 2, \cdots)$.

Proof. We only prove the theorem when $n = \infty$. Let $P \in \mathcal{A}'(T)$ be idempotent and admit the representation

$$P = \begin{bmatrix} P_{11} & P_{12} & P_{13} & \cdots \\ P_{21} & P_{22} & P_{23} & \cdots \\ & \cdots & & \cdots \end{bmatrix}.$$

Since $TP = PT$, $B_3P_{21} = P_{21}B_1$. Since $\ker \tau_{B_3B_1} = \{0\}$, $P_{21} = 0$. Similarly, $P_{lk} = 0$ $(l > k)$. Thus,

$$P = \begin{bmatrix} P_{11} & P_{12} & P_{13} & \cdots \\ & P_{22} & P_{23} & \cdots \\ & & P_{33} & \cdots \\ & 0 & & \ddots \end{bmatrix},$$

and $P_{ll}B_{2l-1} = B_{2l-1}P_{ll}$ and $P_{ll}^2 = P_{ll}$ $(l = 1, 2, \cdots)$. By Proposition 2.28 $B_{2l-1} \in (SI)(\mathcal{H})$ and $P_{ll} = 0$ or 1. Assume that $P_{11} = 0$ (if $P_{11} = 1$ consider $1 - P$). If $P_{22} = 1$ since $P_{12}B_3 = B_1P_{12} + C_2$, $C_2 \in \operatorname{ran} \tau_{B_1B_3}$, contradiction. Thus $P_{22} = 0$ and therefore $P_{12} = 0$. By the same argument, $P_{ll} = 0$ $(l = 1, 2, \cdots)$ and $P = 0$, i.e., $T \in (SI)$.

Observe that T is a compact perturbation of $B^{(\infty)}$, therefore $\sigma(T)\backslash\sigma(B^{(\infty)}) = \sigma(T)\backslash\sigma(B) = \sigma_0(T)$. Since $T \in (SI)$, $\sigma_0(T) = \varnothing$. Thus $\sigma(T) = \sigma(B)$. Since $\ker \tau_{B_iB_j} = \{0\}$ $(i \neq j)$, computation shows that $\ker \tau_{T,B_{2k}} = \ker \tau_{B_{2k}T} = \{0\}$. It is easily seen that $T \in \mathcal{B}_\infty(\Omega)$. □

Corollary 3.14. *Given $T \in \mathcal{L}(\mathcal{H}^{(n)})$ $(1 \leq n \leq \infty)$ with the representation*

$$T = \begin{bmatrix} A_1 & C_2 & C_3 & \cdots & C_n \\ & A_2 & & & \\ & & A_3 & & 0 \\ & & & \ddots & \\ & 0 & & & \\ & & & & A_n \end{bmatrix}$$

satisfying

(i) $A_i \in (SI)(\mathcal{H})$ $(i = 1, \cdots, n)$;

(ii) $\ker \tau_{A_iA_j} = \{0\}$ $(i \neq j,\ i, j \geq 2)$, $\ker \tau_{A_iA_1} = \{0\}$ $(i \geq 2)$;

(iii) $C_i \notin \operatorname{ran} \tau_{A_1A_j}$ $(j = 1, 2, \cdots)$;

then T is strongly irreducible.

The next two results are also about the construction of new (SI) operators in terms of (SI) operators.

Proposition 3.15. *Given $T \in \mathcal{L}(\bigoplus \mathcal{H}_n)$ with the representation*

$$T = \begin{bmatrix} T_1 & T_{12} & & * \\ & T_2 & T_{23} & \\ & & T_3 & \ddots \\ & 0 & & \ddots \end{bmatrix}$$

satisfying

(i) $T_n \in (SI)(\mathcal{H}_n)$;

(ii) $T_{n\,n+1} \notin \operatorname{ran} \tau_{T_n T_{n+1}}$;

(iii) $\ker \tau_{B_n A_n} = \{0\}$ $(n \geq 1)$,

where

$$A_n = \begin{bmatrix} T_1 & T_{12} & \dots & T_{1n} \\ & T_2 & \dots & T_{2n} \\ & & \ddots & \\ 0 & & & T_n \end{bmatrix}, \quad B_n = \begin{bmatrix} T_{n+1} & T_{n+1\,n+2} & \dots \\ & T_{n+2} & \dots \\ 0 & & \ddots \end{bmatrix}$$

then T is strongly irreducible.

Proof. Let $P = (P_{ij})_{i,j=1}^{\infty} \in \mathcal{A}'(T)$ be idempotent. By (iv), $P_{ij} = 0\,(i > j)$. Thus P_{ii} is an idempotent commuting with T_i $(i \geq 1)$. Since $T_i \in (SI)$, $P_{ii} = 0$ or 1. Assume that $P_{11} = 0$ (if $P_{11} = 1$, consider $1 - P$). Since $PT = TP$, $P_{12}T_2 = T_1P_{12} + T_{12}P_{22}$. By (ii), $P_{22} = 0$ and therefore $P_{12} = 0$. Inductively, we can prove that $P_{nn} = 0$ and $P_{nm} = 0$ $(n < m)$, i.e., $P = 0$ and $T \in (SI)$. □

Proposition 3.16. *Given $T \in \mathcal{L}(\mathcal{H})$, where $\mathcal{H} = \bigoplus_{k=1}^{n} \mathcal{H}_k$ $(n < \infty)$, such that*

$$T = \begin{bmatrix} T_1 & T_{12} & \dots & T_{1n} \\ & T_2 & \ddots & \vdots \\ & & \ddots & T_{n-1\,n} \\ 0 & & & T_n \end{bmatrix}$$

and

(i) *each $T_i \in (SI)(\mathcal{H}_i)$,*

(ii) $\ker \tau_{T_i T_j} = \{0\}$ $(i > j)$,

(iii) $T_{i\,i+1} \notin \operatorname{ran} \tau_{T_i T_{i+1}}$ $i = 1, 2, \cdots, n-1$;

then $T \in (SI)(\mathcal{H})$.

Proof. Assume that $P \in \mathcal{A}'(T)$ is an idempotent and admits a block matrix representation $P = (P_{ij})_{i,j=1}^{n}$ with respect to $\mathcal{H} = \bigoplus_{i=1}^{n} \mathcal{H}_i$. Since $P_{n1}T_1 = T_nP_{n1}$, by (ii) $P_{n1} = 0$. Similarly, we can prove $P_{ij} = 0$ $(i > j)$. By the same argument of Proposition 3.15 we can prove $T \in (SI)(\mathcal{H})$. □

Given $B_1,\ B_2 \in \mathcal{B}_1(\Omega)$, $(0 \in \Omega)$, by Lemmas 2.20 and 3.10, B_1 and B_2 admit the following matrix representations

$$B_1 \simeq B_1' = \begin{bmatrix} 0 & b^1_{12} & & & * \\ & 0 & b^1_{23} & & \\ & & 0 & b^1_{34} & \\ & & & 0 & \ddots \\ & 0 & & & \ddots \end{bmatrix} \begin{matrix} e_1 \\ e_2 \\ e_3 \\ e_4 \\ \vdots \end{matrix}, \quad B_2 \simeq B_2' = \begin{bmatrix} 0 & b^2_{12} & & & * \\ & 0 & b^2_{23} & & \\ & & 0 & b^2_{34} & \\ & & & 0 & \ddots \\ & 0 & & & \ddots \end{bmatrix} \begin{matrix} e_1 \\ e_2 \\ e_3 \\ e_4 \\ \vdots \end{matrix},$$

where $\{e_n\}_{n=1}^{\infty}$ is an ONB of $\mathcal{H}$ and $b^i_{n\,n+1} > r > 0$ $(i=1,2;\ n = 1,2,\cdots)$ for some r.

Define $r(B_1, B_2) = \overline{\lim}_{n\to\infty}[\prod_{k=1}^{n} b^1_{k\,k+1}/b^2_{k\,k+1}]^{1/n}$.

Proposition 3.17. (i) *If* $r(B_1, B_2) > 1$, *then* $\ker \tau_{B_2 B_1} = \{0\}$.

(ii) *If* $r(B_1 B_2) = 1$, *then given* $\epsilon > 0$ $(\epsilon < r)$, *there exists a compact operator* K *satisfying*

(a) $\|K\| < \epsilon$;

(b) $\ker \tau_{B_1\ B_2+K} = \ker \tau_{B_2+K\ B_1} = \{0\}$;

(c) $B_2 + K \in \mathcal{B}_1(\Omega)$ *and* $r(B_1, B_2 + K) = 1$.

Proof. (ii) Denote $d_i = 1 - \epsilon/2^i$ $(i = 1,2,\cdots)$. Since

$$\overline{\lim_{n\to\infty}} \left[\prod_{k=1}^{n} \frac{b^1_{k\,k+1}}{b^2_{k\,k+1} d_1}\right]^{1/n} = \frac{1}{d_1} > 1,$$

there exists n_1 such that

$$\prod_{k=1}^{n_1} \frac{b^1_{k\,k+1}}{b^2_{k\,k+1} d_1} > 2.$$

Set $\beta_k = 1 - d_1$ $(1 \le k \le n_1)$. Since

$$\overline{\lim_{n\to\infty}} \left[\left(\prod_{k=1}^{n_1} \frac{b^1_{k\,k+1}}{b^2_{k\,k+1}(1-\beta_k)}\right)\left(\prod_{k=n_1+1}^{n} \frac{b^1_{k\,k+1} d_2}{b^2_{k\,k+1}}\right)\right]^{1/n} = d_2 < 1,$$

we can find $n_2 > n_1$ such that

$$\prod_{k=1}^{n_1} \frac{b^1_{k\,k+1}}{b^2_{k\,k+1}(1-\beta_k)} \cdot \prod_{k=n_1+1}^{n_2} \frac{b^1_{k\,k+1} d_2}{b^2_{k\,k+1}} < \frac{1}{2}.$$

Set $\beta_k = 1 - 1/d_2$ $(n_1 + 1 \le k \le n_2)$. Inductively, we can define

$$\beta_k = \begin{cases} 1 - d_{2l-1}, & n_{2l-2} + 1 \le k \le n_{2l-1}, \\ 1 - \frac{1}{d_{2l}}, & n_{2l-1} + 1 < k \le n_{2l} \end{cases}$$

such that

$$\prod_{k=1}^{n_{2l-1}} \frac{b^1_{k\,k+1}}{b^2_{k\,k+1}(1-\beta_k)} > 2^l$$
$$\prod_{k=1}^{n_{2l}} \frac{b^1_{k\,k+1}}{b^2_{k\,k+1}(1-\beta_k)} < 2^{-l} \qquad (l = 1, 2, \cdots), \tag{3.3}$$

and $\lim_{k\to\infty} \beta_k = 0$ and $\sup_k |\beta_k| < \epsilon/2$.

Define $K'e_k = -b^2_{k\,k+1}\beta_k e_{k-1}$ $(k = 2, 3, \cdots)$ and $K'e_1 = 0$. Then K' is compact and $\|K'\| < \epsilon/2$. It is easily seen that $B'_2 + K' \in \mathcal{B}_1(\Omega)$. If $B'_1 X = X(B'_2 + K')$ for some $X \in \mathcal{L}(\mathcal{H})$, then by the argument of Lemma 3.11 we can prove that

$$X = \begin{bmatrix} x_{11} & x_{12} & \cdots \\ & x_{22} & \cdots \\ 0 & & \ddots \end{bmatrix}$$

with respect to $\{e_n\}$ and

$$x_{nn} = \prod_{k=1}^{n-1} \frac{b^2_{k\,k+1}(1-\beta_k)}{b^1_{k\,k+1}} x_{11} \quad n = 1, 2, \cdots .$$

By (3.3), $x_{nn} = 0$ $(n = 1, 2, \cdots)$. Similarly, computation indicates that

$$x_{n\,n+1} = \frac{b^1_{n\,n+1}}{b^2_{12}(1-\beta_1)} \prod_{k=1}^{n} \frac{b^2_{k\,k+1}(1-\beta_k)}{b^1_{k\,k+1}} x_{12} \quad (k = 2, 3, \cdots).$$

By (3.3), $x_{n\,n+1} = 0$ $(n = 1, 2, \cdots)$. Generally, we can prove that $x_{ij} = 0$ $(i < j)$ and therefore, $\ker \tau_{B'_1\,B'_2+K'} = \{0\}$. By the same argument, $\ker \tau_{B'_2+K'\,B'_1} = \{0\}$. From the definition of $\{\beta_k\}$, it is easy to see that $r(B'_1, B'_2 + K') = 1$. Since $B_1 \simeq B'_1$ and $B_2 \simeq B'_2$, we can find a compact operator K satisfying all the requirements of (ii).

(i) If $r(B_1, B_2) > 1$, then there is a subsequence $\{n_i\}_{i=1}^{\infty}$ of natural numbers such that $n_1 < n_2 < \cdots$ and

$$\prod_{k=1}^{n_k} \frac{b^1_{k\,k+1}}{b^2_{k\,k+1}} > k \quad (k = 1, 2, \cdots).$$

By the same argument as (ii), $\ker \tau_{B_2 B_1} = \{0\}$. □

By the arguments of Proposition 3.17 we have

Proposition 3.18. *Let* $\{B_k\}_{k=1}^{\infty}$ *be a sequence of operators in* $\mathcal{B}_1(\Omega)$ *such that* $r(B_k, B_l) = 1$ $(k \neq l)$; *then for* $\epsilon > 0$, *there exists a sequence* $\{K_k\}_{k=1}^{\infty}$ *of compact operators such that*

(i) $\|K_k\| < \epsilon/2^k$, $(k = 1, 2, \cdots)$;

(ii) $T_k = B_k + K_k \in \mathcal{B}(\Omega)$, $(k = 1, 2, \cdots)$;

(iii) $\ker \tau_{T_k T_l} = \{0\}$, $(k \neq l)$.

Given $B \in \mathcal{B}_1(\Omega)$, by Lemma 2.20 B admits an upper triangular representation $B = (b_{ij})_{i,j=1}^{\infty}$ with $b_{ij} = 0$ $(i > j)$. By Lemma 3.10,

$$0 < \underline{\lim}_{n\to\infty} \left[\prod_{k=1}^{n} |b_{k\,k+1}|\right]^{1/n} < \overline{\lim}_{n\to\infty} \left[\prod_{k=1}^{n} |b_{k\,k+1}|\right]^{1/n} < \|B\|.$$

Theorem 1.44 implies that the spectrum of a $\mathcal{B}_1(\Omega)$ operator is very complicated. From the proof of Proposition 3.17 we know that the necessary condition for $\mathcal{B}_1(\Omega)$ operators B_1 and B_2 to be similar or quasisimilar to each other is

$$r(B_1, B_2) = \overline{\lim_{n\to\infty}} \left[\prod_{k=1}^{n} \left|\frac{b^1_{k\,k+1}}{b^2_{k\,k+1}}\right|\right]^{1/n} = 1,$$

otherwise, by Proposition 3.17 one of $\ker \tau_{B_1 B_2}$ and $\ker \tau_{B_2 B_1}$ must be $\{0\}$. Thus, it is necessary to evaluate $\lim_{n\to\infty} \left[\prod_{k=1}^{n} |b_{k\,k+1}|\right]^{1/n}$ and

$$r(B_1, B_2) = \overline{\lim_{n\to\infty}} \left[\prod_{k=1}^{n} \frac{b^1_{k\,k+1}}{b^2_{k\,k+1}}\right]^{1/n}.$$

Proposition 3.19. *Given* $B \in \mathcal{B}_1(D)$ *with* $\sigma(B) = \overline{D}$, *where* $D = \{\lambda \in \mathbb{C} : |\lambda| < 1\}$, *assume that* B *admits an upper triangular matrix representation* $B = (b_{ij})_{i,j=1}^{\infty}$ *with* $b_{ij} = 0$ $(i \geq j)$ *with respect to an ONB* $\{e_n\}_{n=1}^{\infty}$ *of* $\mathcal{H}$. *Then*

$$\lim_{n\to\infty} \left[\prod_{k=1}^{n} |b_{k\,k+1}|\right]^{1/n} = 1.$$

Proof. By Lemma 3.10, there exists $r > 0$ such that

$$0 < r < \inf\{|b_{k\,k+1}|\} \leq \sup\{|b_{k\,k+1}|\} \leq \|B\|.$$

Let A be the right inverse of B such that $BA = 1$, $AB = 1 - P_{\ker B}$, and

$$A = \begin{bmatrix} * & & & * \\ 1/b_{12} & * & & \\ & 1/b_{23} & * & \\ 0 & & \ddots & \ddots \end{bmatrix} \begin{matrix} e_1 \\ e_2 \\ e_3 \\ \vdots \end{matrix}.$$

For each $\lambda \in D$, $(\lambda - A) = (\lambda B - 1)A = \lambda(B - 1/\lambda)A$. Since $(B - 1/\lambda)$ is invertible, $\lambda \in \rho_{s-F}(A)$ and $\operatorname{ind}(\lambda - A) = -1$. Thus $\sigma(A) \supset \overline{D}$. If $|\lambda| > 1$, since $1/\lambda \in D$, $\lambda - A$ is a Fredholm operator with $\operatorname{ind}(\lambda - A) = 0$. Therefore $\sigma_e(A) = \partial D$ and $\sigma_0(A) \subset \mathbb{C}\backslash\overline{D}$. By Proposition 1.20, there is a compact operator K such that $\sigma(A + K) = \overline{D}$. Given an $\epsilon > 0$, there is a natural number n_0 such that $\sigma(T) \subset D_\epsilon = \{\lambda \in \mathbb{C} : |\lambda| < 1 + \epsilon\}$ where $T = A + P_{n_0}KP_{n_0}$ and P_{n_0} is the orthogonal projection of $\mathcal{H}$ onto $\mathcal{H}_{n_0} = \bigvee\{e_n : n = 1, 2, \cdots, n_0\}$.

Computation shows that the $(n_0 + n + 1, n_0)$ entry of the matrix representation of T^{n+1} with respect to the ONB $\{e_k\}$ is $\left[\prod_{i=1}^{n+1} b_{n_0+i\, n_0+i+1}\right]^{-1}$. Thus

$$\varlimsup_{n\to\infty} \left[\prod_{i=1}^{n+1} |b_{n_0+i\, n_0+i+1}|\right]^{-1/(n+1)} \le \lim_{n\to\infty} \sqrt[n+1]{\|T^{n+1}\|} \le 1 + \epsilon$$

and therefore

$$\varlimsup_{n\to\infty} \left[\prod_{k=1}^{n} |b_{k\, k+1}|\right]^{-1/n} \le 1 + \epsilon$$

or

$$\varliminf_{n\to\infty} \left[\prod_{k=1}^{n} |b_{k\, k+1}|\right]^{1/n} \ge \frac{1}{(1 + \epsilon)}.$$

Since ϵ can be arbitrarily small, $\varliminf_{n\to\infty} \left[\prod_{k=1}^{n} |b_{k\, k+1}|\right]^{1/n} \ge 1$. Since the $(1, n + 1)$ entry of the matrix representation of B^n is $\prod_{k=1}^{n} b_{k\, k+1}$, $\left(\prod_{k=1}^{n} |b_{k\, k+1}|\right)^{1/n} \le \sqrt[n]{\|B^n\|}$. Thus $\varlimsup_{n\to\infty} \left[\prod_{k=1}^{n} |b_{k\, k+1}|\right]^{1/n} \le 1$ and the proof is complete. □

Proposition 3.20. *Let $A, B \in \mathcal{L}(\mathcal{H})$ be quasisimilar ($A \underset{qs}{\sim} B$), $B \in \mathcal{B}_1(\Omega)$ and let $\lambda_0 \in \Omega$; then there exist $r_0 > 0$ and two orthonormal bases $\{e_n\}_{n=1}^\infty$ and $\{f_n\}_{n=1}^\infty$ of $\mathcal{H}$ satisfying:*

(i) *A and B admit upper matrix representations $(a_{ij})_{i,j=1}^\infty$ and $(b_{ij})_{i,j=1}^\infty$ with respect to the ONB $\{e_n\}$ and, respectively, ONB $\{f_n\}$ and $b_{nn} = a_{nn} = \lambda_0$ $(n = 1, 2, \cdots)$;*

(ii) $\lim_{n\to\infty}\left[\prod_{k=1}^{n} b_{k\,k+1}/a_{k\,k+1}\right]^{1/n} = 1$;

(iii) $0 < r_0 < \inf\{|a_{k\,k+1}| : k = 1, 2, \cdots\}$;

(iv) $\mathrm{nul}\,(\lambda - A) = 1$, $\bigvee\{\ker(\lambda - A)^n : n \geq 1\} = \mathcal{H}$ *for each* $\lambda \in \Omega$.

Proof. Without loss of generality, we can assume that $\lambda_0 = 0$. Since $A \underset{qs}{\sim} B$ and $B \in \mathcal{B}_1(\Omega)$, (iv) is true.

(i) Assume that $AX = XB$ and $YA = BY$, where $X, Y \in \mathcal{L}(\mathcal{H})$ have trivial kernel and dense range. By Lemma 3.10, B admits an upper triangular representation $B = (b_{ij})_{i,j=1}^{\infty}$ with respect to an ONB $\{e_n\}_{n=1}^{\infty}$ of $\mathcal{H}$ such that $b_{nn} = 0$ $(n = 1, 2, \cdots)$ and $\inf\{|b_{n\,n+1}| : n = 1, 2, \cdots\} = r > 0$ for some r. Observe that $A^n X e_j = XB^n e_j = 0$ for $j = 1, 2, \cdots, n$ and $n = 1, 2, \cdots$. Since $\bigvee\{\ker B^n : n \geq 1\} = \bigvee\{e_n : n \geq 1\} = \mathcal{H}$, $\bigvee\{\ker A^n : n \geq 1\} = \bigvee\{Xe_n : n \geq 1\} = \mathcal{H}$. Thus we can get an ONB $\{f_n\}_{n=1}^{\infty}$ of $\mathcal{H}$ such that A admits an upper triangular matrix representation $A = (a_{ij})_{i,j=1}^{\infty}$ with respect to $\{f_n\}_{n=1}^{\infty}$ and $a_{nn} = 0$ $(n \geq 1)$.

(ii) Note that $U^*XB = A'U^*X$ and $YUA' = BYU$, where U is the unitary operator: $Ue_n = f_n$ $(n = 1, 2, \cdots)$ and $A' = U^*AU$, A' admits the same matrix representation as A with respect to the ONB $\{e_n\}_{n=1}^{\infty}$.

Computations show that U^*X and YU admit upper matrix representations $(x_{ij})_{i,j=1}^{\infty}$ and $(y_{ij})_{i,j=1}^{\infty}$ respectively, with respect to $\{e_n\}_{n=1}^{\infty}$ and

$$\left.\begin{aligned} x_{n+1\,n+1} &= \prod_{k=1}^{n} \frac{b_{k\,k+1}}{a_{k\,k+1}} x_{11} \\ y_{n+1\,n+1} &= \prod_{k=1}^{n} \frac{a_{k\,k+1}}{b_{k\,k+1}} y_{11} \end{aligned}\right. \qquad (n = 1, 2, \cdots).$$

By the argument of Proposition 3.17

$$\overline{\lim_{n\to\infty}}\left[\prod_{k=1}^{n} \frac{|b_{k\,k+1}|}{|a_{k\,k+1}|}\right]^{1/n} \leq 1 \quad \text{and} \quad \overline{\lim_{n\to\infty}}\left[\prod_{k=1}^{n} \frac{|a_{k\,k+1}|}{|b_{k\,k+1}|}\right]^{1/n} \leq 1.$$

Thus

$$\lim_{n\to\infty}\left[\prod_{k=1}^{n} \frac{|a_{k\,k+1}|}{|b_{k\,k+1}|}\right]^{1/n} = 1.$$

(iii) Note that $YX = (h_{ij})_{i,j=1}^{\infty}$ admits an upper matrix representation with respect to $\{e_n\}_{n=1}^{\infty}$ and $h_{ii} = y_{ii}x_{ii}$ $(i = 1, 2, \cdots)$. From $YXB = YAX = BYX$ we get

$$y_{n+1\,n+1}x_{n+1\,n+1} = \frac{\prod_{k=1}^{n} b_{k\,k+1}}{\prod_{k=1}^{n} b_{k\,k+1}} y_{11}x_{11} = y_{11}x_{11} \quad (n = 1, 2, \cdots).$$

Since X and Y are injective, $y_{nn}x_{nn} = y_{11}x_{11} \neq 0$ $(n = 1, 2, \cdots)$ and $\inf_n\{|x_{nn}|\} > \alpha > 0$. Note that $a_{k\,k+1} = [x_{kk}/x_{k+1\,k+1}]b_{k\,k+1}$ and $\inf\{|b_{k\,k+1}| : k \geq 1\} = r > 0$, thus $\inf\{|a_{k\,k+1}| : k \geq 1\} = r_0 > 0$. □

3.3 The strongly irreducible operators in the class of essentially normal operators

Lemma 3.21. *Let Ω be a bounded, connected, open subset of $\mathbb{C}$. Then there exists an essentially normal operator $B \in \mathcal{B}_1(\Omega)$ such that*

(i) $\sigma(B) = \overline{\Omega}$,

(ii) $\Omega = \rho_F(B) \cap \sigma(B)$.

Proof. Denote $\Phi = (\overline{\Omega})^\circ$, the interior of $\overline{\Omega}$, and $\Gamma = \Phi \backslash \Omega$. Let $\{\lambda_i\}_{i=1}^\infty$ be a dense subset of Γ satisfying $\text{card}\,\{n : \lambda_n = \lambda_k\} = \infty$, $k = 1, 2, \cdots$. Denote $\mathcal{H} = A^2(\Phi^*)$. Let $\{e_n\}_{n=1}^\infty$ be an ONB of $\mathcal{H}$.

Define

$$A = \begin{bmatrix} N_1(\Phi^*)^* & 0 \\ 0 & M \end{bmatrix}$$

on $\mathcal{H} \oplus \mathcal{H}$, where $M \in \mathcal{L}(\mathcal{H})$ such that $Me_n = \lambda_n e_n$ $(n = 1, 2, \cdots)$. Thus A is a quasitriangular operator with $\sigma(A) = \sigma_W(A) = \overline{\Omega}$ and $\text{ind}\,(\lambda - A) = 1$ for $\lambda \in \Omega$. By Theorem 1.44, there exists a compact operator K such that $B = A + K \in \mathcal{B}_1(\Omega)$ and B satisfies (i) and (ii). □

Example 3.22. Define $\gamma_1 = 1$, $\gamma_2 = 1/4$, $\gamma_3 = (\gamma_1\,\gamma_2)^3, \cdots, \gamma_n = (\gamma_1 \cdots \gamma_{n-1})^n$, $\cdots$, and let $\{\alpha_n\}$ be the sequence

$$\gamma_1,\, \gamma_2, \cdots, \gamma_9,\, \gamma_1,\, \gamma_2, \cdots, \gamma_{90},\, \gamma_1,\, \gamma_2, \cdots, \gamma_{900},\, \gamma_1,\, \gamma_2, \cdots, \gamma_{9000},\, \gamma_1, \cdots.$$

Let V be the unilateral weighted shift defined by $Ve_n = \alpha_n e_{n+1}$ $(n \geq 1)$ with respect to an ONB $\{e_n\}_{n=1}^\infty$ of the Hilbert space $\mathcal{H}$. Then V is a quasinilpotent unicellular operator and V^k is not compact for all $k = 1, 2, \cdots$. The interested reader is referred to D.A. Herrero [7].

Combining Example 3.22 with the analysis of Lemma 3.21, we have

Lemma 3.23. *Let Ω be a bounded, connected open subset of $\mathbb{C}$. Then there exists an operator $B \in \mathcal{B}_1(\Omega)$ satisfying*

(i) $\sigma(B) = \overline{\Omega}$, $\Omega = \rho_F(B) \cap \sigma(B)$;

(ii) *For each isolated point* λ_0 *of* $\sigma_e(B)$ *(*λ_0 *must be an isolated point of* $\partial\Omega$ *also),* $\left[(\pi(B)-\lambda_0)E(\lambda_0;\pi(B))\right]^k \neq 0$ *for* $k=1,2,\cdots$*, where* π *is the canonical projection from* $\mathcal{L}(\mathcal{H})$ *onto* $\mathcal{A}(\mathcal{H})=\mathcal{L}(\mathcal{H})/\mathcal{K}(\mathcal{H})$ *and* $E(\lambda_0;\pi(B))$ *is the Riesz idempotent of* $\pi(B)$ *corresponding to* λ_0*.*

Proof. Let Φ, Γ, $\mathcal{H}$, $\{e_n\}_{n=1}^{\infty}$ be given as in Lemma 3.21. Let Γ_1 be the perfect part of Γ and $\Gamma\backslash\Gamma_1=\{\mu_n\}_{n=1}^{l}$ $(0\le l\le\infty)$. Let $\{\lambda_i\}_{i=1}^{\infty}$ be the dense subset of Γ_1 satisfying card$\{n:\lambda_n=\lambda_k\}=\infty$, $k=1,2,\cdots$. The operators $N\in\mathcal{L}(\mathcal{H})$ and $G\in\mathcal{L}(\mathcal{H}^{(l)})$ are given by

$$Ne_n=\lambda_n e_n \quad n=1,2,\cdots$$

$$G=\bigoplus_{n=1}^{l}(\mu_n+V^*).$$

Define

$$A=\begin{bmatrix} N_+(\Phi^*)^* & 0 & 0\\ 0 & N & 0\\ 0 & 0 & G\end{bmatrix}$$

on $\mathcal{H}^{(l+2)}$. By Theorem 1.44, there exists a compact operator K such that $B=A+K\in\mathcal{B}_1(\Omega)$ satisfies (i) and (ii). □

Theorem 3.24. *Let* Ω *be a bounded, connected open subset of* $\mathbb{C}$*, and let* n *be a natural number. Then there exists an essentially normal (SI) operator* B *satisfying*

(i) $B\in\mathcal{B}_n(\Omega)$, $\sigma(B)=\overline{\Omega}$;

(ii) $\Omega=\rho_F(B)\cap\sigma(B)$.

Proof. The theorem is a consequence of Proposition 3.13 and Lemma 3.21. □

Theorem 3.25. *Let* Ω *be a bounded, connected open subset of* $\mathbb{C}$*,* $\epsilon>0$*; then there exists a (SI) operator* B *in* $CalB_n(\Omega)$*,* $(1\le n\le\infty)$ *satisfying*

(i) $\sigma(B)=\overline{\Omega}$, $\rho_F(B)\cap\sigma(B)=\Omega$;

(ii) *For each isolated point* λ_0 *of* $\sigma_e(B)$, $\left[(\pi(B)-\lambda_0)E(\lambda_0;\pi(B))\right]^k\neq 0$, $k=1,2,\cdots$;

(iii) $B=\left[N_+(\Phi^*)^*\right]^{(n)}\bigoplus N\bigoplus\left[\bigoplus_{n=1}^{l}(\mu_n+V^*)\right]+K_\epsilon$*, where* $\{\mu_n\}_{n=1}^{l}$ $(0\le l\le\infty)$ *is the subset of isolated points of* $\partial\Omega$*,* V *is the operator given in Example 3.22,* N *is a normal operator with* $\sigma(N)=\sigma_{lre}(N)=\partial\Omega$*, and* K_ϵ *is compact with* $\|K_\epsilon\|<\epsilon$*.*

Proof. The theorem is a consequence of Proposition 3.13 and Lemma 3.23. □

Theorem 3.26. *Let Ω be a bounded, connected open subset of $\mathbb{C}$. Then there exists an essentially normal and strongly irreducible operator B satisfying*

(i) $\sigma(B) = \overline{\Omega}$;

(ii) $\rho_F(B) \cap \sigma(B) = \Omega$, $\operatorname{ind}(\lambda - B) = 0$, $\lambda \in \Omega$;

(iii) $\operatorname{min\,ind}(\lambda - B) = 1$, $\lambda \in \Omega$.

Proof. By Theorem 3.24, there exist (SI) operators B_1 and B_2 such that $B_1 \in \mathcal{B}_1(\Omega)$, $\sigma(B_1) = \overline{\Omega}$, $\rho_F(B_1) \cap \sigma(B_1) = \Omega$ and $B_2^* \in \mathcal{B}_1(\Omega^*)$, $\sigma(B_2) = \overline{\Omega}$, $\rho_F(B_2) \cap \sigma(B_2) = \Omega$. It follows from $\sigma_r(B_1) \cap \sigma_l(B_2) \neq \varnothing$ and Theorem 1.11 that for $\epsilon > 0$, there exists a compact operator K, $\|K\| < \epsilon$, such that $K \notin \operatorname{ran} \tau_{B_1 B_2}$.

Define

$$B = \begin{bmatrix} B_1 & K \\ 0 & B_2 \end{bmatrix} \in \mathcal{L}(\mathcal{H}^{(2)}).$$

Then B is essentially normal and satisfies (i), (ii) and (iii). We only need prove that $B \in (SI)$. Assume that $P = \begin{bmatrix} P_1 & P_{12} \\ P_{21} & P_2 \end{bmatrix} \in \mathcal{L}(\mathcal{H}^{(2)})$ is an idempotent commuting with B. Thus $P_{21} B_1 = B_2 P_{21}$. Note that $\bigvee\{\ker(\lambda - B_1) : \lambda \in \Omega\} = \mathcal{H}$ and $\sigma_p(B_2) = \varnothing$. By Proposition 1.14, $\ker \tau_{B_2 B_1} = \{0\}$ and $P_{21} = 0$. Observe that $K \notin \operatorname{ran} \tau_{B_1 B_2}$. By the argument of Proposition 3.16, we can get $P = 0$ or 1. □

An operator $T \in \mathcal{L}(\mathcal{H})$ is *almost normal* if T is the form of normal+compact. Theorem 3.8 implies that for a connected perfect subset σ of $\mathbb{C}$, there exists an almost normal (SI) operator T with the nice property such that $\sigma(T) = \sigma$. The following proposition strengthens the result.

Proposition 3.27. *Let σ be a connected perfect subset of $\mathbb{C}$, let $\{u_n\}_{n=1}^{\infty}$ be a countable subset of σ of pairwise distinct numbers and $\epsilon > 0$; then there exists an almost normal operator $A \in (SI)(\mathcal{H})$, where $\mathcal{H}$ is a Hilbert space, and a compact operator K with $\|K\| < \epsilon$ satisfying*

(i) $\sigma(A) = \sigma_{lre}(A) = \sigma$, $\sigma_p(A) \cap \{u_n\}_{n=1}^{\infty} = \varnothing$;

(ii) *There is a countable subset $\{\lambda_i\}_{i=1}^{\infty}$ of $\sigma_p(A)$ such that $\{\lambda_i\}^- = \sigma$ and*

$$\bigvee\{\ker(\lambda_i - A)^k : k \geq 1, i \geq 1\} = \mathcal{H};$$

(iii) $\sigma_p(A^*) = \varnothing$;

(iv) *$N = A - K$ is normal and $\sigma(N) = \sigma(A)$.*

In order to prove Proposition 3.27, we need two lemmas.

Lemma 3.28. *Let σ be a connected perfect subset of $\mathbb{C}$, $\{u_n\}_{n=1}^{\infty}$ and $\{v_k\}_{k=1}^{l}$ $(l<\infty)$ be two subsets of σ and let ϵ be a positive number such that $0<\epsilon<\frac{1}{2}\min\{|v_j-v_k|: j\neq k,\ 1\leq j,k\leq l\}$; then for each $M>0$ there exists a sequence of L pairwise distinct complex numbers $\{\lambda_i\}_{i=1}^{L}$ in $S(v_1,\epsilon)\cap[\sigma\backslash\{u_n\}_{n=1}^{\infty}]$ $(L<\infty)$, such that*

$$|\lambda_j-\lambda_{j+1}|\leq 3\epsilon/2^{j+1}\quad (j=1,2,\cdots,L-1)$$

and

$$\prod_{j=1}^{L-1}\frac{|\lambda_j-v_k|}{|\lambda_j-\lambda_{j+1}|^{1/2}}\geq M\quad (k=2,3,\cdots,l),$$

where $S(v_1,\epsilon)=\{x\in\mathbb{C}: |x-v_1|<\epsilon\}$.

Proof. Since σ is connected and perfect, there exist $\lambda_j\in S(v_1,\epsilon)\cap[\sigma\backslash\{u_n\}_{n=1}^{\infty}]$ such that $\epsilon/2^{j+1}\leq|\lambda_j-v_1|\leq\epsilon/2^j$ $(j=1,2,\cdots)$ and $\{\lambda_j\}_{j=1}^{\infty}$ are pairwise distinct. Thus,

$$|\lambda_j-\lambda_{j+1}|\leq|\lambda_j-v_1|+|v_1-\lambda_{j+1}|\leq\frac{\epsilon}{2^j}+\frac{\epsilon}{2^{j+1}}=\frac{3\epsilon}{2^{j+1}}$$

and

$$|\lambda_j-v_k|>\epsilon\quad (k=2,3,\cdots,l;\ j=1,2,\cdots).$$

Note that

$$\frac{|\lambda_j-v_k|}{|\lambda_j-\lambda_{j+1}|^{1/2}}\geq\left(\frac{\epsilon}{3}\right)^{1/2}2^{(j+1)/2}\qquad (k\geq 2).$$

Thus, for each k $(2\leq k\leq l)$,

$$\prod_{j=1}^{n-1}\frac{|\lambda_j-v_k|}{|\lambda_j-\lambda_{j+1}|^{1/2}}\geq\left(\frac{\epsilon}{3}\right)^{(n-1)/2}\cdot 2^{(n^2+n-2)/4}\to\infty\quad (n\to\infty).$$

Therefore we can find a natural number L such that

$$\prod_{j=1}^{L-1}\frac{|\lambda_j-v_k|}{|\lambda_j-\lambda_{j+1}|^{1/2}}\geq M\quad (k=2,3,\cdots,l).$$

□

Lemma 3.29. *Let σ and $\{u_n\}_{n=1}^{\infty}$ be given as in Lemma 3.28. Then there exists a sequence $\{\lambda_k\}_{k=1}^{\infty}$ of numbers in $\sigma\backslash\{u_n\}_{n=1}^{\infty}$ such that*

(i) $\{\lambda_k\}^{-}=\sigma$;

(ii) $\overline{\lim_{n\to\infty}}\prod_{k=1}^{n}\frac{|\lambda_k-u_m|}{|\lambda_k-\lambda_{k+1}|^{1/2}}=\infty\quad (m=1,2,\cdots);$

(iii) $\operatorname{card}\{k : \lambda_k = \lambda_n\} = \infty \quad (n = 1, 2, \cdots)$;
(iv) $\lambda_n \neq \lambda_{n+1} \quad (n = 1, 2, \cdots)$;
(v) $\lim_{n\to\infty} |\lambda_n - \lambda_{n+1}| = 0$.

Proof. Using Lemma 3.28 with respect to $u_1, u_2, \cdots, u_{10}$, we can find $\lambda_1, \lambda_2, \cdots, \lambda_{L_1}$ in a neighbourhood of u_1 such that $\{\lambda_i\}_{i=1}^{L_1} \subset \sigma\backslash\{u_n\}_{n=1}^{\infty}$, $\lambda_j \neq \lambda_{j+1}$ $(1 \leq j \leq L_1 - 1)$, $\epsilon_1 := |\lambda_{L_1-1} - \lambda_{L_1}| < 1/2$ and

$$\prod_{j=1}^{L_1-1} \frac{|\lambda_j - u_m|}{|\lambda_j - \lambda_{j+1}|^{1/2}} \geq 1 \qquad (m = 2, 3, \cdots, 10).$$

Choose $\lambda_{L_1+1}, \cdots, \lambda_{L_2}$ in $\sigma\backslash\{u_n\}_{n=1}^{\infty}$ such that

(i) $0 < |\lambda_j - \lambda_{j+1}| < \epsilon_1 \quad (L_1 \leq j \leq L_2 - 1)$;
(ii) $\sigma \subset [\{\lambda_j\}_{j=1}^{L_2}]_{\epsilon_1} := \{x \in \mathbb{C} : \operatorname{dist}(x, \{\lambda_j\}_{j=1}^{L_2}) < \epsilon_1\}$;
(iii) $|\lambda_{L_2} - u_2| < \epsilon_1$;

Using Lemma 3.28 with respect to $\{u_n\}_{n=1}^{100}$, find $\{\lambda_j\}_{j=L_2+1}^{L_3}$ in a neighbourhood of u_2 such that $\{\lambda_j\}_{j=L_2+1}^{L_3} \subset \sigma\backslash\{u_n\}_{n=1}^{\infty}$, $\lambda_j \neq \lambda_{j+1}$ $(L_2 \leq j \leq L_3 - 1)$, $\epsilon_2 := |\lambda_{L_3-1} - \lambda_{L_3}| < 1/2^2$ and $\prod_{j=1}^{L_3-1} |\lambda_j - u_m|/|\lambda_j - \lambda_{j+1}|^{1/2} \geq 2$ $(1 \leq m \leq 10^2$ and $m \neq 2)$.

Choose $\{\lambda_j\}_{j=L_3+1}^{L_4} \subset \sigma\backslash\{u_n\}_{n=1}^{\infty}$ such that

(i) $\{\lambda_j\}_{j=1}^{L_3} \subset \{\lambda_j\}_{j=L_3+1}^{L_4}$;
(ii) $0 < |\lambda_j - \lambda_{j+1}| < \epsilon_2 \quad (L_3 \leq j \leq L_4 - 1)$;
(iii) $\sigma \subset [\{\lambda_j\}_{j=1}^{L_4}]_{\epsilon_2}$;
(iv) $|\lambda_{L_4} - u_3| < \epsilon_2$.

In general, for each k find $\{\lambda_j\}_{j=L_{2k}+1}^{L_{2k+1}}$ in a neighbourhood of u_{k+1} such that

$$\{\lambda_j\}_{j=L_{2k}+1}^{L_{2k+1}} \subset \sigma\backslash\{u_n\}_{n=1}^{\infty}; \quad \lambda_j \neq \lambda_{j+1} \; (L_{2k} \leq j \leq L_{2k+1} - 1),$$

$\epsilon_{k+1} := |\lambda_{L_{2k+1}-1} - \lambda_{L_{2k+1}}| < 1/2^{k+1}$ and

$$\prod_{j=1}^{L_{2k+1}-1} \frac{|\lambda_j - u_m|}{|\lambda_j - \lambda_{j+1}|^{1/2}} \geq k + 1 \quad (1 \leq m \leq 10^{k+1} \text{ and } m \neq k + 1).$$

Choose $\{\lambda_j\}_{j=L_{2k+1}+1}^{L_{2(k+1)}} \subset \sigma\backslash\{u_n\}_{n=1}^{\infty}$ such that

(i) $\{\lambda_j\}_{j=1}^{L_{2k+1}} \subset \{\lambda_j\}_{j=L_{2k+1}+1}^{L_{2(k+1)}}$;

(ii) $0 < |\lambda_j - \lambda_{j+1}| < \epsilon_{k+1} \quad (L_{2k+1} \le j \le L_{2(k+1)} - 1)$;

(iii) $\sigma \subset \left[\{\lambda_j\}_{j=1}^{L_{2(k+1)}}\right]_{\epsilon_{k+1}}$;

(iv) $|\lambda_{L_{2(k+1)}} - u_{k+2}| < \epsilon_{k+1}$.

It is easily seen that $\{\lambda_n\}_{n=1}^{\infty}$ satisfies all the requirements of the lemma. $\square$

Proof of Proposition 3.27. Let $\{e_n\}_{n=1}^{\infty}$ be an ONB of a Hilbert space $\mathcal{H}$ and $\{\lambda_j\}_{j=1}^{\infty}$ be given as in Lemma 3.29. Since $\lim_{j\to\infty} |\lambda_j - \lambda_{j+1}|^{1/2} = 0$, there exists n_0 such that $|\lambda_n - \lambda_{n+1}|^{1/2} < \epsilon$ $(n \ge n_0)$. Thus without loss of generality, we can assume that $|\lambda_n - \lambda_{n+1}|^{1/2} < \epsilon$ $(n \ge 1)$ (otherwise, eliminate the first $n_0 - 1$ terms from the sequence $\{\lambda_n\}_{n=1}^{\infty}$).

Define $A, N, K \in \mathcal{L}(\mathcal{H})$ by $Ne_n = \lambda_n e_n$ $(n \ge 1)$; $Ke_1 = 0$ and $Ke_n = \alpha_{n-1}e_{n-1}$, where $\alpha_{n-1} = |\lambda_{n-1} - \lambda_n|^{1/2}$ $(n = 2, 3, \cdots)$; and $A = N + K$. Since $\lim_{n\to\infty} \alpha_n = 0$, K is compact with $\|K\| < \epsilon$ and A is almost normal. If P is an idempotent commuting with A, then $(A - \lambda_1)Pe_1 = P(A - \lambda_1)e_1 = 0$. Computation shows that $\ker(A - \lambda_1) = \bigvee\{e_1\}$. Thus $Pe_1 = p_{11}e_1$ for some complex number p_{11}. If $Pe_i = \sum_{j=1}^{i} p_{ji}e_j$ $(i = 1, 2, \cdots, n-1)$, then

$$(A - \lambda_n)Pe_n = P(A - \lambda_n)e_n = \alpha_{n-1}Pe_{n-1} = \alpha_{n-1}\sum_{j=1}^{n-1} p_{j\,n-1}e_j.$$

Computations show that if $(A - \lambda_n)f \in \bigvee\{e_1, \cdots, e_{n-1}\}$, then $f \in \bigvee\{e_1, \cdots, e_n\}$. Thus P admits an upper matrix representation

$$P = \begin{bmatrix} p_{11} & & * \\ & p_{22} & \\ & 0 & \ddots \end{bmatrix}$$

with respect to $\{e_n\}_{n=1}^{\infty}$. Since $P^2 = P$, $p_{nn} = 1$ or 0 $(n = 1, 2, \cdots)$. Comparing the $(n, n+1)$ entries of both sides of $AP = PA$, we have

$$\lambda_n p_{n\,n+1} + \alpha_n p_{n+1\,n+1} = p_{nn}\alpha_n + p_{n\,n+1}\lambda_{n+1}$$

or

$$|p_{n\,n+1}| = \frac{|p_{n+1\,n+1} - p_{nn}|\alpha_n}{|\lambda_n - \lambda_{n+1}|} = \frac{|p_{n+1\,n+1} - p_{nn}|}{|\lambda_n - \lambda_{n+1}|^{1/2}} \quad (n = 1, 2, \cdots).$$

Since $|\lambda_n - \lambda_{n+1}|^{1/2} \to 0$ $(n \to \infty)$, there exists a natural number m such that $p_{n+1\,n+1} = p_{nn}$ $(n > m)$. Assume that $p_{nn} \equiv 0$ $(n > m)$ and m is the largest number satisfying $p_{mm}=1$ (if $p_{nn}\equiv 1$ $(n>m)$, consider $1-P$). Thus,

$$\begin{bmatrix} p_{mm} & p_{m\,m+1} & \cdots \\ 0 & 0 & \cdots \\ \cdots & \cdots & \end{bmatrix}\begin{bmatrix} \lambda_m & \alpha_m & 0 \\ & \lambda_{m+1} & \ddots \\ 0 & & \ddots \end{bmatrix} =$$

$$\begin{bmatrix} \lambda_m & \alpha_m & 0 \\ & \lambda_{m+1} & \ddots \\ 0 & & \ddots \end{bmatrix}\begin{bmatrix} p_{mm} & p_{m\,m+1} & \cdots \\ 0 & 0 & \cdots \\ \cdots & \cdots & \end{bmatrix}$$

implies that

$$p_{mm}\alpha_m + p_{m\,m+1}\lambda_{m+1} = \lambda_m p_{m\,m+1}$$

and

$$p_{m\,m+n}\alpha_{m+n} + p_{m\,m+n+1}\lambda_{m+n+1} = \lambda_m p_{m\,m+n+1} \qquad (n \geq 1).$$

Since $\operatorname{card}\{n : \lambda_m = \lambda_n\} = \infty$, there exists $n > 0$ such that $\lambda_{m+n+1} = \lambda_m$. Thus $p_{m\,m+n} = 0$. Since $p_{m\,m+n} = 0$, $p_{m\,m+n-1} = 0, \cdots$, finally, $p_{mm} = 0$, a contradiction. Thus $P = 0$ and $A \in (SI)(\mathcal{H})$.

(i) (iv) Since $A =$ Normal+Compact and since $\sigma(A)$ is connected,

$$\sigma(A) = \sigma_{lre}(A) = \sigma(N) = \{\lambda_n\}^- = \sigma.$$

If $(A - u_m)x = 0$ for some m and $x = \sum x_i e_i$, then computations indicate that

$$x_n = -\frac{(\lambda_{n-1} - u_m)}{\alpha_{n-1}}\, x_{n-1} \quad (n = 2, 3, \cdots),$$

i.e.

$$|x_n| = \prod_{k=1}^{n-1} \frac{|\lambda_k - u_m|}{|\lambda_k - \lambda_{k+1}|^{1/2}}\, |x_1| \quad (n = 2, 3, \cdots).$$

Since

$$\overline{\lim_{n\to\infty}} \prod_{k=1}^{n-1} \frac{|\lambda_k - u_m|}{|\lambda_k - \lambda_{k+1}|^{1/2}} = \infty,$$

$x_1 = 0$ and therefore $x = 0$ and $\sigma_p(A) \cap \{u_m\}_{m=1}^{\infty} = \varnothing$.

(ii) From the definition of A, it is easily seen that

$$\bigvee \{\ker(\lambda_i - A)^k : i \geq 1,\ k \geq 1\} = \mathcal{H}.$$

(iii) Assume that $(A-\lambda)^*x = 0$ for some λ and $x = \sum x_i e_i$. Thus $(\lambda_1^* - \lambda^*)x_1 = 0$, $(\lambda_n^* - \lambda^*)x_n = -\alpha_{n-1}x_{n-1}$ $(n > 1)$. If $\lambda \notin \{\lambda_n\}_{n=1}^{\infty}$, then $x = 0$. If $\lambda \in \{\lambda_n\}_{n=1}^{\infty}$, since card $\{k, \lambda_k = \lambda\} = \infty$, a simple computation shows that $x = 0$. Thus $\sigma_p(A^*) = \varnothing$. □

3.4 The norm-closure of strongly irreducible operators and the strongly irreducible operators in some other classes of operators

The Riesz decomposition theorem asserts that if $\sigma(T) = \sigma_1 \cup \sigma_2$, where σ_1, σ_2 are disjoint compact subsets of $\mathbb{C}$, then $T = T_1 \dot{+} T_2$ and $\sigma(T_i) = \sigma_i$ $(i = 1, 2)$. The following theorem characterizes the norm-closure $(SI)(\mathcal{H})^-$ of the class $(SI)(\mathcal{H})$. In a certain sense, this characterization can be considered as an "approximate inverse" of the Riesz decomposition theorem.

Theorem 3.30. $(SI)(\mathcal{H})^- = \{T \in \mathcal{L}(\mathcal{H}) : \sigma(T)$ *is connected*$\}$.

It is not difficult to see that $\sigma(T)$ is necessarily connected for each T in $(SI)(\mathcal{H})^-$. Thus, we must show only that every T in $\mathcal{L}(\mathcal{H})$ with a connected spectrum can be uniformly approximated by (SI) operators. In order to prove this we need the following lemmas.

Lemma 3.31. *Suppose that*

$$R = \begin{bmatrix} A_1 & & & & Q_{11} & Q_{12} & \cdots & Q_{1m} \\ & A_2 & & 0 & Q_{21} & Q_{22} & \cdots & Q_{2m} \\ & & \ddots & & \vdots & \vdots & \ddots & \vdots \\ & & & A_n & Q_{n1} & Q_{n2} & \cdots & Q_{nm} \\ & & & & C_1 & & & \\ & & & & & C_2 & & 0 \\ & & 0 & & & & \ddots & \\ & & & & & & & C_m \end{bmatrix}$$

with respect to the orthogonal direct sum decomposition

$$\mathcal{H} = \Big(\bigoplus_{j=1}^{n} \mathcal{H}_j^1\Big) \oplus \Big(\bigoplus_{i=1}^{m} \mathcal{H}_i^2\Big) \qquad (n, m < \infty),$$

where (i) $\{\sigma(A_j)\}_{j=1}^n$ *and* $\{\sigma(C_i)\}_{i=1}^m$ *are two families of pairwise disjoint compact sets such that*

$$\sigma(R) = \Big[\bigcup_{j=1}^{n} \sigma(A_j)\Big] \cup \Big[\bigcup_{i=1}^{m} \sigma(C_i)\Big]$$

and $\sigma(R)$ *is connected;*

(ii) $A_j,\ C_i \in (SI)$ *and* $\ker \tau_{C_i A_j} = \{0\}$ $(1 \le j \le n,\ 1 \le i \le m)$;

(iii) *Either* $\sigma(A_j) \cap \sigma(C_i) = \varnothing$ *and* $Q_{ji} = 0$, *or* $\sigma(A_j) \cap \sigma(C_i) \neq 0$ *and* $Q_{ji} \notin \operatorname{ran} \tau_{A_j C_i}$.

Then $R \in (SI)(\mathcal{H})$.

Proof. Let P be an idempotent commuting with R and assume that P admits the block matrix representation $P = (P_{ij})_{i,j=1}^{n+m}$ with respect to the decomposition of the space $\mathcal{H}$. Since $\ker \tau_{C_i A_j} = \{0\}$ $(1 \le j \le n,\ 1 \le i \le m)$, computations indicate that $P_{ji} = 0$ $(n+1 \le j \le n+m$ and $1 \le i \le n)$. Since $\{\sigma(A_j)\}_{j=1}^n$ and $\{\sigma(C_i)\}_{i=1}^m$ are two families of pairwise disjoint compact sets, $\ker \tau_{A_i A_j} = 0$, $\ker \tau_{C_i C_j} = \{0\}$ $(i \neq j)$. Thus computation shows that $P_{ji} = 0$ $(n \ge j > i \ge 1$ or $n+m \ge j > i \ge n+1)$, i.e., P has an upper triangular matrix representation with respect to the given decomposition. Moreover, the same arguments indicate that $P_{ji} = 0$ $(1 \le j < i \le n$ or $n+1 \le j < i \le n+m)$. Since P_{jj} is an idempotent and $A_j, C_i \in (SI)$, $P_{jj} = 0$ or 1 $(j = 1, 2, \cdots, n+m)$. Without loss of generality, we can suppose that $P_{11} = 0$. Since $\sigma(R)$ is connected, there must exist an integer i $(1 \le i \le m)$ such that $\sigma(A_1) \cap \sigma(C_i) \neq \varnothing$. It follows from $PR = RP$ that

$$P_{1\,n+i} C_i = A_1 P_{1\,n+i} + Q_{1i} P_{n+i\,n+i}$$

or

$$Q_{1i} P_{n+i\,n+i} = \tau_{A_1 C_i}(P_{1\,n+i}).$$

Since $Q_{1i} \notin \operatorname{ran} \tau_{A_1 C_i}$, $P_{n+i\,n+i} = 0$. If $\sigma(A_j) \cap \sigma(C_i) \neq \varnothing$, by the same argument, $P_{jj} = 0$. Since $\sigma(R)$ is connected, after a finite number of steps $P_{jj} = 0$ $(1 \le j \le n+m)$. It follows from $P^2 = P$ that $P = 0$, i.e. $R \in (SI)$. □

Lemma 3.32. *Let* T *be in* $\mathcal{L}(\mathcal{H})$ *with connected spectrum* $\sigma(T)$, *if* $\sigma_{lre}(T)$ *is the closure of an analytic Cauchy domain* Ω. *Then there exists a* (SI) *operator* A *such that the spectral picture* $\Lambda(A)$ *of* A *is the same as the spectral picture* $\Lambda(T)$ *of* T *and*

$$\min\operatorname{ind}(A-\lambda)^k \le \min\operatorname{ind}(T-\lambda)^k, \quad (k \ge 1,\ \lambda \in \rho_{s-F}(A)).$$

Proof. Let (Ω_1, k_1), (Ω_2, k_2), $\cdots$, (Ω_n, k_n) be the finitely many components of $\sigma(T)\backslash\overline{\Omega}$, where $k_i = \operatorname{ind}(\lambda - T)$, $\lambda \in \Omega_i$ $(i = 1, 2, \cdots, n)$. Thus $\{\overline{\Omega}_i\}$ are pairwise disjoint and each Ω_i is a connected Cauchy domain.

If $|k_i| < \infty$, by Theorem 3.5, Theorems 3.25 and 3.26, there is a (SI) operator $A_i = A(\Omega_i, k_i) \in \mathcal{L}(\mathcal{H})$ such that $\sigma(A_i) = \overline{\Omega}_i$, $\operatorname{ind}(A_i - \lambda) = k_i$ $(\lambda \in \Omega_i)$ and either $\min\operatorname{ind}(A_i - \lambda) = 0$ (if $k_i \neq 0$), or $\min\operatorname{ind}(A_i - \lambda) = 1$ (if $k_i = 0$).

If $|k_i| = \infty$, using the Bergman operator $N_+(\Omega_i)$ and by Lemma 3.12, Proposition 3.13 we can find an operator $A_i = A(\Omega_i, k_i) \in \mathcal{L}(\mathcal{H})$ such that $\sigma(A_i) = \overline{\Omega}_i$, $\operatorname{ind}(A_i - \lambda) = k_i$ and $\min\operatorname{ind}(A_i - \lambda) = 0$ $(\lambda \in \Omega_i)$.

Let Φ_1, $\Phi_2, \cdots, \Phi_m$ be all the components of $\overline{\Omega}$. By Theorem 3.7, we can find (SI) operators $C_1, C_2, \cdots, C_m$ such that $\sigma(C_i) = \sigma_{lre}(C_i) = \Phi_i$.

Set

$$A = \begin{bmatrix} A_1 & & & & Q_{11} & Q_{12} & \cdots & Q_{1m} \\ & A_2 & & 0 & Q_{21} & Q_{22} & \cdots & Q_{2m} \\ & & \ddots & & \vdots & \vdots & \ddots & \vdots \\ & & & A_n & Q_{n1} & Q_{n2} & \cdots & Q_{nm} \\ & & & & C_1 & & & \\ & & & & & C_2 & & 0 \\ & & 0 & & & & \ddots & \\ & & & & & & & C_m \end{bmatrix} \in \mathcal{L}(\mathcal{H}^{(n+m)}),$$

where $\{Q_{ij} : 1 \leq i \leq n;\ 1 \leq j \leq m\}$ are defined as in Lemma 3.31. By Lemma 3.31, $A \in (SI)$ and satisfies all the requirements of the lemma. $\square$

We are now in a position to prove Theorem 3.30.

Proof of Theorem 3.30. Let T be in $\mathcal{L}(\mathcal{H})$ with connected $\sigma(T)$; then by Theorem 1.27, there is a sequence $\{T_n\}_{n=1}^{\infty}$ of operators such that $\sigma_{lre}(T_n)$ is the closure of an analytic Cauchy domain, $\sigma(T_n)$ is connected and $\|T_n - T\| \to 0$ $(n \to \infty)$. Thus it suffices to prove that each operator T with connected $\sigma(T)$ whose $\sigma_{lre}(T)$ is the closure of an analytic Cauchy domain is in $(SI)(\mathcal{H})^-$. By Lemma 3.32 and the Similarity orbit theorem there exists a (SI) operator A such that $T \in S(A)^-$. This implies that $(SI)(\mathcal{H})^- \supset \{T \in \mathcal{L}(\mathcal{H}) : \sigma(T) \text{ is connected}\}$. The inverse inclusion comes from the upper continuity of spectrum. $\square$

In 3.3, we constructed a family of (SI) Cowen–Douglas operators. Proposition 2.28 asserts that $\mathcal{B}_1(\Omega) \subset (SI)(\mathcal{H})$. In fact we have the following

Theorem 3.33. *$B \in \mathcal{B}_n(\Omega)$ is strongly irreducible if and only if $P|_{\ker(\lambda_0 - B)} = 0$ or 1 for each idempotent $P \in \mathcal{A}'(B)$ and some point $\lambda_0 \in \Omega$.*

Proof. By the proof of Lemma 2.20,

$$B = \begin{bmatrix} \lambda_0 & B_{12} & & * \\ & \lambda_0 & B_{23} & \\ & & \lambda_0 & \ddots \\ & 0 & & \ddots \end{bmatrix} \in \mathcal{L}(\bigoplus \mathbb{C}^n),$$

where $\lambda_0 \in \Omega$ and $B_{k\,k+1}$ is invertible $(k = 1, 2, \cdots)$. Calculations indicate that

$$P = \begin{bmatrix} P_{11} & P_{12} & P_{13} & \cdots \\ 0 & P_{22} & P_{23} & \cdots \\ 0 & 0 & P_{33} & \cdots \\ & & & \ddots \end{bmatrix} \begin{matrix} \mathbb{C}^n \\ \mathbb{C}^n \\ \mathbb{C}^n \\ \vdots \end{matrix},$$

and P_{kk} is idempotent $(k = 1, 2, \cdots)$. $PB = BP$ implies that

$$P_{kk}B_{k\,k+1} = B_{k\,k+1}P_{k+1\,k+1} \qquad (k = 1, 2, \cdots),$$

i.e., $P_{kk} \sim P_{11}$. Thus $B \in (SI)$ if and only if $P|_{\ker(\lambda_0 - B)} = P_{11} = 0$ or 1. □

Consider the analytic Toeplitz operator T_ϕ with symbol $\phi \in H^\infty$. If $\rho_F(T_\phi) \cap \sigma(T_\phi) \neq \varnothing$, then T_ϕ^* is a Cowen–Douglas operator, therefore, we have

Corollary 3.34. *Let T_ϕ be an analytic Toeplitz operator. If $\rho_F(T_\phi) \cap \sigma(T_\phi) \neq \varnothing$, then $T_\phi \in (SI)$ if and only if $P|_{\operatorname{ran}(T_\phi - \lambda)} = 0$ or 1 for each idempotent P commuting with T_ϕ and some point $\lambda \in \rho_F(T_\phi) \cap \sigma(T_\phi)$.*

Denote $A(D) = \{\phi \in H^\infty : \phi \text{ is continuous in } \overline{D}\}$. $A(D)$ is called a *disc algebra.* Then the strong irreducibility and irreducibility of T_ϕ $(\phi \in A(D))$ are closely related.

Theorem 3.35. *Let $\phi \in A(D)$ and $\phi(D)^- \neq \phi(\partial D)$. Then T_ϕ is strongly irreducible if and only if T_ϕ is irreducible.*

Proof. We only need prove the sufficiency part. Since $\phi(D)^- \neq \phi(\partial D)$, $\sigma(T_\phi)\backslash\sigma_e(T_\phi) \neq \varnothing$. Without loss of generality, we can assume that $0 \in \sigma(T_\phi)\backslash\sigma_e(T_\phi)$, i.e., T_ϕ is a Fredholm operator, and $H^2 \ominus \phi H^2$ is a finite dimensional subspace. By the Factorization Theorem, $\phi = \chi \cdot F$, where χ is a finite Blaschke product and F is an outer function; and there exists a finite Blaschke product g such that $\phi(z) = F(g(z))$ (C.C.Cowen [1]). By T.E. Thomson [1] we have

$$\mathcal{A}'(T_\phi) = \mathcal{A}'(T_z) \cap \mathcal{A}'(T_F) = \mathcal{A}'(T_g).$$

Note that T_g is a partial isometry and it follows from the fact that T_ϕ is irreducible that $\operatorname{codim} \operatorname{ran} T_g = 1$. Since T_ϕ is irreducible and g is a finite Blaschke product, $\mathcal{A}'(T_g) = \mathcal{A}'(T_z)$. Thus the fact that $\mathcal{A}'(T_z)$ has no non-trivial idempotent implies that $T_\phi \in (SI)$. □

Proposition 3.36. *Let $\phi \in H^\infty$ and χ and F be the inner and outer factors of ϕ respectively, i.e., $\phi = \chi \cdot F$. If $\chi(z) = z^n$ $(n \geq 1)$ and*

$$F(z) = a_0 + a_1 z^{p_1} + \cdots + a_k z^{p_k} + \cdots \quad (a_k \neq 0,\ k = 1, 2, \cdots),$$

then $T_\phi \in (SI)(\mathcal{H})$ if and only if g.c.d$(n, p_1, p_2, \cdots, p_{k_0}) = 1$ *for some natural number k_0, where* g.c.d$(n, p_1, p_2, \cdots, p_{k_0})$ *denotes the greatest common divisor of $n, p_1, \cdots,$ and p_{k_0}.*

Proof. If $T_\phi \in (SI)$ and g.c.d$(n, p_1, \cdots, p_k) \geq 2$ for all $k = 1, 2, \cdots$, then there exists a natural number s $(s > 1)$, such that s is the common factor of n and p_k $(k = 1, 2, \cdots)$. Thus ϕ is a function of z^s. It is easily seen that $\mathcal{A}'(T_{z^s}) \subset \mathcal{A}'(T_\phi)$. Since $T_\phi \in (SI)$, $T_{z^s} \in (SI)$ and $s = 1$, a contradiction.

If g.c.d $(n, p_1, \cdots, p_{k_0}) = 1$ for some k_0, then $\mathcal{A}'(T_\phi) = \mathcal{A}'(T_{z^n}) \cap \mathcal{A}'(T_F)$ (T.E. Thomson [1]). A simple computation shows that $T_\phi \in (SI)$. □

Proposition 3.37. *Given an entire function $f(z) = a_1 z^{p_1} + \cdots + a_k z^{p_k} + \cdots$, $T_f \in (SI)$ if and only if there exists an integer k_0 such that* g.c.d$(p_1, p_2, \cdots, p_{k_0}) = 1$.

Proof. By I.N. Baker, J.A. Deddens and J.L. Ullman [1], we can find a natural number s such that $\mathcal{A}'(T_f) = \mathcal{A}'(T_{z^s})$. Thus, if $T_f \in (SI)$, then $T_{z^s} \in (SI)$ and $s = 1$. Therefore, there exists a natural number k_0 such that g.c.d$(p_1, p_2, \cdots, p_{k_0}) = 1$.

If g.c.d$(p_1, p_2, \cdots, p_{k_0}) = 1$ for some k_0, since $\mathcal{A}'(T_f) = \mathcal{A}'(T_{z^s})$, it suffices to prove that $s = 1$. If $s > 1$, then there exists a natural number p_m such that s is not a divisor of p_m. Let $m_0 = \min\{m : s \text{ is not a divisor of } p_m\}$, and set $p(z) = \sum_{k=0}^{m_0 - 1} a_k z^{p_k}$. Then $p(z)$ is a function of z^s and $\mathcal{A}'(T_{z^s}) = \mathcal{A}'(T_{f-p})$. Since $f(z) - p(z) = \sum_{k=m_0}^{\infty} a_k z^{p_k}$, by J.A. Deddens and T.K Wong [1], s is a divisor of p_{m_0}, a contradiction. □

Proposition 2.31 asserts that the bilateral weighted shift with an annulus spectrum is a (SI) operator. The next several results discuss the bilateral weighted shift with circle spectrum.

Let $T \in \mathcal{L}(\mathcal{H})$ be the bilateral weighted shift given by $Te_n = \omega_n e_{n+1}$ $(n = 0, \pm 1, \pm 2, \cdots)$, where $\{e_n\}_{n=-\infty}^{\infty}$ is an ONB of $\mathcal{H}$ and $\{\omega_n\}_{n=-\infty}^{\infty}$ is a bounded sequence of complex numbers such that $\omega_n \neq 0$ $(n = 0, \pm 1, \pm 2, \cdots)$. In the sense of unitary equivalence, we can assume that $\omega_n > 0$ $(n = 0, \pm 1, \pm 2, \cdots)$.

Denote

$$\begin{aligned} &\beta(0) = 1, \\ &\beta(n) = \omega_0 \omega_1 \cdots \omega_{n-1} \qquad (n > 0), \\ &\beta(n) = (\omega_{-1}\omega_{-2} \cdots \omega_{-n})^{-1} \quad (n < 0), \end{aligned}$$

and

$$r_1(T) = \min\left\{ \lim_{n\to\infty} \left[\inf \frac{\beta(n+j)}{\beta(j)}\right]^{1/n}, \lim_{n\to\infty} \left[\inf \frac{\beta(-n+j)}{\beta(j)}\right]^{1/n} \right\}.$$

It is easily seen that if $\sigma(T)$ is an annulus, then $r_1(T) < r(T)$, where $r(T)$ is the spectral radius of T; if $\sigma(T)$ is a circle, then $r_1(T) = r(T)$. We always consider T as a multiplication on the Hilbert space $\mathcal{H}$:

$$\mathcal{H} = L^2(\beta) = \left\{ f = \{f(n)\} : \sum_{n=-\infty}^{\infty} |f(n)|^2 |\beta(n)|^2 < \infty \right\},$$

and its commutant is $L^\infty(\beta)$, where $L^\infty(\beta) = l^\infty \cap L^2(\beta)$. Let $L^\infty(C)$ denote the set of essentially bounded measurable functions on the unit circle C, and $\phi_F(z)$ denote the characteristic function of the subset F of C.

Lemma 3.38. *Let T be a bilateral weighted shift with $r_1(T) = r(T)$. Then $T \in (SI)(\mathcal{H})$ if and only if there does not exist any measurable subset F of C such that $m(F) \neq 0$, 2π, and $\chi_F(z) \in L^\infty(\beta)$.*

Proof. By Proposition 8 and Theorem 3 of A.L. Shilds [1], $T \in (SI)(\mathcal{H})$ if and only if $\phi(z) = 0$ or 1 for $\phi(z) \in L^\infty(\beta)$ such that $\phi(z)^2 = \phi(z)$. Thus $T \in (SI)(\mathcal{H})$ if and only if there does not exist $F \subset C$ such that $m(F) \neq 0$, 2π and $\chi_F(z) \in L^\infty(\beta)$.

□

Lemma 3.39. *Let T_1, T_2 be two bilateral weighted shifts with $r_1(T_1) = r(T_1) = r_1(T_2) = r(T_2) = 1$ and let $\beta_1(n)$ and $\beta_2(n)$ denote the β sequences of T_1 and, respectively, T_2. If*

$$m\beta_1(n) \leq \beta_2(n) \leq M\beta_1(n) \qquad \textit{for some } M \geq m > 0,$$

then the strong irreducibilities of T_1 and T_2 are same.

Proof. The condition implies that $T_1 \sim T_2$. □

Lemma 3.40. *Let $M \subset \partial D$, $m(M) \neq 0,\ 2\pi$ and let $\chi_M(z) = \sum_{n=-\infty}^{\infty} b_n z^n$. Then $|b_n| = |b_{-n}|$ $(n = 1, 2, \cdots)$ and $\sum |b_n|$ diverges, where $\{b_n\}$ are the coefficients of the formal Laurent series of $\chi_M(z)$.*

Proof. Let $z = e^{i\theta}$, $b_n = c_n + id_n$; then

$$\begin{aligned}\chi_M(e^{i\theta}) &= \sum_{n=-\infty}^{\infty} b_n e^{in\theta} = \sum_{n=-\infty}^{\infty} (c_n + id_n)(\cos n\theta + i\ \sin n\theta)\\ &= c_0 \sum_{n=1}^{\infty} \left[(c_n + c_{-n})\ \cos n\theta + (d_{-n} - d_n)\ \sin n\theta\right]\\ &\quad + i\left(\sum_{n=1}^{\infty} \left[(c_n - c_{-n})\ \sin n\theta + (d_n + d_{-n})\ \cos n\theta\right] + d_0\right).\end{aligned}$$

Since χ_M is a real-valued function, $c_n = c_{-n}$, $d_n = -d_{-n}$ $(n = 1, 2, \cdots)$ and $d_0 = 0$, i.e., $b_n = b_{-n}^*$ $(n = 1, 2, \cdots)$. If $\sum_{-\infty}^{\infty} |b_n| < \infty$, then $\sum_{n=-\infty}^{\infty} b_n z^n$ converges to a continuous function on $C = \partial D$. Thus by the uniqueness of the Laurent expansion of a function, $\chi_M(e^{i\theta})$ equals a continuous function a.e.. But this is impossible, since $m(M) \neq 0,\ 2\pi$. □

Theorem 3.41. *Let T be a bilateral weighted shift with $r_1(T) = r(T) = 1$. Then $T \in (SI)$ if $\{\beta(n)\}$ satisfies one of the following conditions:*

(i) $\sum_{n=0}^{\infty} 1/\beta^2(n) < \infty$;

(ii) $\sum_{n=0}^{\infty} 1/\beta^2(-n) < \infty$;

(iii) $\sum_{n=0}^{\infty} \beta^2(n) < \infty$;

(iv) $\sum_{n=0}^{\infty} \beta^2(-n) < \infty$;

(v) $\sum_{n=0}^{\infty} \inf_k \left\{\min\left(\beta(n+k)/\beta(k),\ \beta(k)/\beta(n+k)\right)\right\} < \infty$.

Proof. (i) If $T \notin (SI)(\mathcal{H})$, then by Lemma 3.38 there is a measurable subset M of C, such that $m(M) \neq 0,\ 2\pi$ and $\chi_M \in L^\infty(\beta) \subset L^2(\beta)$. Assume that $\chi_M(z) = \sum_{n=-\infty}^{\infty} b_n z^n$; then $\sum_{n=0}^{\infty} b_n^2 \beta^2(n)$ converges. Since $\sum_{n=0}^{\infty} 1/\beta^2(n) < \infty$,

$$\left(\sum_{n=0}^{\infty} |b_n|\right)^2 \leq \left(\sum_{n=0}^{\infty} |b_n|^2 \beta^2(n)\right)\left(\sum_{n=0}^{\infty} \frac{1}{\beta^2(n)}\right) < \infty,$$

contradicting Lemma 3.40.

By the same argument, we can prove (ii). Note that since the β sequence of T^* is $\{1/\beta(-n)\}$, (iii) and (iv) can be proved by (i) and (ii).

(v) If $T \notin (SI)(\mathcal{H})$, then there is a subset $M \subset C$ such that $m(M) \neq 0,\ 2\pi$ and $\chi_M(z) \in L^\infty(\beta)$. Assume that $\chi_M(z) = \sum_{n=-\infty}^{\infty} b_n z^n$. If the norm of the "multiplication by $\chi_M(z)$" operator on $L^2(\beta)$ is R, then for each integer k, we have

$$\left(\chi_M(z)\, z^k,\ z^{n+k}\right) \leq R\|z^k\|\,\|z^{n+k}\|.$$

Thus

$$|b_n|\,\beta^2(n+k) \leq R\,\beta(k)\,\beta(n+k) \quad \text{or} \quad |b_n| \leq R\,\frac{\beta(k)}{\beta(n+k)}.$$

Similarly, $|b_{-n}| \leq R\big(\beta(n+k)/\beta(k)\big)$. By Lemma 3.40, $|b_n| = |b_{-n}|$. Thus

$$|b_n| \leq R \inf_k \min\left(\frac{\beta(n+k)}{\beta(k)},\ \frac{\beta(k)}{\beta(n+k)}\right)$$

and $\sum_{n=-\infty}^{\infty} |b_n| < \infty$, which contradicts Lemma 3.40. □

To close this chapter, we consider the strong irreducibility of a class of operators with abelian commutants.

Proposition 3.42. *Let $T \in \mathcal{L}(\mathcal{H})$. Then*

(i) *if $\mathcal{A}^a(T)$ is strictly cyclic, then $T \in (SI)(\mathcal{H})$ if and only if $\sigma(T)$ is connected;*

(ii) *if $\mathcal{A}'(T)$ is a strictly cyclic abelian algebra, then $T \in (SI)(\mathcal{H})$ if and only if the maximal ideal space Σ is connected.*

Proof. (i) Let Σ denote the maximal ideal space of $\mathcal{A}^a(T)$. Note that $\sigma(T) = \sigma_{\mathcal{A}^a(T)}(T)$. Consider the Gelfand transform: $\Sigma \to \mathbb{C}$ defined by $\hat{T}(\phi) = \phi(T)$. Thus $\sigma(T) = \{\hat{T}(\phi) : \phi \in \Sigma\}$. The mapping ˆ is injective. In fact if $\phi_1(T) = \phi_2(T)$ for $\phi_1, \phi_2 \in \Sigma$, then $\phi_1((\lambda - T)^{-1}) = \phi_2((\lambda - T)^{-1})$ for all $\lambda \notin \sigma(T)$. Thus

$\phi_1(A) = \phi_2(A)$ for all $A \in \mathcal{A}^a(T)$, and $\phi_1 = \phi_2$. Therefore Σ and $\sigma(T)$ are isomorphic. Since $\mathcal{A}^a(T)$ is strictly cyclic, $\mathcal{A}^a(T) = \mathcal{A}'(T) = \mathcal{A}''(T)$. It follows from Shilov's Idempotent Theorem (D.A. Herrero [5]) that the Gelfand transform $\hat{P}$ of each idempotent P in $\mathcal{A}^a(T) = \mathcal{A}'(T)$ is in $C(\Sigma) \simeq C(\sigma(T))$. Since $\hat{P}$ is a characteristic function, $T \in (SI)(\mathcal{H})$ if and only if $\sigma(T)$ is connected.

(ii) By the same argument, we can prove (ii). □

3.5 Open problems

Problem 3.1. *Are the strong irreducibility and irreducibility of analytic Toeplitz operators same?*

Problem 3.2. *Can we find a necessary and sufficient condition expressed in terms of the weight sequence for a bilateral weighted shift to be strongly irreducible?*

Problem 3.3. *Can we find a (SI) Cowen–Douglas operator $B \in \mathcal{B}_n(\Omega)$ with the nice property, where Ω is an arbitrary connected bounded open set Ω and n is a natural number?*

3.6 Remarks

In 1988, in personal correspondence with C.L. Jiang, D.A. Herrero first gave the concept of a strongly irreducible operator with the nice property. In the meantime he asked about the possibility of the existence of the (SI) operator with the nice property. It is J. Dixmier and C. Foiaş [1] who first used Sobolev space theory in operator theory. Later, D.A. Herrero, T.J. Taylor and Z.Y. Wang [1], and Z.Y. Wang [1] pursued deep work on this aspect. C.L. Jiang and Z.Y. Wang solved the questions of D.A. Herrero about the (SI) operator with the nice property in terms of the multiplication operator on a Sobolev space (C.L. Jiang and Z.Y. Wang [2]). The results in 3.2 and 3.3 are basically due to C.L. Jiang and Z.Y. Wang. Y.Q. Ji, S. Power and S.H. Sun also made some contributions (C.L. Jiang, S.H. Sun and Z.Y. Wang [1], C.L. Jiang [1], Y.Q. Ji, C.L. Jiang and Z.Y. Wang [2] [3] [4] and C.L. Jiang, S. Power and Z.Y. Wang [1]). D.A. Herrero and C.L. Jiang [1] proved Theorem 3.30. Theorem 3.41 is due to G.H. Gong. Theorem 3.35, Propositions 3.36 and Lemma 3.38 were contributed by C.L. Jiang, G.F. Cao and S.L. Sun [1]. The other work in 3.4 is due to Y.Q. Ji, C.L.Jiang and Z.Y.Wang (C.L. Jiang [1], Y.Q. Ji, C.L. Jiang and Z.Y. Wang [2] [4]).

4. The spectral pictures of strongly irreducible operators and the approximate Jordan decomposition theorem

4.1 The closure of the direct sums of (SI) operators

People have founded a variety of "Jordan decomposition theorems" with respect to different kinds of classes of operators. L.J. Gray [1] first proved that a class of nilpotent operators can be uniquely expressed as the orthogonal direct sum of countably many Jordan nilpotents in the sense of similarity. C. Apostol, R. Douglas and C. Foiaş [1] proved that the nilpotent operators can be classified up to quasisimilarity by their "Jordan forms". K.R. Davidson and D.A. Herrero [1] extended this result to bitriangular operators. (An operator T is a *bitriangular operator* if both T and T^* are triangular operators.) These results can be considered as "Jordan decomposition theorems" for nilpotent operators or bitriangular operators. If we consider the (SI) operators on a Hilbert space $\mathcal{H}$ as the natural generalization of Jordan blocks, it is natural to ask the following question: can every operator in $\mathcal{L}(\mathcal{H})$ be expressed as a direct (not necessary orthogonal) sum of (SI) operators; or does every operator in $\mathcal{L}(\mathcal{H})$ have a strongly irreducible decomposition? The next proposition gives a negative answer to this question.

Proposition 4.1. *Let $A \in \mathcal{L}(\mathcal{H})$ be a self-adjoint operator with empty point spectrum. Then A cannot be represented as a direct sum of countably many (SI) operators.*

Proof. Assume that $A = \sum \dot{+} A_i$ with respect to the direct sum decomposition of the space $\mathcal{H} = \sum \dot{+} \mathcal{H}_i$, where $A_i = A|_{\mathcal{H}_i} \in (SI)(\mathcal{H}_i)$. Since A is self-adjoint, each A_i is self-adjoint. Thus $\sigma(A_i)$ is a singleton, $\sigma(A_i) = \{\lambda_i\}$, and λ_i is an eigenvalue of A_i and therefore $\lambda_i \in \sigma_p(A)$, a contradiction. □

Proposition 4.1 implies that if Jordan blocks are replaced by (SI) operators, we can only have an "approximate Jordan theorem".

Theorem 4.2. *Given $T \in \mathcal{L}(\mathcal{H})$ and $\epsilon > 0$, there exist a natural number n and n (SI) operators $T_1, T_2, \cdots, T_n$ such that*

$$\left\| T - \sum_{i=1}^{n} \dot{+} T_i \right\| < \epsilon.$$

Proof. By Theorem 1.27, there exists $T_\epsilon \in \mathcal{L}(\mathcal{H})$ such that $\sigma(T_\epsilon)$ has finitely many components $\sigma(T_\epsilon) = \{\sigma_i\}_{i=1}^n$ and $\|T - T_\epsilon\| < \epsilon/2$. By the Riesz decomposition theorem $T_\epsilon = \sum_{i=1}^n \dot{+} T_i'$, where $T_i' = T_\epsilon|_{\mathcal{H}_i}$, $\mathcal{H} = \sum_{i=1}^n \dot{+} \mathcal{H}_i$ and $\sigma(T_i') = \sigma_i$ $(i = 1, 2, \cdots, n)$. By Theorem 3.30 we can find (SI) operators $T_1, T_2, \cdots, T_n$ such that $\|T_\epsilon - \sum_{i=1}^n \dot{+} T_i\| < \epsilon/2$, therefore $\|T - \sum_{i=1}^n \dot{+} T_i\| < \epsilon$. □

It seems noticeable that there are two classes of operators in $\mathcal{L}(\mathcal{H})$ which are more similar to the Jordan blocks. One is the unicellular operators and another is $\{\lambda + \mu S, \lambda + \mu S^* : \lambda \in \mathbb{C},\ \mu > 0,\ S \text{ is the unilateral shift}\}$. Therefore it is necessary to find the norm closures of the following two classes of operators:

$$DU(\mathcal{H}) := \left\{ T \in \mathcal{L}(\mathcal{H}) : T \sim \bigoplus_{i=1}^{m} T_i,\ T_i \text{ is unicellular},\ 1 \le i \le m,\ 1 \le m \le \infty \right\};$$

$$DJ(\mathcal{H}) := \left\{ T \in \mathcal{L}(\mathcal{H}) : \begin{array}{l} T \sim \bigoplus_{i=1}^{m} (\lambda_i + \mu_i S_i),\ \lambda_i \in \mathbb{C},\ \mu_i > 0, \\ S_i = S \text{ or } S^*,\ i = 1, 2, \cdots, m,\ 1 \le m \le \infty \end{array} \right\}.$$

Theorem 4.3. $DU(\mathcal{H})^- = \{A \in \mathcal{L}(\mathcal{H}) : \operatorname{ind}(\lambda - A) = 0$ *for* $\lambda \in \rho_{s-F}(A)\}$.

Proof. Note that $\sigma(T) = \sigma_{lre}(T)$ for all unicellular operators T (otherwise, there is a non-empty open set $\Omega \subset \sigma_p(T)$ or $\Omega \subset \sigma_p(T^*)$, which contradicts the unicellularity of T). For the same reason, if A is similar to an orthogonal direct sum $\bigoplus_{i=1}^m T_i$ of unicellular operators T_i, then $\sigma(A) = \sigma_{lre}(A)$. Suppose that $\{A_n\}_{n=1}^\infty \subset DU(\mathcal{H})$ such that $\|A_n - A\| \to 0$ $(n \to \infty)$. If there is a non-empty open set $\Omega \subset \rho_{s-F}^{\pm}(A) \cap \sigma(A)$, then by the stability of semi-Fredholm operators, when n is big enough there is a non-empty open set $\Omega_1 \subset \Omega$ such that $\Omega_1 \subset \rho_{s-F}(A_n) \cap \sigma(A_n)$ and $\operatorname{ind}(\lambda - A_n) \neq 0$ for $\lambda \in \Omega_1$. This implies that

$$DU(\mathcal{H})^- \subset \{A \in \mathcal{L}(\mathcal{H}) : \operatorname{ind}(\lambda - A) = 0 \text{ for } \lambda \in \rho_{s-F}(A)\}.$$

Conversely, if $\text{ind}(\lambda - B) = 0$ for all $\lambda \in \rho_{s-F}(B)$, then for $\epsilon > 0$, by Proposition 1.22 and the argument of Theorem 1.23, there exists a compact operator K with $\|K\| < \epsilon/2$ such that $\sigma(B+K) = \sigma_{lre}(B+K) \cup \sigma_0(B+K)$ and $\sigma_0(B+K) = \sigma_0(B)$. By Theorem 1.27, there exists an operator B_ϵ such that $\|B_\epsilon - (B+K)\| < \epsilon/2$ and $\sigma(B_\epsilon) = \sigma_{lre}(B_\epsilon) \cup \sigma_0(B_\epsilon)$ consists of finitely many components. Since ϵ can be chosen arbitrarily small, it suffices to prove that $B_\epsilon \in DU(\mathcal{H})^-$. Let $\{\sigma_i\}_{i=1}^n$ be the components of $\sigma(B_\epsilon)$, and let $E(\sigma_i, B_\epsilon)$ be the Riesz idempotent of B_ϵ corresponding to σ_i and set $B_i = B_\epsilon|_{E(\sigma_i,B_\epsilon)\mathcal{H}}$ $(i = 1, 2, \cdots, n)$. If $\sigma_i \subset \sigma_{lre}(B_\epsilon)$, then $\sigma(B_i) = \sigma_{lre}(B_i) = \sigma_i$. Set $T_i = \lambda_i + V$, where λ_i is in σ_i and V is given in Example 3.22. If $\sigma_i = \{\lambda_i\} \in \sigma_0(B_\epsilon)$, then $\dim E(\sigma_i, B_\epsilon)\mathcal{H} = m_i < \infty$. Set $T_i = \lambda_i + J_{m_i}$, where J_{m_i} is the $m_i \times m_i$ Jordan nilpotent. Define $T = \bigoplus_{i=1}^n T_i$. By the similarity orbit theorem $B_\epsilon \in S(T)^-$, i.e., there is a sequence $\{X_i\}_{i=1}^\infty$ of invertible operators such that $\lim_{i\to\infty} X_i^{-1} T X_i = B_\epsilon$. Thus $B_\epsilon \in DU(\mathcal{H})^-$. □

An operator T is a *biquasitriangular operator* ($T \in (BQT)$) if $\text{ind}(\lambda - T) = 0$ for all $\lambda \in \rho_{s-F}(T)$. Thus Theorem 4.3 indicates that $DU(\mathcal{H})^- = (BQT)$. Theorem 4.3 implies that it is inadequate to consider the unicellular operators as the replacement of Jordan blocks in $\mathcal{L}(\mathcal{H})$, since $DU(\mathcal{H})^- \neq \mathcal{L}(\mathcal{H})$.

Let

$$T \sim J = \Big(\bigoplus_k J_k\Big) \oplus \Big(\bigoplus_l J_l'\Big),$$

where $J_k = \lambda_k + \mu_k S$, $J_l' = (\lambda_l' + \mu_l' S^*)$, $\lambda_k, \lambda_l' \in \mathbb{C}$, $\mu_k, \mu_l' > 0$ and S is the unilateral shift. Let D_k (D_l') denote the open disc centred at λ_k (λ_l') with radius μ_n (μ_l'). Then we have

$$\sigma(J_k) = D_k^-, \quad \sigma_e(J_k) = \partial D_k;$$
$$\sigma(J_l') = D_l'^-, \quad \sigma_e(J_l') = \partial D_l'.$$

Lemma 4.4. *Let T, J be given as above. Then*

(i) $\sigma(T) = [\bigcup_k D_k]^- \cup [\bigcup_l D_l']^-$;

(ii) $\lambda \in \rho_{s-F}(T)$ *if and only if* $\lambda \notin [\bigcup_k \partial D_k]^- \cup [\bigcup_l \partial D_l']^-$ *and at least one of the following is finite*

$$\text{nul}(\lambda - T)^* = \text{card}\{k : \lambda \in D_k\};$$
$$\text{nul}(\lambda - T) = \text{card}\{l : \lambda \in D_l'\};$$

and

$$\text{ind}(\lambda - T) = \text{card}\{l : \lambda \in D'_l\} - \text{card}\{k : \lambda \in D_k\}.$$

Proof. Since $J_k, J_l'^*$ are subnormal operators,

$$\|(\lambda - J_k)^{-1}\| = [\text{dist}(\lambda, \sigma(J_k))]^{-1} \text{ and } \|(\lambda - J'_l)^{-1}\| = [\text{dist}(\lambda, \sigma(J'_l))]^{-1}$$

for all k, l and $\lambda \notin [\bigcup_k D_k]^- \cup [\bigcup_l D'_l]^-$. Thus $\{\|(\lambda - J_k)^{-1}\|\}_k$ and $\{\|(\lambda - J'_l)^{-1}\|\}_l$ are uniformly bounded for $\lambda \notin [\bigcup_k D_k]^- \cup [\bigcup_l D'_l]^-$ and $\lambda \in \rho(T)$, i.e., $\sigma(T) \subset [\bigcup_k D_k]^- \cup [\bigcup_l D'_l]^-$. It is easily seen that $\sigma(T) \supset [\bigcup_k D_k]^- \cup [\bigcup_l D'_l]^-$. Thus (i) is true.

Since $\sigma_p(T) = \sigma_p(J) = \bigcup_l D'_l$ and $\sigma_p(T^*)^* = \sigma_p(J^*)^* = \bigcup_k D_k$, $\text{nul}(\lambda - T)^* = \text{card}\{k : \lambda \in D_k\}$ and $\text{nul}(\lambda - T) = \text{card}\{l : \lambda \in D'_l\}$. Since $\partial D_k = \sigma_{lre}(J_k) \subset \sigma_{lre}(T)$ and $\partial D'_l = \sigma_{lre}(J'_l) \subset \sigma_{lre}(T)$, if $\lambda \in \rho_{s-F}(T)$ then $\lambda \notin [\bigcup_k \partial D_k]^- \cup [\bigcup_l \partial D'_l]^-$. Thus at least one of $\text{nul}(\lambda - T)$ and $\text{nul}(\lambda - T)^*$ is finite and

$$\text{ind}(\lambda - T) = \text{card}\{l : \lambda \in D'_l\} - \text{card}\{k : \lambda \in D_k\}. \qquad \square$$

Let $T_n \in DJ(\mathcal{H})$ and $T_n \sim J_n = \left(\bigoplus_k J_{n,k}\right) \oplus \left(\bigoplus_l J'_{n,l}\right)$ $(n = 1, 2, \cdots)$, where $J_{n,k} = \lambda_{n,k} + \mu_{n,k} S$ and $J'_{n,l} = \lambda'_{n,l} + \mu'_{n,l} S^*$. Let $\{D(n,k); D'(n,l)\}$ denote the open discs related to the operator T_n, where $D(n,k) = \{z \in \mathbb{C} : |z - \lambda_{n,k}| < \mu_{n,k}\}$, $D'(n,l) = \{z \in \mathbb{C} : |z - \lambda'_{n,l}| < \mu'_{n,l}\}$.

Theorem 4.5. *$T \in DJ(\mathcal{H})^-$ if and only if at least one of the following are satisfied.*

(i) *$T \in (BQT)$ and $\sigma_0(T) = \varnothing$;*

(ii) *For each $\epsilon > 0$, there exist finitely many discs $\Sigma := \{D_j\}_{j=1}^p$ and related pairs of integers $\{(N_j, M_j)\}_{j=1}^p$, $0 \le N_j, M_j \le \infty$ $(j = 1, 2, \cdots, p)$ satisfying*

  (a) *$\left(\bigcup_{j=1}^p D_j\right)^- \subset \sigma(T)$, $\left(\bigcup_{j=1}^p \partial D_j\right) \subset \sigma_{lre}(T)$;*

  (b) *$\sum_{\lambda \in D_j} M_j \le \text{nul}(\lambda - T)$, $\sum_{\lambda \in D_j} N_j \le \text{nul}(\lambda - T)^*$ and $\sum_{\lambda \in D_j}(M_j - N_j) = \text{ind}(\lambda - T)$ for all $\lambda \in \left[\sigma(T) \cap \rho_{s-F}(T)\right] \backslash \sigma_{lre}(T)_\epsilon$;*

  (c) *$\sigma_0(T) = \varnothing$.*

Note that if λ is not in any D_j, then $\sum_{\lambda \in D_j} M_j = \sum_{\lambda \in D_j} N_j = 0$. In order to prove Theorem 4.5, we need several lemmas.

Lemma 4.6. *If $T \in DJ(\mathcal{H})^-$ then $\sigma_0(T) = \varnothing$.*

Proof. Assume that $J_n \sim T_n \to T$ and $\lambda_0 \in \sigma_0(T)$. Let $D = \{z \in \mathbb{C} : |z - \lambda_0| < \epsilon\}$ and $\overline{D} \cap \sigma(T) = \{\lambda_0\}$, let $E(\lambda_0, T)$ be the Riesz idempotent of T corresponding to λ_0.

Thus $\dim \operatorname{ran} E(\lambda_0, T) < \infty$. It follows from the upper continuity of the spectrum that when n is big enough, $\sigma(T_n) \cap D \neq \varnothing$ and $\sigma(T_n) \cap \partial D = \varnothing$. Since $T_n \to T$, $(\lambda - T_n)^{-1} \to (\lambda - T)^{-1}$ $(n \to \infty)$ for all $\lambda \in \partial D$. Thus $E(\lambda_0, T_n) \to E(\lambda_0, T)$ $(n \to \infty)$. This implies that $\dim \operatorname{ran} E(\lambda_0, T_n) < \infty$ when n is big enough, i.e., $\sigma_0(T_n) \neq \varnothing$, contradicting Lemma 4.4. □

Given two discs D_1 and D_2, if $D_i = \{z \in \mathbb{C} : |z - z_i| < r_i\}$ $(i = 1, 2)$, define $d(D_1, D_2) = |z_1 - z_2| + |r_1 - r_2|$. Then it is obvious that d is a distance in the set of discs.

Definition 4.7. Let $T_n \in DJ(\mathcal{H})$ $(n = 1, 2, \cdots)$ and $T_n \to T$ $(n \to \infty)$. An open disc D is called a *convergent disc of $\{T_n\}$ related to T* if the following are satisfied:

(i) $\partial D \subset \sigma_{lre}(T)$ and there is a component U of $\rho_{s-F}(T)$ such that $D \supset U$;

(ii) there exist a subsequence $\{T_{n_j}\}$ of $\{T_n\}$ and a sequence of related discs $\{D(n_j, k_j)\}$ or $\{D'(n_j, l_j)\}$ such that

$$\lim_{j \to \infty} d\big(D(n_j, k_j), D\big) = 0 \quad \text{or} \quad \lim_{j \to \infty} d\big(D'(n_j, l_j), D\big) = 0,$$

where $D(n_j, k_j)$ or $D'(n_j, l_j)$ is a disc of T_{n_j} $(j = 1, 2, \cdots)$.

Lemma 4.8. *Assume that $T_n \in DJ(\mathcal{H})$ $(n = 1, 2, \cdots)$ such that $T_n \to T$ $(n \to \infty)$ and V is a component of $\rho^{\pm}_{s-F}(T)$. Then there exists a convergent disc D of $\{T_n\}$ related to T such that $D \supset V$.*

Proof. Pick a $z_0 \in V$. Without loss of generality we can assume that $\operatorname{ind}(z_0 - T) < 0$. It follows from the stability of semi-Fredholm operators that there is an $N > 0$ such that $\operatorname{ind}(z_0 - T_n) < 0$ and therefore $\operatorname{nul}(z_0 - T_n)^* > 0$ for $n \geq N$. By Lemma 4.4 there exist at least one $D(n, k_n)$ containing z_0 and $z_0 \notin [\bigcup_k \partial D(n, k)]^- \cup [\bigcup_l \partial D'(n, l)]^-$. By λ_n and μ_n we shall denote the centre and radius of $D(n, k_n)$ respectively. Then there exists a subsequence $\{(\lambda_{n_j}, \mu_{n_j})\}$ of $\{(\lambda_n, \mu_n)\}$ such that $(\lambda_{n_j}, \mu_{n_j}) \to (\lambda_0, \mu_0) \in \mathbb{C} \times \mathbb{R}_+$ $(j \to \infty)$ according to the $\mathbb{C}^2$ distance. If $\mu_0 = 0$ then $\lambda_0 = z_0$ and λ_0 is a limit point of the points in the boundaries of the discs related to T_{n_j} which are in $\sigma_{lre}(T_{n_j})$. This contradicts $z_0 \in \rho_{s-F}(T)$. Thus $\mu_0 > 0$. Set $D = \{z \in \mathbb{C} : |z - \lambda_0| < \mu_0\}$. Then for the same reason $\partial D \subset \sigma_{lre}(T)$.

If $(\mathbb{C} \backslash D) \cap V \neq \varnothing$, since $\lambda_0 \in D(n, k_n)$ $(n \geq N)$ and D is the limit of a subsequence of $\{D(n, k_n)\}$, $\lambda_0 \in \overline{D}$. Since V is connected, $\partial D \cap V \neq \varnothing$. This contradicts $\partial D \subset \sigma_{lre}(T)$. Thus $D \supset V$. □

Lemma 4.9. *Let $T_n \in DJ(\mathcal{H})$ $(n = 1, 2, \cdots)$, $T_n \to T$ $(n \to \infty)$ and let D be a convergent disc of $\{T_n\}$ related to T. Then $\overline{D} \subset \sigma(T)$.*

Proof. Without loss of generality we can assume that

$$\lim d\big(D(n, k_n), D\big) = 0 \qquad (n \to \infty).$$

If $\lambda_0 \in D \cap \rho(T)$, then there exists an N such that $\lambda_0 \in D(n, k_n) \subset \sigma(T_n)$ $(n \geq N)$, which contradicts $\lambda_0 \in \rho(T)$. □

Lemma 4.10. *Let $T_n \in DJ(\mathcal{H})$ $(n = 1, 2, \cdots)$, $T_n \to T$ $(n \to \infty)$ and let V be a component of $\rho_F(T)$. Then there exist a subsequence $\{T_{n_m}\}$ of $\{T_n\}$ and a finite set $\Sigma = \{D_j\}_{j=1}^{p}$ of convergent discs of $\{T_n\}$ related to T such that*

(i) *$D_i \supset V$ $(i = 1, 2, \cdots, p)$;*

(ii) *for each j $(1 \leq j \leq p)$ and each $\epsilon > 0$, there exist natural numbers M and N such that $1/2^M < \epsilon$ and*

$$\operatorname{card}\big\{k : d\big(D(n_m, k), D_j\big) < 1/2^M\big\},$$
$$\operatorname{card}\big\{l : d\big(D'(n_m, l), D_j\big) < 1/2^M\big\}$$

are two non-negative integers independent of m provided $m > N$ and at least one of them is non-zero;

(iii) *if D is a convergent disc of $\{T_{n_m}\}$ related to T containing V, then $D \in \Sigma$.*

Proof. Without loss of generality we can assume that $V \subset \rho_F^-(T)$. By Lemma 4.8, there exist a disc D_1, a subsequence of $\{T_n\}$ and a sequence of related discs which converges to D_1. For convenience we shall denote them by $\{T_j^1\}_{j=1}^{\infty}$ and $D^1(j, k_j)$ respectively, where $D^1(j, k_j)$ is the k_j-th disc of T_j^1. We can always assume, after a unitary transformation if necessary, that $k_j = 1$ $(j = 1, 2, \cdots)$, i.e.,

$$\lim_{j \to \infty} d\big(D^1(j, 1),\, D_1\big) = 0.$$

Find a subsequence $\{T_m^2\}$ of $\{T_j^1\}$ and a sequence $\{D^2(m, k_m)\}$ of related discs with $k_m > 1$ such that $d\big(D^2(m, k_m), D_1\big) \to 0$ $(m \to \infty)$. Similarly, we assume that $k_m = 2$, i.e., $\lim_{m \to \infty} \big(D^2(m, k), D_1\big) = 0$ $(k = 1, 2)$.

Inductively, find a subsequence $\{T_j^q\}$ of $\{T_m^{q-1}\}$ and a sequence $\{D^q(j, q)\}$ of related discs such that

$$\lim_{j \to \infty} \big(D^q(j, k),\, D_1\big) = 0 \qquad (k = 1, 2, \cdots, q).$$

This procedure must finish after a finite number of steps. Otherwise, if $q = 2\,\mathrm{nul}\,(\lambda_0 - T)^*$ and $0 < \epsilon < \mathrm{dist}\,(\lambda_0, \partial D_1)$, where λ_0 is a fixed point in V, then there are M, N such that $1/2^M < (1/2)\epsilon$ and $d\big(D^q(j,k), D_1\big) < 1/2^M$ $(k = 1, 2, \cdots, q)$ when $j > N$. Thus computation indicates that $\lambda_0 \in D^q(j,k)\,(k=1,2,\cdots,q)$ when $j > N$. By Lemma 4.4 $\mathrm{nul}\,(\lambda_0 - T_j^q)^* \geq q > \mathrm{nul}\,(\lambda_0 - T)^*$ $(j > N)$. But according to the stability of the semi-Fredholm operator $\mathrm{nul}\,(\lambda_0 - T_j^q)^* \leq \mathrm{nul}\,(\lambda_0 - T)^*$ when j is large enough, a contradiction. Thus there exists a subsequence $\{T_j^Q\}$ of $\{T_n\}$ such that for each $\epsilon > 0$, there are M and N satisfying $\epsilon < 1/2^M$ and $\mathrm{card}\,\{k : d(D^Q(n,k), D_1) < 1/2^M\}$ is a constant independent of n when $n > N$.

If $\{T_j^Q\}_{j=1}^{\infty}$ has no subsequence $\{T_j^{Q+1}\}_{j=1}^{\infty}$ which has a sequence $\{D'^{\,Q+1}(j, l_j)\}_{j=1}^{\infty}$ such that $\lim_{j\to\infty} d(D'^{\,Q+1}(j, l_j), D_1) = 0$, then for each $\epsilon > 0$, there are M and N such that $\epsilon < 1/2^M$ and $\mathrm{card}\,\{k : d(D'^{\,Q}(n,k), D_1) < 1/2^M\} = 0$ provided $n > N$, and D_1 satisfies the requirements of the lemma. Otherwise, by the same argument we can find a subsequence $\{T_m^{Q+R}\}_{m=1}^{\infty}$ of $\{T_j^Q\}_{j=1}^{\infty}$ $(R < \infty)$ such that for each $\epsilon > 0$, there are M and N satisfying $\epsilon < 1/2^M$ and $\mathrm{card}\,\{l : d(D'^{\,Q+p}(j,l), D_1) < 1/2^M\}$ is a constant independent of j provided $j > N$.

If the subsequence $\{T_m^{Q+R}\}$ has no other convergent discs containing V and if we recover $\{T_m^{Q+R}\}_{m=1}^{\infty}$ as $\{T_{n_m}\}_{m=1}^{\infty}$, then $\{T_{n_m}\}$ and $\Sigma = \{D_1\}$ satisfy the requirements of the lemma. Otherwise, we can repeat the procedure and get a subsequence $\{T_{n_m}\}$ and a set of convergent discs $\Sigma = \{D_1, \cdots, D_j\}_{j=1}^{p}$ satisfying (i) and (ii). p must be finite, i.e., after a finite number of steps this procedure must stop. In fact, if $p = 4[\mathrm{nul}\,(\lambda_0 - T) + \mathrm{nul}\,(\lambda_0 - T)^*]$, and $0 < \epsilon < \min\{\mathrm{dist}\,(\lambda_0, \partial D_j) : j = 1, 2, \cdots, p\}$, then there are M and N such that $1/2^M < \epsilon/2$ and

$$\mathrm{card}\,\big\{k :\, d\big(D(n_m, k), D_j\big) < 1/2^M\big\} = a_j,$$
$$\mathrm{card}\,\big\{l :\, d\big(D'(n_m, l), D_j\big) < 1/2^M\big\} = b_j,$$

$(j = 1, 2, \cdots, p)$ and $m > N$. Thus

$$\sum_{j=1}^{p} a_j + \sum_{j=1}^{p} b_j \geq p > 2[\mathrm{nul}\,(\lambda_0 - T) + \mathrm{nul}\,(\lambda_0 - T)^*]$$

and

$$\sum a_j > [\mathrm{nul}\,(\lambda - T) + \mathrm{nul}\,(\lambda - T)^*] \text{ or } \sum b_j > [\mathrm{nul}\,(\lambda - T) + \mathrm{nul}\,(\lambda - T)^*].$$

In either case we have a contradiction. □

Lemma 4.11. *Let $T_n \in DJ(\mathcal{H})$ $(n = 1, 2, \cdots)$, $T_n \to T$ $(n \to \infty)$ and let V be a component of $\rho_{s-F}^{\infty}(T)$ (or $\rho_{s-F}^{-\infty}(T)$). Then there exist a subsequence $\{T_{n_m}\}$ of $\{T_n\}$ and a convergent disc D of $\{T_n\}$ related to T such that*

(i) $D \supset V$;

(ii) $D \cap (\rho_F(T) \cup \rho_{s-F}^{-\infty}(T)) = \varnothing$ ($D \cap (\rho_F(T) \cup \rho_{s-F}^{\infty}(T)) = \varnothing$, *respectively*);

(iii) *for each $\epsilon > 0$, there exist natural numbers M and N such that $1/2^M < \epsilon$ and*

$$\operatorname{card}\{k : d(D(n_m, k), D) < 1/2^M) = \text{a finite constant } (\infty, \text{ respectively}),$$
$$\operatorname{card}\{l : d(D'(n_m, l), D) < 1/2^M) = \infty \text{ (a finite constant, respectively)}$$

provided $m > N$.

Proof. Denote $F_k = V \backslash (\partial V)_{\epsilon/2^k}$, where ϵ is a fixed positive number and is chosen so small that each F_k $(k = 1, 2, \cdots)$ is a connected compact subset of V. Thus $F_1 \subset F_2 \subset \cdots$ and $\bigcup_{i=1}^{\infty} F_i = V$. For each k, there is an $n_k \in \mathbb{N}$ such that $D'(n, l) \supset F_k$ for all $n \geq n_k$ and $l = 1, 2, \cdots$. Otherwise, by the connectedness of F_k, there are two subsequences $\{m_j\}_{j=1}^{\infty}$, $\{l_j\}_{j=1}^{\infty}$ of $\mathbb{N}$ and $\{\lambda_j\}_{j=1}^{\infty} \subset \mathbb{C}$ such that

$$\lambda_j \in \partial D'(m_j, l_j) \cap F_k \qquad j = 1, 2, \cdots.$$

Since $\{\lambda_j\}$ is bounded, $\{\lambda_j\}$ has a limit point $\mu \in F_k \subset V$. Note that $\lambda_j \in \sigma_{lre}(T_{m_j})$. This is contradictory to the stability of the semi-Fredholm operator. Thus we can assume (by considering the subsequence $\{T_{n_k}\}_{k=1}^{\infty}$ only) that $D'(n, l) \supset F_k$ for all $n \geq k$ and all $l = 1, 2, \cdots$.

By the Weierstrass theorem, there are subsequences $\{l_j^n\}_{j=1}^{\infty}$ $(n = 1, 2, \cdots)$ and a sequence $\{D_n\}_{n=1}^{\infty}$ of discs such that

$$\mathbb{N} \supset \{l_j^1\} \supset \{l_j^2\} \supset \cdots \quad \text{and} \quad \lim_{j \to \infty} d(D'(n, l_j^n), D_n) = 0 \quad (n = 1, 2, \cdots).$$

For the same reason, a subsequence $\{D_{n_k}\}_{k=1}^{\infty}$ of $\{D_n\}$ has a limit D, i.e., $\lim_{k \to \infty} d(D_{n_k}, D) = 0$. It is easily seen that $D_n \supset F_n$ $(n = 1, 2, \cdots)$ and

$$\lim_{k \to \infty} d(D'(n_k, l_k^k), D) = 0.$$

Thus D is a convergent disc of $\{T_n\}$ related to T, $D \supset V$ and computation indicates that for each $\epsilon > 0$, there is M, $N \in \mathbb{N}$ such that $1/2^M < \epsilon$ and

$$\operatorname{card}\{l : \ d(D'(n_k, l), \ D) < 1/2^M\} = \infty \quad \text{for all } k > N.$$

By the argument of Lemma 4.10, we can find a subsequence of $\{T_{n_k}\}_{k=1}^{\infty}$ (for simplicity it is still denoted by $\{T_{n_k}\}$) such that card$\{j : d(D(n_k, j), D) < 1/2^M\}$ is a constant when k is big enough. Since $V \subset \rho_{s-F}^{\infty}(T)$, the constant is finite. Thus (i) and (iii) is proved.

If $\lambda_0 \in D \cap \rho_F(T)$, we have nul$(\lambda_0 - T) < \infty$ and nul$(\lambda_0 - T)^* < \infty$. If $0 < \epsilon <$ dist$(\lambda_0, \partial D)$, then when $k > N$, card$\{l : \lambda_0 \in D'(n_k, l)\} = \infty$ i.e. nul$(\lambda_0 - T_{n_k}) = \infty$, which contradicts nul$(\lambda_0 - T) < \infty$. Thus $D \cap \rho_F(T) = \varnothing$. Similarly, we can prove that $D \cap \rho_{s-F}^{-\infty}(T) = \varnothing$.

By the same arguments, we can prove the lemma when $V \subset \rho_{s-F}^{-\infty}(T)$. □

Let $\{T_n\}, \{T_{n_m}\}_{m=1}^{\infty}, T, V \subset \rho_F(T)$, and $\Sigma = \{D_j\}_{j=1}^{p}$ be given as in Lemma 4.10. Define

$$N_{jn} = \overline{\lim_{m \to \infty}} \text{ card}\{k : d(D(n_m, k), D_j) < 1/2^n\} \tag{4.1}$$

$$M_{jn} = \overline{\lim_{m \to \infty}} \text{ card}\{l : d(D'(n_m, l), D_j) < 1/2^n\} \tag{4.2}$$

$$N_j = \overline{\lim_{n \to \infty}} N_{jn} \tag{4.3}$$

$$M_j = \overline{\lim_{n \to \infty}} M_{jn} \tag{4.4}$$

$(j = 1, 2, \cdots, p)$.

Lemma 4.12. *Let $\{T_n\}_{n=1}^{\infty} \subset DJ(\mathcal{H})$, T, $\{T_{n_m}\}_{m=1}^{\infty}$, $V \subset \rho_F(T)$ and $\Sigma = \{D_j\}_{j=1}^{p}$ be given as in Lemma 4.10. Then for each $\lambda_0 \in V$, there exists $N \in \mathbb{N}$ such that*

$$\begin{aligned} \text{nul}(\lambda_0 - T_{n_m}) &= \sum_j M_j \le \text{nul}(\lambda_0 - T), \\ \text{nul}(\lambda_0 - T_{n_m})^* &= \sum_j N_j \le \text{nul}(\lambda_0 - T)^* \end{aligned} \tag{4.5}$$

and

$$\text{ind}(\lambda_0 - T) = \sum_j N_j - \sum_j M_j.$$

when $m > N$.

Proof. By Lemma 4.10, when n, m are big enough,

$$N_j = \text{card}\left\{k : d(D(n_m, k), D_j) < \frac{1}{2^n}\right\}$$

and

$$M_j = \text{card}\left\{l: \ d(D'(n_m,l),D_j) < \frac{1}{2^n}\right\}.$$

By the proof of Lemma 4.10, when m is big enough, there are at least $R := \sum_j N_j + \sum_j M_j$ discs related to T_{n_m} each of which contains λ_0.

Claim. When m is big enough, there are only R discs related to T_{n_m} containing λ_0. Otherwise, there is a subsequence of $\{T_{n_m}\}$ (for simplicity we can consider it is $\{T_{n_m}\}$ itself) and a sequence $\{D_{n_m}\}$ of related discs (type $D(n_m,k)$ or type $D'(n_m,l)$) such that $\lambda_0 \in D_{n_m}$ and

$$D_{n_m} \notin \left\{D(n_m,k): \ d(D(n_m,k),D_j) < \frac{1}{2^n}\right\} \cup \left\{D'(n_m,l): \ d(D'(n_m,l),D_j) < \frac{1}{2^n}\right\}$$
$$j = 1,2,\cdots,p; \ n \text{ is big enough.} \quad (4.6)$$

Since $\{D_{n_m}\}$ is a bounded set, it has at least one limit point D. By the same argument as Lemma 4.8, D is a convergent disc of $\{T_{n_m}\}$ related to T and $D \supset V$. But by (4.6) $D \notin \Sigma$. This contradicts Lemma 4.10 and therefore there are R and only R discs related to T_{n_m}, each of which contains λ_0 when m is big enough.

It is obvious by Lemma 4.4 that

$$\text{nul}\,(\lambda_0 - T_{n_m}) = \sum_j M_j \leq \text{nul}\,(\lambda_0 - T)$$

and

$$\text{nul}\,(\lambda_0 - T_{n_m})^* = \sum_j N_j \leq \text{nul}\,(\lambda_0 - T)^*$$

when m is big enough. By the stability of the semi-Fredholm operator,

$$\text{ind}\,(\lambda_0 - T) = \sum_j M_j - \sum_j N_j. \qquad \square$$

Now we are in a position to prove Theorem 4.5.

Proof of Theorem 4.5. "Sufficient" part. If (ii) is satisfied, by Theorem 1.27 and Proposition 1.22, for each $\epsilon > 0$ there exists an operator $T_\epsilon \in \mathcal{L}(\mathcal{H})$ satisfying

(i) $\|T - T_\epsilon\| < \epsilon$;
(ii) $\sigma_{lre}(T_\epsilon) = \sigma_{lre}(T)_\epsilon$;
(iii) $\rho^r_{s-F}(T_\epsilon) = \rho_{s-F}(T)\backslash\sigma_{lre}(T_\epsilon) = \rho_{s-F}(T_\epsilon)$;

(iv) $\forall\lambda \in \rho_{s-F}(T_\epsilon)$, $\min\operatorname{ind}(\lambda - T_\epsilon)^k \le \min\operatorname{ind}(\lambda - T)^k$ $(k = 1, 2, \cdots)$ and $\operatorname{ind}(\lambda - T_\epsilon) = \operatorname{ind}(\lambda - T)$.

Let Δ be a set of countably many discs contained in $\sigma_{lre}(T)_\epsilon$ such that $\bigcup\{\partial D : D \in \Delta\}$ is dense in $\sigma_{lre}(T)_\epsilon$.

If $D = \{z \in \mathbb{C} : |z - c| < d\}$, then denote $J_D = c + rS$ and $J'_D = c + rS^*$. Set

$$J_1 = \bigoplus_{j=1}^{p} J_{D_j}^{(N_j)},\ J_2 = \bigoplus_{j=1}^{p} J'^{(M_j)}_{D_j}\ J_3 = \bigoplus_{D\in\Delta} J_D \text{ and } J_\epsilon = \bigoplus_{i=1}^{3} J_i.$$

Then by the similarity orbit theorem, $T_\epsilon \in S(J_\epsilon)^-$. Since ϵ is arbitrary, $T \in DJ(\mathcal{H})^-$.

If (i) is satisfied, define $J_\epsilon = J_3 = \bigoplus_{D\in\Delta} J_D$; then by the same arguments we can prove $T \in DJ(\mathcal{H})^-$.

"Necessary " part. Assume that $T \in DJ(\mathcal{H})^-$ and $T_n \to T$ $(n \to \infty)$, where $T_n \in DJ(\mathcal{H})$. By Lemma 4.6, $\sigma_0(T) = \varnothing$. If for each $\epsilon > 0$, $\sigma(T)\backslash\sigma_{lre}(T)_\epsilon \subset \rho^\circ_{s-F}(T)$, then $T \in (BQT)$ and T satisfies (i). Otherwise, when ϵ is small enough, there is $\lambda_0 \in \sigma(T)\backslash\sigma_{lre}(T)_\epsilon$ and $\operatorname{ind}(\lambda_0 - T) \ne 0$. Note that $\sigma(T)\backslash\sigma_{lre}(T)_\epsilon$ consists of finitely many components. Suppose that they are denoted by $\{V_i\}_{i=1}^R \cup \{U_j\}_{j=1}^Q$, where $\{V_i\}$ are components of $\left[\sigma(T)\cap\rho_F(T)\right]\backslash\sigma_{lre}(T)_\epsilon$ and $\{U_j\}$ are components of $\rho^{\pm\infty}_{s-F}(T)\backslash\sigma_{lre}(T)_\epsilon$.

By Lemma 4.10, we can find a subsequence $\{T_{n_m}\}_{m=1}^\infty$ of $\{T_n\}$, finitely many convergent discs $\{D_j\}_{j=1}^{l_1}$ of $\{T_{n_m}\}$ related to T and corresponding pairs of integers $\{(N_j, M_j)\}_{j=1}^{l_1}$, where $0 \le N_j, M_j < \infty$ $(j = 1, 2, \cdots, l_1)$ such that each V_j is contained in some D_k, $\bigcup_{j=1}^{l_1} D_j^- \subset \sigma(T)$, $\bigcup_{j=1}^{l_1} \partial D_j \subset \sigma_{lre}(T)$ and for each $\lambda \in \left[\sigma(T)\cap\rho_F(T)\right]\backslash\sigma_{lre}(T)_\epsilon$,

$$\sum_{\lambda\in D_j} N_j \le \operatorname{nul}(\lambda - T)^*,\ \sum_{\lambda\in D_j} M_j \le \operatorname{nul}(\lambda - T) \text{ and } \sum_{\lambda\in D_j}(M_j - N_j) = \operatorname{ind}(\lambda - T).$$

By Lemma 4.11, there is a subsequence of $\{T_{n_m}\}_{m=1}^\infty$, for convenience we still denote it by $\{T_{n_m}\}_{m=1}^\infty$, and Q convergent discs $\{D_j\}_{j=l_1+1}^{Q}$ of $\{T_{n_m}\}_{m=1}^\infty$ related to T and Q pairs of integers $\{(N_j, M_j)\}_{j=l_1+1}^{l_1+Q}$, where $0 \le M_j, N_j \le \infty$, but $(N_j, M_j) \ne (\infty, \infty)$ $(j = l_1 + 1, \cdots, l_1 + Q)$ such that $U_j \subset D_j$, $\bigcup_{j=l_1+1}^{l_1+Q} D_j^- \subset \sigma(T)$, $\bigcup_{j=l_1+1}^{l_1+Q} \partial D_j \subset \sigma_{lre}(T)$ and for each $\lambda \in U_j$, $N_j = \infty$, $M_j = \operatorname{nul}(\lambda - T)$, if $U_j \subset \rho^{-\infty}_{s-F}(T)$ or $N_j = \operatorname{nul}(\lambda - T)^*$, $M_j = \infty$ if $U_j \subset \rho^{\infty}_{s-F}(T)$ $(j = l_1 + 1, \cdots, l_1 + Q)$.

Denote $p = l_1 + Q$, then $\Sigma = \{D_j\}_{j=1}^p$ and $\{(N_j, M_j)\}_{j=1}^p$ satisfy (ii) and the proof of Theorem 4.5 is now complete. □

Example 4.13. Assume that Ω is a rectangle in $\mathbb{C}$ with width 3/2 and length 9/2, and B is an operator in $B_n(\Omega)$ with $\sigma(B) = \overline{\Omega}$. Then by Theorem 4.5 $B \notin DJ(\mathcal{H})^-$.

Theorem 4.5 implies that it is also inadequate to consider the operator class $\{\lambda + \mu S, \lambda + \mu S^* : \lambda \in \mathbb{C}, \mu > 0, \ S$ is the unilateral shift$\}$ as the replacement of Jordan blocks in $\mathcal{L}(\mathcal{H})$, since $DJ(\mathcal{H})^-$ is only a rather small part of $\mathcal{L}(\mathcal{H})$. Comparing Theorem 4.2 with Theorems 4.3, 4.5, we can find that the class of strongly irreducible operators is an adequate replacement of Jordan blocks in $\mathcal{L}(\mathcal{H})$.

4.2 The uniqueness of the strongly irreducible decompositions of operators

In this section we will discuss the relation between the semisimplicity of the commutant of an operator and the uniqueness of the strongly irreducible decomposition of the operator. Also, we will prove that the class of operators which have a unique representation of a direct sum of strongly irreducible operators with the nice property is dense in $\mathcal{L}(\mathcal{H})$.

A Banach algebra $\mathcal{B}$ is *semisimple* if $\operatorname{rad}\mathcal{B} = \{0\}$, i.e., $A \in \mathcal{B}$ and $\sigma_{\mathcal{B}}(A) = \{0\}$ imply $A = 0$. It is easily seen that for each operator $T \in \mathcal{L}(\mathcal{H})$, the Banach algebra $\mathcal{A}'(T)$ is semisimple if $\mathcal{A}'(T)$ does not contain any non-zero quasinilpotent. Proposition 4.1 asserts that not every operator has a strongly irreducible decomposition and Theorem 4.2 implies that the class of operators which have strongly irreducible decompositions is dense in $\mathcal{L}(\mathcal{H})$. Therefore, we need to discuss the following two questions.

Question a. *If an operator has a strongly irreducible decomposition, when is this decomposition unique?*

Question b. *What is the norm-closure of the class of operators which have unique strongly irreducible decompositions?*

Definition 4.14. Given an operator $T \in \mathcal{L}(\mathcal{H})$, if a sequence of uniformly bounded idempotents $\mathcal{P} = \{P_n\}_{n=1}^{l}$ $(1 \le l \le \infty)$ satisfies the following conditions

(i) $\mathcal{P} \subset \mathcal{A}'(T)$;

(ii) $P_m P_n = 0$ $(n \neq m)$;

(iii) $\sum_{i=1}^{n} P_i \to I$ $(n \to l)$ in the strong operator topology;

(iv) $T_n = T|_{P_n\mathcal{H}} \in (SI)$ $(n = 1, 2, \cdots)$;

the $\mathcal{P}$ is called a *strongly irreducible decomposition* of T, or briefly, an (SI) decomposition of T. Furthermore, if $\mathcal{P}'$ is a strongly irreducible decomposition of T implies $\mathcal{P}' = \mathcal{P}$, then $\mathcal{P}$ is called the *unique strongly irreducible decomposition of* T.

Example 4.15. S is the unilateral shift on $\mathcal{H}$, $T = S \oplus S \in \mathcal{L}(\mathcal{H} \oplus \mathcal{H})$. Set $P_1 = \begin{bmatrix} 1 & 0 \\ 0 & 0 \end{bmatrix}$, $P_2 = \begin{bmatrix} 0 & 0 \\ 0 & 1 \end{bmatrix}$, $Q_1 = \frac{1}{2}\begin{bmatrix} 1 & 1 \\ 1 & 1 \end{bmatrix}$, $Q_2 = \frac{1}{2}\begin{bmatrix} 1 & -1 \\ -1 & 1 \end{bmatrix}$. Then (P_1, P_2) and (Q_1, Q_2) are two strongly irreducible decompositions of T.

Proposition 4.16. *Let* $\mathcal{P} = \{P_n\}_{n=1}^{l}$ *be a strongly irreducible decomposition of* $T \in \mathcal{L}(\mathcal{H})$, $T_n = T|_{P_n\mathcal{H}}$ $(n = 1, 2, \cdots, l)$, *then the following are equivalent:*

(i) $\mathcal{P}$ *is the unique strongly irreducible decomposition of* T;

(ii) $\ker \tau_{T_n T_m} = \{0\}$ $(n \neq m)$;

(iii) $\mathcal{A}'(T) = \sum_{n=1}^{l} \dot{+} \mathcal{A}'(T_n) := \{B = \sum_{n=1}^{l} \dot{+} B_n : B_n \in \mathcal{A}'(T_n)\}$.

Proof. (i)$\Rightarrow$(ii). Without loss of generality, we can assume that $\ker \tau_{T_1 T_2} \neq \{0\}$, i.e., there exists $C \in \mathcal{L}(\mathcal{H}_2, \mathcal{H}_1)$ such that $C \neq 0$ and $T_1 C = C T_2$, where $\mathcal{H}_n = \operatorname{ran} P_n$. Define $Q_1, Q_2 \in \mathcal{L}(\mathcal{H})$ by

$$Q_1\Big(\sum_{n=1}^{l} \dot{+} x_n\Big) = x_1 + Cx_2$$

$$Q_2\Big(\sum_{n=1}^{l} \dot{+} x_n\Big) = x_2 - Cx_2$$

where $x = \sum_{n=1}^{l} \dot{+} x_n \in \mathcal{H}$ and $x_n \in \mathcal{H}_n$ $(n = 1, 2, \cdots, l)$. Then Q_1, Q_2 are idempotents commuting with T and $Q_1 Q_2 = Q_2 Q_1 = 0$.

Since $\operatorname{ran} Q_1 = \operatorname{ran} P_1 = \mathcal{H}_1$, $T|_{\operatorname{ran} Q_1} \in (SI)$. Computation shows that $T_2 = W T_2' W^{-1}$, where W, $T_2' \in \mathcal{L}(\mathcal{H}_1 \dot{+} \mathcal{H}_2)$, $W(x_1 \dot{+} x_2) = (x_1 + Cx_2) \dot{+} x_2$, $W^{-1}(x_1 \dot{+} x_2) = (x_1 - Cx_2) \dot{+} x_2$, $T_2'(x_1 \dot{+} x_2) = -CT_2 x_2 + T_2 x_2$ for $x_1 \in \mathcal{H}_1$ and $x_2 \in \mathcal{H}_2$.

Since $T_2 \in (SI)$, $T_2' = T|_{\operatorname{ran} Q_2} \in (SI)$. Thus $\mathcal{P}_1 = \{Q_1, Q_2, P_i \ (i = 3, 4, \cdots)\}$ is a strongly irreducible decomposition of T different from $\mathcal{P}$, a contradiction.

(ii)$\Rightarrow$(iii). We can get the conclusion directly from $\ker \tau_{T_n T_m} = \{0\}$ $(n \neq m)$ and calculation.

(iii)$\Rightarrow$(i). Let $P \in \mathcal{A}'(T)$ be an idempotent such that $T|_{\operatorname{ran} P} \in (SI)$. By (iii) $P = \sum_{n=1}^{l} \dot{+} P_n'$. Since $T = \sum_{n=1}^{l} \dot{+} T_n$ and each P_n' is an idempotent, $P_n' = 0$ or 1 $(n = 1, 2, \cdots, l)$. Thus $P \in \mathcal{P}$ and P is the unique strongly irreducible decomposition of T. □

In what follows, Ω denotes a connected, bounded open subset of $\mathbb{C}$.

Theorem 4.17. *Let $T \in \mathcal{B}_n(\Omega)$. Then*

(i) *If $n = 1$, then $\mathcal{A}'(T)$ is semisimple;*

(ii) *T can be expressed as a direct sum of at most n (SI) Cowen–Douglas operators.*

Proof. (i) Choose an $x_\lambda \in \ker(\lambda - T)$ $(\lambda \in \Omega)$, $x_\lambda \neq 0$. Thus $\bigvee\{x_\lambda : \lambda \in \Omega\} = \mathcal{H}$. If $A \in \mathcal{A}'(T)$ is a quasinilpotent, then $TAx_\lambda = ATx_\lambda = \lambda Ax_\lambda$ and $Ax_\lambda \in \ker(\lambda - T)$. Since $\operatorname{nul}(\lambda - T) = 1$, $Ax_\lambda = t(\lambda)x_\lambda$ for some $t(\lambda) \in \mathbb{C}$. Thus $t(\lambda) \in \sigma(A) = \{0\}$ and $t(\lambda) = 0$ or $Ax_\lambda = 0$ $(\lambda \in \Omega)$. Since $\{x_\lambda : \lambda \in \Omega\}$ spans $\mathcal{H}$, $A = 0$ and $\mathcal{A}'(T)$ is semisimple.

(ii) If $T \in (SI)$, then (ii) is true. If T is not an (SI) operator, then $T = T_1 \dot{+} T_2$. Observe that $\sigma_p(T) \cap \Omega = [\sigma_p(T_1) \cap \Omega] \cup [\sigma_p(T_2) \cap \Omega]$ and $\sigma_p(T_i) \cap \Omega \neq \varnothing$ $(i = 1, 2)$. Since $\rho_F(T) \supset \Omega$, $\rho_F(T_i) \supset \Omega$ $(i = 1, 2)$. Denote $n_i = \operatorname{ind}(\lambda - T_i)$ $\lambda \in \Omega$. Then $T_i \in \mathcal{B}_{n_i}(\Omega)$ $(i = 1, 2)$ and $n_1 + n_2 = n$. If at least one of T_1, T_2 is not an (SI) operator, repeat this argument. Since $\mathcal{B}_1(\Omega)$ operators are in (SI) and $n < \infty$, after a finite number of steps T can be written as a direct sum of at most n Cowen–Douglas (SI) operators. □

Theorem 4.18. (i) *If $T \in \mathcal{B}_n(\Omega)$ and $\mathcal{A}'(T)$ is semisimple, then T has a unique strongly irreducible decomposition.*

(ii) *If $T \in \mathcal{B}_n(\Omega)$ can be expressed as a direct sum of n (SI) operators, then the following are equivalent:*

(a) *$\mathcal{A}'(T)$ is an abelian algebra;*

(b) *$\mathcal{A}'(T)$ is semisimple;*

(c) *T can be uniquely expressed as a direct sum of (SI) operators.*

Proof. (i) By Theorem 4.17, T can be expressed as a direct sum of at most n (SI) operators. Let $\mathcal{P} = \{P_k\}_{k=1}^{l}$ $(1 \leq l \leq n)$ be an (SI) decomposition of T and $T = \sum_{k=1}^{l} \dot{+} T_k$, where $T_k = T|_{\operatorname{ran} P_k}$. If $\ker \tau_{T_k T_m} \neq \{0\}$ for some $k \neq m$, then there exists an operator C such that $T_k C = C T_m$ and $C \neq 0$. Define $B \in \mathcal{L}(\mathcal{H})$ by $B(\sum_{i=1}^{l} x_i) = Cx_m$, where $x_i \in \operatorname{ran} P_i$ $(i = 1, 2, \cdots, l)$. Thus $B^2 = 0$, i.e., B is a nilpotent. Since $T_k C = C T_m$, $B \in \mathcal{A}'(T)$. It follows from the semisimplicity of $\mathcal{A}'(T)$ that $B = 0$, or $C = 0$, a contradiction. Thus $\ker \tau_{T_k T_m} = \{0\}$ $(k \neq m)$. By Proposition 4.16, $\mathcal{P}$ is the unique (SI) decomposition of T.

(ii) (a)$\Rightarrow$(b). If there exists an operator C such that $C \neq 0$ and $C \in \ker \tau_{T_k T_m}$ for some $k \neq m$, define $B \in \mathcal{L}(\mathcal{H})$ by $B(\sum_{i=1}^{n} x_i) = Cx_m$, where $x_i \in \operatorname{ran} P_i$ $(i = 1, 2, \cdots, n)$. Thus $B \in \mathcal{A}'(T)$. Since $\mathcal{A}'(T)$ is abelian, $BP_k = P_k B$. Thus for

$x_m \in \operatorname{ran} P_m$, we have $Cx_m = BP_m x_m = P_m Bx_m = 0$, i.e., $C = 0$. By Theorem 4.17, $T = \sum_{k=1}^{n} \dot{+} T_k$ and $T_k \in \mathcal{B}_1(\Omega)$ $(k = 1, 2, \cdots, n)$. By Proposition 4.16, $\mathcal{A}'(T) = \sum_{i=1}^{n} \dot{+} \mathcal{A}'(T_i)$. If $A \in \mathcal{A}'(T)$ and $\sigma(A) = \{0\}$, then $A = \sum_{i=1}^{n} \dot{+} A_i$, where $A_i \in \mathcal{A}'(T_i)$. By Theorem 4.17, $A_i = 0$ $(i = 1, 2, \cdots, n)$, or $A = 0$. Thus $\mathcal{A}'(T)$ is semisimple.

By (i) we can get (c) from (b).

(c)$\Rightarrow$(a) By Proposition 4.16, $\mathcal{A}'(T) = \sum_{i=1}^{n} \dot{+} \mathcal{A}'(T_i)$. Since $T_i \in \mathcal{B}_1(\Omega)$ $(i = 1, 2, \cdots, n)$, it suffices to prove that the commutant of each $\mathcal{B}_1(\Omega)$ operator is abelian.

Assume that $A \in \mathcal{B}_1(\Omega)$ and $D, F \in \mathcal{A}'(A)$. Since $\operatorname{nul}(\lambda - A) = 1$ for all $\lambda \in \Omega$, $Dx_\lambda = d(\lambda)x_\lambda$, $Fx_\lambda = f(\lambda)x_\lambda$ for all $x_\lambda \in \ker(\lambda - A)$, where $d(\lambda), f(\lambda) \in \mathbb{C}$, $\lambda \in \Omega$. Since $\{x_\lambda\}$ spans $\mathcal{H}$, $DF = FD$ and $\mathcal{A}'(A)$ is abelian. $\square$

It is worth noting that the semisimplicity of $\mathcal{A}'(T)$ in Theorem 4.18 is only a sufficient condition for the uniqueness of the (SI) decomposition of the $\mathcal{B}_n(\Omega)$ operator T. The next theorem and example indicate that it is not necessary.

Theorem 4.19. *Assume that $T \in \mathcal{L}(\mathcal{H})$ has an (SI) decomposition and $\mathcal{A}'(T)$ is abelian. Then T has a unique (SI) decomposition.*

Proof. It suffices to prove that if P_1, P_2 are idempotents in $\mathcal{A}'(T)$ such that $T|_{\operatorname{ran} P_1}$ and $T|_{\operatorname{ran} P_2} \in (SI)$, then $P_1 = P_2$ or $P_1P_2 = 0$. If $P_1 \neq P_2$ and $P_1P_2 \neq 0$, P_1P_2 is an idempotent in $\mathcal{A}'(T)$ and $P_1P_2|_{\operatorname{ran} P_i}$ commutes with $T|_{\operatorname{ran} P_i}$ $(i = 1, 2)$. This contradicts $T|_{\operatorname{ran} P_i} \in (SI)$. $\square$

Corollary 4.20. *If $T \in \mathcal{L}(\mathcal{H})$ has the nice property and T has an (SI) decomposition, then this decomposition is unique.*

Example 4.21. Let $T = M_n(\Omega)$ be given as in Theorem 3.5. Then $T^* \in \mathcal{B}_n(\Omega^*)$. By Theorem 3.5 $\mathcal{A}'(T^*)$ is abelian. By Theorem 4.19, T^* has an (SI) decomposition and therefore the (SI) decomposition is unique. Note that $\mathcal{A}'(T^*)$ is not semisimple.

Proposition 4.22. *Let $T \in \mathcal{B}_n(\Omega)$ with semisimple $\mathcal{A}'(T)$. Then for each $\epsilon > 0$, there exists a compact operator K, $\|K\| < \epsilon$, such that $T + K \in (SI)$.*

Proof. It is easily seen that there exists an invertible operator X such that $A = X^{-1}TX = \bigoplus_{k=1}^{l} A_k \in \mathcal{L}(\mathcal{H}^{(l)})$ $(1 \leq l \leq n)$, where $A_k \in (SI)$ and by Theorem 4.17, $A_k \in B_{m_k}(\Omega)$, $0 < m_k \leq n$, $1 \leq k \leq l$ and $\sum_{k=1}^{l} m_k = n$. Therefore, $\sigma_r(A_1) \cap \sigma_l(A_k) \neq \varnothing$ $(k = 2, 3, \cdots, l)$. By Theorem 1.11 there exist compact operators $C_2, C_3, \cdots, C_l$ such that $C_k \notin \operatorname{ran} \tau_{A_1 A_k}$ and

$$\|C_k\| < \frac{\epsilon}{2^k \|X\| \|X^{-1}\|} \qquad k = 2, 3, \cdots, l.$$

Set $C = \begin{bmatrix} 0 & C_2 & C_3 & \dots & C_l \\ & & 0 & & \end{bmatrix} \in \mathcal{L}(\mathcal{H}^{(l)})$. Then C is compact with $\|C\| < \epsilon/(2\|X\|\|X^{-1}\|)$ and therefore, $\|XCX^{-1}\| < \epsilon/2$. By Corollary 3.14 $A + C \in (SI)$. Hence $T + K \in (SI)$ where $K = XCX^{-1}$ and $\|K\| < \epsilon$. □

Theorem 4.23. *The class of operators each of which can be uniquely expressed as a direct sum of finitely many (SI) operators with the nice property is dense in $\mathcal{L}(\mathcal{H})$.*

Proof. By the argument of Theorem 1.23, we can prove that there is a $K \in \mathcal{K}(\mathcal{H})$, $\|K\| < \epsilon/4$, such that $\sigma_0(T + K) = \sigma_0(T)$ and $\min\operatorname{ind}(T + K - \lambda) = 0$ for all $\lambda \in \rho_{s-F}(T)\backslash\sigma_0(T)$. By Theorem 1.27, there exists a $G \in \mathcal{L}(\mathcal{H})$, $\|G\| < \epsilon/4$, such that $\sigma_0(T + K + G) \subset \sigma_0(T)$, $\min\operatorname{ind}(T + K + G - \lambda) = 0$ for all $\lambda \in \rho_{s-F}(T + K + G)\backslash\sigma_0(T)$ and $\sigma_{lre}(T + K + G)$ is the closure of an analytic Cauchy domain Φ. Set $T_\epsilon = T + K + G$; thus it suffices to find $A \in \mathcal{L}(\mathcal{H})$ such that A is the direct sum of finitely many (SI) operators with the nice property and $\|A - T_\epsilon\| < \epsilon/2$.

Since Φ is an analytic Cauchy domain, $\rho_{s-F}(T_\epsilon) \cap \sigma(T_\epsilon)$ consists of finitely many components $\{\Omega_i\}_{i=1}^{l}$. Denote $n_i = \operatorname{ind}(\lambda - T_\epsilon)$ $(\lambda \in \Omega_i)$; then $n_i \neq 0$ $(i = 1, 2, \cdots, l)$. Since $l < \infty$, there exists an $\eta > 0$ such that when $|n_i| = \infty$, $\bigwedge(\Omega_i, \eta) := \bigcup\{\overline{\Omega}_i + r\eta : 0 \leq r \leq 1\}$ satisfies

(i) $\mathbb{C}\backslash\overline{\Omega}_i$ and $\mathbb{C}\backslash\bigwedge(\Omega_i, \eta)$ have the same number of components;

(ii) $\bigwedge(\Omega_i, \eta) \subset \sigma(T_\epsilon)$;

(iii) $\left[\sigma_{lre}(T_\epsilon)\backslash\bigcup\{\bigwedge(\Omega_i, \eta) : |n_i| = \infty\}\right]^\circ = \Psi$ is an analytic Cauchy domain and therefore $\{\bigwedge(\Omega_i, \eta) : |n_i| = \infty\} \cup \{\Omega_j : |n_j| < \infty\}$ are pairwise disjoint. Let $\{\Psi_i\}_{i=l+1}^{m}$ $(m < \infty)$ be the set of components of Ψ. Then there is a sequence $\{\sigma_i\}_{i=l+1}^{m}$ of compact sets such that $\sigma_i \subset \Psi_i$ $(i = l + 1, \cdots, m)$. Assume that $\sigma_0(T_\epsilon) = \{\lambda_i\}_{i=m+1}^{n}$ $(n < \infty)$ and let $p_i = \dim E(\lambda_i, T_\epsilon)$ $(i = m + 1, \cdots, n)$.

- For each Ω_i $(0 < n_i < \infty)$, construct (SI) operator $A_i \in \mathcal{B}_{n_i}(\Omega_i)$ by Theorem 3.5 such that $\sigma(A_i) = \overline{\Omega}_i$ and $\mathcal{A}'(A_i^*)$ is strictly cyclic.
- For each Ω_i $(0 > n_i > -\infty)$, construct (SI) operator $A_i^* \in \mathcal{B}_{|n_i|}(\Omega_i^*)$ by Theorem 3.5 such that $\sigma(A_i) = \overline{\Omega}_i$ and $\mathcal{A}'(A_i)$ is strictly cyclic.
- For each Ω_i $(n_i = \infty)$, construct an (SI) operator A_i by Proposition 3.6 such that $\sigma(A_i) = \bigwedge(\Omega_i, \eta)$, $\sigma_{lre}(A_i) = \bigcup\{\partial\Omega_i + r\eta : 0 \leq r \leq 1\}$, $\ker(\lambda - A_i)^* = \{0\}$ and $\operatorname{ind}(\lambda - A_i) = \infty$ for all $\lambda \in \Omega_i\backslash\sigma_{lre}(A_i)$ and $\mathcal{A}'(A_i)$ is strictly cyclic.
- For each Ω_i $(n_i = -\infty)$, construct an (SI) operator A_i by Proposition 3.6 such that $\sigma(A_i) = \bigwedge(\Omega_i, \eta)$, $\sigma_{lre}(A_i) = \bigcup\{\partial\Omega_i + r\eta : 0 \leq r \leq 1\}$, $\ker(\lambda - A_i) = \{0\}$ and $\operatorname{ind}(\lambda - A_i) = -\infty$ for all $\lambda \in \Omega_i\backslash\sigma_{lre}(A_i)$ and $\mathcal{A}'(A_i^*)$ is strictly cyclic.

- For each σ_i $(l+1 \le i \le m)$, construct an (SI) operator A_i by Theorem 3.7 such that $\sigma(A_i) = \sigma_{lre}(A_i) = \sigma_i$, and $\mathcal{A}'(A_i)$ is strictly cyclic.
- Set $A_i = \lambda_i + q_{p_i}$ $(m+1 \le i \le n)$, where q_{p_i} is $p_i \times p_i$ Jordan nilpotent.

Since $\{\sigma(A_i) : i = 1, 2, \cdots, n\}$ are pairwise disjoint, by Corollary 1.8 $\ker \tau_{A_i A_j} = \{0\}$ $(i \ne j)$. Set $B = \bigoplus_{i=1}^{n} A_i$. Then by Proposition 4.16, the (SI) decomposition of B is unique. By the similarity theorem, $T_\epsilon \in S(B)^-$. Thus there exists an invertible operator X such that $\|XBX^{-1} - T_\epsilon\| < \epsilon/2$ and $A := XBX^{-1}$ has a unique nice (SI) decomposition. □

4.3 The spectral picture of (SI) operators

The realization of strong irreducibility for some special spectral pictures was discussed in Chapter 3. In this section we will analyse the following question:

Given an operator $T \in \mathcal{L}(\mathcal{H})$ with connected spectrum $\sigma(T)$, does there exist an (SI) operator A such that $\Lambda(A) = \Lambda(T)$, $\min\operatorname{ind}(\lambda - A)^k \le \min\operatorname{ind}(\lambda - T)^k$ for $k = 1, 2, \cdots$ and $\lambda \in \rho_{s-F}(T)$ and $T \in S(A)^-$? Here $\Lambda(T)$ denotes the *spectral picture* of T, i.e., the compact set $\sigma_{lre}(T)$, plus the data corresponding to the indices of $\lambda - T$ for λ in the bounded components of $\rho_{s-F}(T)$.

In order to give an affirmative answer to this question, we need the following lemma.

Lemma 4.24. *Let*

$$R = \begin{bmatrix} A & Q_1 & 0 \\ 0 & B & Q_2 \\ 0 & 0 & C \end{bmatrix} \in \mathcal{L}(\mathcal{H})$$

with respect to the orthogonal direct sum decomposition

$$\mathcal{H} = \Big(\bigoplus_{i=1}^{l_1} \mathcal{H}_i^1\Big) \oplus \Big(\bigoplus_{j=1}^{l_2} \mathcal{H}_j^2\Big) \oplus \Big(\bigoplus_{k=1}^{l_3} \mathcal{H}_k^3\Big),$$

where each $\mathcal{H}_m^i$ is a Hilbert space and $0 \le l_1, l_2, l_3 \le \infty$. Assume that

(i) $A = \bigoplus_{i=1}^{l_1} A_i$, $B = \bigoplus_{j=1}^{l_2} B_j$ *and* $C = \bigoplus_{k=1}^{l_3} C_k$;

(ii) *For arbitrary $n, m \in \mathbb{N}$ $(1 \le n, m \le l_2)$, there exist a sequence $\{j_s\}_{s=1}^{r}$ of finitely many integers and a sequence $\{D_s\}_{s=1}^{r+1}$ of operators in $\{A_i\}_{i=1}^{l_1} \cup \{C_k\}_{k=1}^{l_3}$ such* *that*

$$\sigma_l(B_{j_s})\cap\sigma_r(D_{s+1})\neq\varnothing \text{ and } \sigma_l(B_{j_{s+1}})\cap\sigma_r(D_{s+1})\neq\varnothing, \quad \text{if } D_{s+1}\in\{A_i\}_{i=1}^{l_1},$$
$$\sigma_r(B_{j_s})\cap\sigma_l(D_{s+1})\neq\varnothing \text{ and } \sigma_r(B_{j_{s+1}})\cap\sigma_l(D_{s+1})\neq\varnothing, \quad \text{if } D_{s+1}\in\{C_k\}_{k=1}^{l_3},$$

$(0\le s\le r,\ j_0=n,\ j_{r+1}=m)$;

(iii) *For each* A_i (*or* C_k), *there exists at least one* B_j *satisfying*

$$\sigma_r(A_i)\cap\sigma_l(B_j)\neq\varnothing \quad (\text{or } \sigma_l(C_k)\cap\sigma_r(B_j)\neq\varnothing);$$

(iv)

$$Q_1=\begin{bmatrix} q^1_{11} & \cdots & q^1_{1l_2}\\ \cdots & & \cdots\\ q^1_{l_11} & \cdots & q^1_{l_1l_2}\end{bmatrix}\in\mathcal{L}\Big(\bigoplus_{j=1}^{l_2}\mathcal{H}^2_j,\bigoplus_{i=1}^{l_1}\mathcal{H}^1_i\Big)$$

and

$$Q_2=\begin{bmatrix} q^2_{11} & \cdots & q^2_{1l_3}\\ \cdots & & \cdots\\ q^2_{l_21} & \cdots & q^2_{l_2l_3}\end{bmatrix}\in\mathcal{L}\Big(\bigoplus_{k=1}^{l_3}\mathcal{H}^3_k,\bigoplus_{j=1}^{l_2}\mathcal{H}^2_j\Big),$$

satisfying

$$q^1_{ij}\notin\operatorname{ran}\tau_{A_iB_j} \text{ if } \sigma_r(A_i)\cap\sigma_l(B_j)\neq\varnothing; \quad \text{otherwise } q^1_{ij}=0,$$
$$q^2_{jk}\notin\operatorname{ran}\tau_{B_jC_k} \text{ if } \sigma_r(B_j)\cap\sigma_l(C_k)\neq\varnothing; \quad \text{otherwise } q^2_{jk}=0;$$

(v) A_i, B_j, C_k *are* (SI) *operators and*

$$\ker\tau_{AB}=\ker\tau_{BC}=\ker\tau_{AC}=\{0\}$$

and

$$\ker\tau_{A_iA_j}=\ker\tau_{B_iB_j}=\ker\tau_{C_iC_j}=\{0\}\ (i\neq j).$$

Then $R\in(SI)$.

Proof. Let $P\in\mathcal{A}'(R)$ be idempotent. Then it follows from $RP=PR$, (v) and A_i, B_j, $C_k\in(SI)$ that

$$P=\begin{bmatrix} \bigoplus\epsilon_i1^1_i & P_{12} & P_{13}\\ 0 & \bigoplus\delta_j1^2_j & P_{23}\\ 0 & 0 & \bigoplus\gamma_k1^3_k\end{bmatrix}$$

with respect to the decomposition of the space, where ϵ_i, δ_j, γ_k are 0 or 1, and 1_i^1, 1_j^2, 1_k^3 are the identity operators on $\mathcal{H}_i^1$, $\mathcal{H}_j^2$, $\mathcal{H}_k^3$ respectively ($1 \leq i \leq l_1$, $1 \leq j \leq l_2$, $1 \leq k \leq l_3$).

$PR = RP$ implies that

$$AP_{12} - P_{12}B = P_{11}Q_1 - Q_1P_{22} \tag{4.7}$$

$$BP_{23} - P_{23}C = P_{22}Q_2 - Q_2P_{33} \tag{4.8}$$

where $P_{11} = \bigoplus \epsilon_i 1_i^1$, $P_{22} = \bigoplus \delta_j 1_j^2$ and $P_{33} = \bigoplus \gamma_k 1_k^3$.

Assume that $\delta_1 = 0$ (if $\delta_1 = 1$, consider $1 - P$) and assume that $\{j_s\}_{s=1}^r$ and $\{D_s\}_{s=1}^{r+1}$ satisfy (ii) with respect to $n = 1$ and m ($1 \leq m \leq l_2$). If $D_1 = A_h$ and if P_{12}, P_{23} admit the following matrix representations

$$P_{12} = \begin{bmatrix} f_{11} & \cdots & f_{1l_2} \\ \cdots & & \cdots \\ f_{l_11} & \cdots & f_{l_1l_2} \end{bmatrix}, \quad P_{23} = \begin{bmatrix} g_{11} & \cdots & g_{1l_3} \\ \cdots & & \cdots \\ g_{l_21} & \cdots & g_{l_2l_3} \end{bmatrix}$$

then it follows from (4.7), that

$$A_h f_{h1} - f_{h1} B_1 = \epsilon_h q_{h1}^1.$$

Since $q_{h1}^1 \notin \operatorname{ran} \tau_{A_h B_1}$, $\epsilon_h = 0$. From (4.7) again

$$A_h f_{hj_1} - f_{hj_1} B_{j_1} = -\delta_{j_1} q_{hj_1}^1.$$

Since $q_{hj_1}^1 \notin \operatorname{ran} \tau_{A_h B_{j_1}}$, $\delta_{j_1} = 0$.

If $D_1 = C_k$, then (4.8) implies that

$$B_1 g_{1k} - g_{1k} C_k = -\gamma_k q_{1k}^2.$$

Since $q_{1k}^2 \notin \operatorname{ran} \tau_{B_1 C_k}$, $\gamma_k = 0$. From (4.8) again

$$B_{j_1} g_{j_1k} - g_{j_1k} C_k = \delta_{j_1} q_{j_1k}^2.$$

Since $q_{j_1k}^2 \notin \operatorname{ran} \tau_{B_{j_1} C_k}$, $\delta_{j_1} = 0$.

Repeating this argument, after $r+1$ steps we reach $\delta_m = 0$. Since m is arbitrary, $P_{22} = 0$.

For each i $(1 \leq i \leq l_1)$ and each k $(1 \leq k \leq l_3)$, by (iii) there exist B_{j_1} and B_{j_2} such that

$$\sigma_r(A_i) \cap \sigma_l(B_{j_1}) \neq \varnothing \quad \text{and} \quad \sigma_r(B_{j_2}) \cap \sigma_l(C_k) \neq \varnothing.$$

Then it follows from (4.7) and (4.8) that

$$A_i f_{ij_1} - f_{1j_1} B_{j_1} = \epsilon_i q^1_{ij_1} \quad \text{and} \quad B_{j_2} g_{j_2 k} - g_{j_2 k} C_k = -\gamma_k g_{j_2 k}.$$

Therefore $\epsilon_i = \gamma_k = 0$, $P = 0$ and $R \in (SI)(\mathcal{H})$. □

By the same argument as Lemma 4.24, we have the following lemma.

Lemma 4.25. *Let* $R = \begin{bmatrix} A & Q_1 & 0 \\ 0 & B & Q_2 \\ 0 & 0 & C \end{bmatrix} \in \mathcal{L}(\mathcal{H})$ *satisfy* (i), (ii), (iii) *and* (iv) *of Lemma 4.24 and*

(v) *If* $P \in \mathcal{A}'(R)$ *is idempotent and*

$$P = \begin{bmatrix} \bigoplus \epsilon_i 1^1_i & P_{12} & P_{13} \\ 0 & \bigoplus \delta_j 1^2_j & P_{23} \\ 0 & 0 & \bigoplus \gamma_k 1^3_k \end{bmatrix}$$

with respect to the decomposition of the space, where ϵ_i, δ_j, γ_k *are* 0 *or* 1 *and* 1^1_i, 1^2_j *and* 1^3_k *are identity operators on* $\mathcal{H}^1_i$, $\mathcal{H}^2_j$ *and* $\mathcal{H}^3_k$ *respectively* $(1 \leq i \leq l_1,\ 1 \leq j \leq l_2,\ 1 \leq k \leq l_3)$, *then* $P = 0$ *or* 1, *where* 1 *is the identity on* $\mathcal{H}$.

Theorem 4.26. *Let* T *be in* $\mathcal{L}(\mathcal{H})$ *with connected spectrum* $\sigma(T)$. *Then there exists a strongly irreducible operator* L *satisfying*

(i) $\Lambda(L) = \Lambda(T)$;

(ii) $T \in S(L)^-$;

(iii) *If there is another strongly irreducible operator* L_1 *satisfying (i), then* $L_1 \in S(L)^-$.

Proof. Suppose that

$$\rho^+_{s-F}(T) = \{\Omega_{1i}\}_{i \in \Lambda_1},$$
$$\sigma(T) \cap \rho^\circ_{s-F}(T) = \{\Omega_{2j}\}_{j \in \Lambda_2},$$
$$\rho^-_{s-F}(T) = \{\Omega_{3k}\}_{k \in \Lambda_3}.$$

According to Theorem 3.25 construct (SI) operators A_{1i} $(i \in \Lambda_1)$, A_{2j}, A'_{2j} $(j \in \Lambda_2)$ and A_{3k} $(k \in \Lambda_3)$ satisfying

(i) $A_{1i} \in \mathcal{B}_{n_i}(\Omega_{1i})$, $n_i = \text{ind}(\lambda - A_{1i}) = \text{ind}(\lambda - T)$ $(\lambda \in \Omega_{1i})$ and $A_{1i} = [N_+(\Phi_{1i}^*)^*]^{(n_i)} \bigoplus N_{1i} \oplus [\bigoplus_n(\mu_n^i + V^*)] + K_{1i}$, where $\Phi_{1i} = (\overline{\Omega}_{1i})^\circ$, N_{1i} is a normal operator with $\sigma(N_{1i}) = \sigma_{lre}(N_{1i}) = \partial\Omega_{1i}$, $\{\mu_n^i\}$ are the isolated points of $\partial\Omega_{1i}$, K_{1i} is compact with $\|K_{1i}\| < 1/2^i$, and $\sigma(A_{1i}) = \overline{\Omega}_{1i}$;

(ii) $A_{2j} \in \mathcal{B}_1(\Omega_{2j})$ and $A_{2j} = N_+(\Phi_{2j}^*)^* \bigoplus N_{2j} \bigoplus [\bigoplus_n(\nu_n^j + V^*)] + K_{2j}$, where $\Phi_{2j} = (\overline{\Omega}_{2j})^\circ$, N_{2j} is a normal operator with $\sigma(N_{2j}) = \sigma_{lre}(N_{2j}) = \partial\Omega_{2j}$. $\{\nu_n^j\}$ are the isolated points of $\partial\Omega_{2j}$, K_{2j} is compact with $\|K_{2j}\| < 1/2^j$, and $\sigma(A_{2j}) = \overline{\Omega}_{2j}$;

(iii) $A'_{2j} \in \mathcal{B}_1(\Omega_{2j}^*)$ and $A'_{2j} = N_+(\Phi_{2j})^* \bigoplus N_{2j}^* \oplus [\bigoplus(\nu_n^j + V)^*] + K'_{2j}$ where K'_{2j} is compact with $\|K'_{2j}\| < 1/2^j$, and $\sigma(A_{2j}^{\prime *}) = \overline{\Omega}_{2j}$;

(iv) $A_{3k} \in \mathcal{B}_{m_k}(\Omega_{3k}^*)$, $m_k = \text{ind}(\lambda - A_{3k}) = -\text{ind}(\lambda - T)$ $(\lambda \in \Omega_{3k})$ and

$$A_{3k} = [N_+(\Phi_{3k})^*]^{(m_k)} \bigoplus N_{3k}^* \bigoplus [\bigoplus_n (\lambda_n^k + V)^*] + K_{3k},$$

where $\Phi_{3k} = (\overline{\Omega}_{3k})^\circ$, N_{3k} is a normal operator with $\sigma(N_{3k}) = \sigma_{lre}(N_{3k}) = \partial\Omega_{3k}$, $\{\lambda_n^k\}$ are the isolated points of $\partial\Omega_{3k}$, K_{3k} is compact with $\|K_{3k}\| < 1/2^k$ and $\sigma(A_{3k}^*) = \overline{\Omega}_{3k}$;

Rearrange $\{A_{1i} : i \in \Lambda_1\} \cup \{A_{2j} : j \in \Lambda_2\} = \{A_i\}_{i=1}^{l_1}$. Rearrange $\{A_{2j}^{\prime *} : j \in \Lambda_2\} \cup \{A_{3k}^* : k \in \Lambda_3\} = \{C_k\}_{k=1}^{l_3}$. By Proposition 1.14

$$\ker \tau_{A_i A_j} = \ker \tau_{C_i C_j} = \{0\} \qquad (i \neq j). \tag{4.9}$$

Denote $\sigma = \sigma(T) \backslash [\rho_{s-F}(T)^-]^\circ$ and assume that $\{\sigma_j\}_{j=1}^{l_2}$ is a sequence of pairwise disjoint connected subsets of σ satisfying

(i) each σ_j contains some $\partial\sigma(A_i)$ or some $\partial\sigma(C_k)$, or σ_j consists of more than one point and $\partial\sigma_j \cap ([\bigcup \partial\sigma(A_i)] \cup [\bigcup \partial\sigma(C_k)])^- \neq \varnothing$;

(ii) $[\bigcup_{j=1}^{l_2} \sigma_j]^- = \sigma$.

According to Proposition 3.27 construct an (SI) almost normal operator $B_j = N_j + K_j$ $(1 \leq j \leq l_2)$ such that $\sigma(B_j) = \sigma(N_j) = \overline{\sigma}_j$, N_j is normal, K_j is compact with $\|K_j\| < 1/2^j$ and there exists a countable subset $\{\lambda_n^j : 1 \leq n < \infty\}$ of σ_j such that $\{\lambda_n^j : n = 1, 2, \cdots\}^- = \overline{\sigma}_j$, $\{\lambda_n^j : n = 1, 2, \cdots\} \cap \{\sigma_p(B_k) = \varnothing$ $(j \neq k)$ and

$$\bigvee \ker \{(\lambda_n^j - B_j)^m : n, m \geq 1\} = \text{the acting space of } B_j.$$

By Proposition 1.14

$$\ker \tau_{B_j B_k} = \{0\} \qquad (j \neq k). \tag{4.10}$$

Set $A = \bigoplus_{i=1}^{l_1} A_i$, $B = \bigoplus_{j=1}^{l_2} B_j$ and $C = \bigoplus_{k=1}^{l_3} C_k$; by Proposition 1.14, $\ker \tau_{AB} = \ker \tau_{BC} = \ker \tau_{AC} = \{0\}$.

Define

$$L = \begin{bmatrix} A & Q_1 & 0 \\ 0 & B & Q_2 \\ 0 & 0 & C \end{bmatrix} \in \mathcal{L}(\mathcal{H}_1 \oplus \mathcal{H}_2 \oplus \mathcal{H}_3),$$

where $\mathcal{H}_1 = \bigoplus_{i=1}^{l_1} \mathcal{H}_i^1$, $\mathcal{H}_2 = \bigoplus_{j=1}^{l_2} \mathcal{H}_j^2$, $\mathcal{H}_3 = \bigoplus_{k=1}^{l_3} \mathcal{H}_k^3$ and $A_i \in \mathcal{L}(\mathcal{H}_i^1)$, $B_j \in \mathcal{L}(\mathcal{H}_j^2)$, $C_k \in \mathcal{L}(\mathcal{H}_k^3)$ $(1 \le i \le l_1,\ 1 \le j \le l_2,\ 1 \le k \le l_3)$.

In what follows, if $Q \in \mathcal{L}(Y, Z)$, where Y, Z are Hilbert spaces and if Y_1, Z_1 are subspaces of Y and, respectively, Z, denote $Q(Y_1, Z_1) = P_{Z_1} Q|_{Y_1}$, where P_{Z_1} is the orthogonal projection onto Z_1. In order to define Q_1 and Q_2 we need to define $Q_1(\mathcal{H}_j^2, \mathcal{H}_i^1)$ and $Q_2(\mathcal{H}_k^3, \mathcal{H}_j^2)$ $(1 \le i \le l_1,\ 1 \le j \le l_2,\ 1 \le k \le l_3)$. Denote $\Sigma_i = \{1, 2, \cdots, l_i\}$ $(i = 1, 2, 3)$. In Σ_2 define an equivalence relation as follows.

$j \sim j'$ if and only if there exist $\{j_1, \cdots, j_n\} \subset \Sigma_2$ and $\{D_s\}_{s=1}^{n+1} \subset \{A_i, C_k : 1 \le i \le l_1, 1 \le k \le l_3\}$ such that

$$\sigma(B_{j_m}) \cap \sigma(D_m) \ne \varnothing \text{ and } \sigma(B_{j_{m+1}}) \cap \sigma(D_m) \ne \varnothing \quad (m = 0, 1, \cdots, n),$$

where $j_0 = j$ and $j_{n+1} = j'$.

Let J_r^1 denote the r-th equivalence class of Σ_2 and set

$$L_r^1 = \{i \in \Sigma_1 : \sigma(A_i) \cap \sigma(B_j) \ne \varnothing \text{ for some } j \in J_r^1\},$$
$$M_r^1 = \{k \in \Sigma_3 : \sigma(C_k) \cap \sigma(B_j) \ne \varnothing \text{ for some } j \in J_r^1\}.$$

For each r, if $i \in L_r^1$, $j \in J_r^1$ and $k \in M_r^1$, find compact operators $Q_1(\mathcal{H}_j^2, \mathcal{H}_i^1)$ and $Q_2(\mathcal{H}_k^3, \mathcal{H}_j^2)$ by Theorem 1.11 such that

$$Q_1(\mathcal{H}_j^2, \mathcal{H}_i^1) \notin \operatorname{ran} \tau_{A_i B_j} \text{ and } \|Q_1(\mathcal{H}_j^2, \mathcal{H}_i^1)\| < \frac{1}{2^{ij}} \text{ if } \sigma(A_i) \cap \sigma(B_j) \ne \varnothing,$$
$$Q_1(\mathcal{H}_j^2, \mathcal{H}_i^1) = 0 \qquad \text{if } \sigma(A_i) \cap \sigma(B_j) = \varnothing;$$
$$Q_2(\mathcal{H}_k^3, \mathcal{H}_j^2) \notin \operatorname{ran} \tau_{B_j C_k} \text{ and } \|Q_2(\mathcal{H}_k^3, \mathcal{H}_j^2)\| < \frac{1}{2^{k_j}} \text{ if } \sigma(B_j) \cap \sigma(C_k) \ne \varnothing,$$
$$Q_2(\mathcal{H}_k^3, \mathcal{H}_j^2) = 0 \qquad \text{if } \sigma(B_j) \cap \sigma(C_k) = \varnothing.$$

In Σ_2 define a second equivalence relation as follows.

$j \sim j'$ if and only if j and $j' \in J_r^1$ for some r or $j \in J_r^1$, $j' \in J_{r'}^1$ $(r \ne r')$ and there exist a finite set $\{r_1, \cdots, r_n\}$ of numbers and

$$\{D_s\}_{s=1}^{n+1} \subset \{\bigoplus\{A_i : i \in L_t^1\}, \bigoplus\{C_k : k \in M_s^1\} : t, s = 1, 2, \cdots\}$$

such that

$$\begin{aligned}&\sigma\big(\bigoplus\{B_j : j \in J^1_{r_m}\}\big) \cap \sigma(D_m) \neq \varnothing,\\ &\sigma\big(\bigoplus\{B_j : j \in J^1_{r_{m+1}}\}\big) \cap \sigma(D_m) \neq \varnothing\end{aligned} \qquad (m = 0, 1, \cdots, n),$$

where $r_0 = r$ and $r_{n+1} = r'$.

Let J^2_r denote the r-th equivalence class of Σ_2 and set $L^2_r = \bigcup\{L^1_{r'} : J^1_{r'} \subset J^2_r\}$ and $M^2_r = \bigcup\{M^1_{r'} : J^1_{r'} \subset J^2_r\}$. For each r, if $L^1_n \subset L^2_r$, $J^1_m \subset J^2_r$, $M^1_p \subset M^2_r$ and $n \neq m \neq p$, define

$$Q_1\big(\bigoplus\{\mathcal{H}^2_j;\ j \in J^1_m\}, \bigoplus\{\mathcal{H}^1_i : i \in L^1_n\}\big) \notin \operatorname{ran} \tau_{\bigoplus\{A_i : i\in L^1_n\}, \bigoplus\{B_j : j \in J^1_m\}}$$

and

$$\big\|Q_1(\bigoplus\{\mathcal{H}^2_j : j \in J^1_m\}, \bigoplus\{\mathcal{H}^1_i : i \in L^1_n\})\big\| < \frac{1}{2^{nm}}$$

if $\sigma\big(\bigoplus\{A_i : i \in L^1_n\}\big) \cap \sigma\big(\bigoplus\{B_j : j \in J^1_m\}\big) \neq \varnothing$;

$$Q_1(\bigoplus\{\mathcal{H}^2_j : j \in J^1_m\}, \bigoplus\{\mathcal{H}^1_i : i \in L^1_n\}) = 0$$

if $\sigma\big(\bigoplus\{A_i : i \in L^1_n\}\big) \cap \sigma\big(\bigoplus\{B_j : j \in J^1_m\}\big) = \varnothing$;

$$Q_2(\bigoplus\{\mathcal{H}^3_k;\ k \in M^1_p\}, \bigoplus\{\mathcal{H}^2_j : j \in J^1_m\}) \notin \operatorname{ran} \tau_{\bigoplus\{B_j : j\in J^1_m\}, \bigoplus\{C_k : k \in M^1_p\}}$$

and

$$\big\|Q_2(\bigoplus\{\mathcal{H}^3_k : k \in M^1_p\}, \bigoplus\{\mathcal{H}^2_j : j \in J^1_m\})\big\| < \frac{1}{2^{pm}},$$

if $\sigma\big(\bigoplus\{B_j : j \in J^1_m\}\big) \cap \sigma\big(\bigoplus\{C_k : k \in M^1_p\}\big) \neq \varnothing$;

$$Q_2(\bigoplus\{\mathcal{H}^3_k : k \in M^1_p\}, \bigoplus\{\mathcal{H}^2_j : j \in J^1_m\}) = 0$$

if $\sigma\big(\bigoplus\{B_j : j \in J^1_m\}\big) \cap \sigma\big(\bigoplus\{C_k : k \in M^1_p\}\big) = \varnothing$.

In general, we define the n-th equivalence relation as follows.

$j \sim j'$ if and only if $j, j' \in J^{n-1}_r$ for some r or $j \in J^{n-1}_r$, $j' \in J^{n-1}_{r'}$ $(r \neq r')$ and there exist a set $\{r_1, \cdots, r_n\}$ of numbers and

$$\{D_s\}_{s=1}^{n+1} \subset \{\bigoplus\{A_i : i \in L^{n-1}_t\}, \bigoplus\{C_k : k \in M^{n-1}_s\} : t,\ s = 1, 2, \cdots\}$$

such that

$$\begin{aligned}&\sigma(\bigoplus\{B_j : j \in J^{n-1}_{r_m}\}) \cap \sigma(D_m) \neq \varnothing,\\ &\sigma(\bigoplus\{B_j : j \in J^{n-1}_{r_{m+1}}\}) \cap \sigma(D_m) \neq \varnothing\end{aligned} \qquad (m = 0, 1, \cdots, n),$$

where $r_0 = r$ and $r_{n+1} = r'$.

Let J_r^n denote the r-th equivalence class of Σ_2 and set

$$L_r^n = \bigcup\{L_{r'}^{n-1} : J_{r'}^{n-1} \subset J_r^n\} \quad \text{and} \quad M_r^n = \bigcup\{M_{r'}^{n-1} : J_{r'}^{n-1} \subset J_r^n\}.$$

If ω is the first infinite ordinal, define the ω-th equivalence relation as follows.

$j \sim j'$ if and only if $j,\ j' \in J_r^n$ for some natural numbers n and r. Denote by J_r^ω the r-th equivalent class and set

$$L_r^\omega = \bigcup\{L_{r'}^n : J_{r'}^n \subset J_r^\omega,\ n \text{ is finite}\} \text{ and } M_r^\omega = \bigcup\{M_{r'}^n : J_{r'}^n \subset J_r^\omega,\ n \text{ is finite}\}.$$

For the $(\omega+1)$-th equivalence relation, $j \sim j'$ if and only if $j,\ j' \in J_r^\omega$ for some r or $j \in J_r^\omega,\ j' \in J_{r'}^\omega\ (r \neq r')$ and there exist a finite set $\{r_1, \cdots, r_n\}$ of numbers and $\{D_s\}_{s=1}^{n+1} \subset \big\{\bigoplus\{A_i : i \in L_t^\omega\},\ \bigoplus\{C_k : k \in M_s^\omega\} : t, s = 1, 2, \cdots\big\}$ such that

$$\begin{aligned} &\sigma\big(\textstyle\bigoplus\{B_j : j \in J_{r_m}^\omega\}\big) \cap \sigma(D_m) \neq \varnothing, \\ &\sigma\big(\textstyle\bigoplus\{B_j : j \in J_{r_{m+1}}^\omega\}\big) \cap \sigma(D_m) \neq \varnothing \end{aligned} \qquad (m = 0, 1, \cdots, n),$$

where $r_0 = r$ and $r_{n+1} = r'$.

Repeat the procedure and define the suitable parts of Q_1 and Q_2 in the same way, until they are defined on the entire space. It is obvious that Q_1, Q_2 are compact operators.

Assume that $P \in \mathcal{A}'(L)$ is idempotent and admits the representation

$$P = \begin{bmatrix} P_{11} & P_{12} & P_{13} \\ P_{21} & P_{22} & P_{23} \\ P_{31} & P_{32} & P_{33} \end{bmatrix} \begin{matrix} \mathcal{H}_1 \\ \mathcal{H}_2 \\ \mathcal{H}_3. \end{matrix}$$

By Proposition 1.14 $P_{ij} = 0\ (i > j)$. Computation implies that $P_{11} = \bigoplus \epsilon_i 1_i^1$, $P_{22} = \bigoplus \delta_j 1_j^2$ and $P_{33} = \bigoplus \gamma_k 1_k^3$ where ϵ_i, δ_j, γ_k are 0 or 1 and 1_j^i is the identity on $\mathcal{H}_j^i$.

For each r, denote

$$\mathcal{H}_{1r} = \Big(\bigoplus_{L_r^1} \mathcal{H}_i^1\Big) \oplus \Big(\bigoplus_{J_r^1} \mathcal{H}_j^2\Big) \oplus \Big(\bigoplus_{M_r^1} \mathcal{H}_k^3\Big).$$

Applying Lemma 4.24 to $L(\mathcal{H}_{1r}, \mathcal{H}_{1r})$ and $P(\mathcal{H}_{1r}, \mathcal{H}_{1r})$, we have $\epsilon_i = \delta_j = \gamma_k = 0$ or 1 $(i \in L_r^1,\ j \in J_r^1,\ k \in M_r^1)$. Using Lemma 4.24 repeatedly for each r and each ordinal β, we get $\epsilon_i = \delta_j = \gamma_k = 0$ or 1 ($i \in L_r^\beta$, $j \in J_r^\beta$, $k \in M_r^\beta$). Thus $\epsilon_i = \delta_j = \gamma_k = 0$ or 1 for all $1 \le i \le l_1$, $1 \le j \le l_2$, $1 \le k \le l_3$, i.e., $P = 0$ or 1 and $L \in (SI)$.

It follows from the construction of L that L is the sum of the block diagonal operator D and the compact operator K and each entry of D is subnormal or co-subnormal. It is easily seen from the properties of subnormal and co-subnormal operators that $\Lambda(D) = \Lambda(T)$. Since $L = D + K \in (SI)$, $\sigma_0(L) = \varnothing$ and $\Lambda(L) = \Lambda(D) = \Lambda(T)$,

$$\min\operatorname{ind}(\lambda - L) = \begin{cases} 1, & \lambda \in \rho_{s-F}^{\circ}(L) \cap \sigma(L), \\ 0, & \lambda \in \rho_{s-F}^{\pm}(L) \end{cases}$$

and $\left[(\lambda_0 - \pi(L))|_{E(\lambda_0, \pi(L))}\right]^k \neq 0$ for each isolated point in $\sigma_e(L)$ and $k = 1, 2, \cdots$. Therefore $T \in S(L)^-$ by the similarity orbit theorem. For the same reason, $L_1 \in S(L)^-$ for all L_1 with $\Lambda(L_1) = \Lambda(T)$, and therefore the proof of the theorem is now complete. □

Corollary 4.27. *Given $T \in \mathcal{L}(\mathcal{H})$, then there exists an operator $A \in \mathcal{L}(\mathcal{H})$ with a unique (SI) decomposition such that $\Lambda(A) = \Lambda(T)$ and $T \in S(A)^-$. Furthermore, $A_1 \in S(A)^-$ for all $A_1 \in \mathcal{L}(\mathcal{H})$ with $\Lambda(A_1) = \Lambda(T)$.*

4.4 Open problems

Problem 4.1. *Does each Cowen–Douglas operator have a unique (SI) decomposition in the sense of similarity?*

Problem 4.2. *What is the necessary and sufficient condition for an operator to have a unique (SI) decomposition?*

4.5 Remarks

Proposition 4.1 was proved by Z.J. Jiang and S.L. Sun [1]. Theorem 4.2 is due to D.A. Herrero and C.L. Jiang [1]. Theorems 4.3 and 4.5 were contributed by C.K. Fong and C.L. Jiang (C.K. Fong and C.L. Jiang [3] and C.L. Jiang [3]).

The results of 4.2 were basically contributed by Y.Q.Ji, C.L.Jiang and Z.Y.Wang [2]. Theorem 4.23 and the results of 4.3 are due to C.L.Jiang and Z.Y.Wang (C.L.Jiang [1], C.L. Jiang and Z.Y. Wang [1], [2]).

5. Compact perturbations of strongly irreducible operators

5.1 $(\mathcal{U}+\mathcal{K})$-orbits and compact perturbations

Denote

$$(\mathcal{U}+\mathcal{K})(\mathcal{H}) := \left\{ R \in \mathcal{L}(\mathcal{H}) : \begin{array}{l} R \text{ is invertible and of the form unitary} \\ \text{operator plus compact operator} \end{array} \right\}.$$

Given $T \in \mathcal{L}(\mathcal{H})$, we will call $(\mathcal{U}+\mathcal{K})(T) := \{R^{-1}TR : R \in (\mathcal{U}+\mathcal{K})(\mathcal{H})\}$ *the* $(\mathcal{U}+\mathcal{K})$-*orbit of* T. $A \sim_{\mathcal{U}+\mathcal{K}} T$ and $T \to_{\mathcal{U}+\mathcal{K}} B$ imply $A \in (\mathcal{U}+\mathcal{K})(T)$ and, respectively, $B \in (\mathcal{U}+\mathcal{K})(T)^-$. While $\sim_{\mathcal{U}+\mathcal{K}}$ defines an equivalence relation, $\to_{\mathcal{U}+\mathcal{K}}$ does not. Let $\mathcal{U}(T)$ denote the *unitary orbit of* T, i.e., $\mathcal{U}(T) := \{U^*TU : U \text{ is unitary}\}$; then we have $\mathcal{U}(T) \subset (\mathcal{U}+\mathcal{K})(T) \subset S(T)$. If $A \in (\mathcal{U}+\mathcal{K})(T)$, it is easily seen that $\pi(A)$ is unitarily equivalent to $\pi(T)$. The following theorem describes the norm closure of the $(\mathcal{U}+\mathcal{K})$-orbit of normal operators. The interested reader can find the proof in P.S. Guinand and L. Marcoux [1].

Theorem 5.1. *If* $N \in \mathcal{L}(\mathcal{H})$ *is normal, then* $(\mathcal{U}+\mathcal{K})(N)^-$ *consists of all operators* A *satisfying*

(i) *A is almost normal, i.e., A is of the form normal plus compact;*

(ii) $\sigma_e(A) = \sigma_e(N)$, $\sigma(A) \supset \sigma(N)$;

(iii) $\text{nul}(\lambda - A) \geq \text{nul}(\lambda - N)$ *for all* $\lambda \in \rho_F(A)$;

(iv) *if* $\{\lambda\}$ *is isolated in* $\sigma(A)$, *then* $(\lambda - A)|_{\mathcal{H}(\lambda;A)} = 0$ *and* $\dim \mathcal{H}(\lambda; A) = \dim \mathcal{H}(\lambda; N)$.

The next several results are trying to find the closure of the $(\mathcal{U}+\mathcal{K})$-orbit of essentially normal operators. The reader can refer to Al-Musallan [1] for the proof of Lemma 5.2.

Lemma 5.2. *Suppose that* $T \in \mathcal{L}(\mathcal{H})$ *is essentially normal and* $\sigma(T) = \sigma_e(T) \cup \sigma_0(T)$. *Assume, moreover, that* $C(\sigma(T)) = \text{Rat}(\sigma(T))^-$. *Then* $T \to_{\mathcal{U}+\mathcal{K}} N$, *where* N *is a normal operator such that* $\sigma(N) = \sigma(T)$, $\sigma_0(T) = \sigma_0(N)$, *and* $\text{nul}(\lambda - N) =$

$\dim \mathcal{H}(\lambda; T)$ *for all* $\lambda \in \sigma_0(N)$, $\mathrm{Rat}\,(\sigma(T))^-$ *is the uniform closure of rational functions with poles outside* $\sigma(T)$.

Lemma 5.3. *Suppose* $T \in \mathcal{L}(\mathcal{H})$ *is almost normal,* $\sigma(T)$ *is a perfect set with Lebesgue measure* 0, N *is normal with* $\sigma(N) = \sigma(T)$; *then for each* $\epsilon > 0$, *there exists* $X \in (\mathcal{U} + \mathcal{K})(\mathcal{H})$ *such that*

(i) $XTX^{-1} - N \in \mathcal{K}(\mathcal{H})$;

(ii) $\|XTX^{-1} - N\| < \epsilon$.

Proof. Since $m(\sigma(T)) = 0$, it follows from H. Hartogs and A. Rosenthal [1] that $\mathrm{Rat}(\sigma(T))^- = C(\sigma(T))$. By Lemma 5.2, for each $\delta > 0$ there is an $X_1 \in (\mathcal{U}+\mathcal{K})(\mathcal{H})$ such that $\|X_1TX_1^{-1} - N\| < \delta$. Since T is almost normal, $X_1TX_1^{-1}$ is also almost normal. Therefore there is a normal operator M and a compact operator K such that $X_1TX_1^{-1} = M + K$. Since $\|M + K - N\| < \delta$, $\|[(M + K)^*, (M + K)]\| < 4\|N\|\delta + 2\delta^2$. Then I.D. Berg and K.R. Davidson [1] assert that there exists a positive valued continuous function on $[0, \infty)$, $f(0) = 0$, such that for each almost normal operator Q, there is a compact operator $K(Q)$ with $\|K(Q)\| \leq f(\|Q^*Q - QQ^*\|^{1/2})$ and $Q + K(Q)$ is normal. According to this theorem, we can find a δ and a compact operator K_1 such that

(i) $\|K_1\| \leq \epsilon/4$;

(ii) $X_1TX_1^{-1} + K_1 = M + K + K_1$ is normal;

(iii) $\sigma(M + K + K_1) \subset \sigma(T)_{\epsilon/4}$.

Since T is almost normal and $\sigma(T)$ is a perfect set with $m(\sigma(T)) = 0$, $\sigma(T) = \sigma_e(T)$. Thus $\sigma(M) = \sigma(T) \cup \sigma_0(M)$. Since $M + K + K_1$ is normal, $\sigma(M + K + K_1) = \sigma(T) \cup \sigma_0(M + K + K_1)$. Thus we can find a compact operator K_2, $\|K_2\| < \epsilon/2$, such that $\sigma(M + K + K_1 + K_2) = \sigma(T) = \sigma(N)$ and $M + K + K_1 + K_2$ is normal. It follows from Voiculescu's theorem that there is a unitary operator U and a compact operator K_3 with $\|K_3\| < \epsilon/2$ such that $U(X_1TX_1^{-1} + K_1 + K_2)U^* = U(M + K + K_1 + K_2)U^* = N + K_3$. Thus $UX_1TX_1^{-1}U^* - N \in \mathcal{K}(\mathcal{H})$, $\|UX_1TX_1^{-1}U^* - N\| < \epsilon$, and $X := UX_1$ satisfies the requirements of the lemma. □

Let Ω be an analytic and simply connected Cauchy domain in $\mathbb{C}$ and $n \in \mathbb{N}$. Then $\mathcal{A}(\Omega, n)$ will denote the class of operators T, each of which satisfies

(i) T is essentially normal;

(ii) $\sigma(T) = \overline{\Omega}$, $\rho_F(T) \cap \sigma(T) = \Omega$;

(iii) $\mathrm{ind}\,(\lambda - T) = -n$, $\mathrm{nul}\,(\lambda - T) = 0$ for all $\lambda \in \Omega$.

The next two results, Theorems 5.4 and 5.5, are the main results of this section. They describe the closures of $(\mathcal{U}+\mathcal{K})$-orbits of $\mathcal{A}(\Omega, n)$ operators.

Theorem 5.4. *Given $T_1, T_2 \in \mathcal{A}(\Omega, n)$ and $\epsilon > 0$, there exists a compact operator K with $\|K\| < \epsilon$ such that $T_1 + K \sim_{\mathcal{U}+\mathcal{K}} T_2$.*

Theorem 5.5. *Let T be in $\mathcal{A}(\Omega, n)$. Then $(\mathcal{U}+\mathcal{K})(T)^-$ consists of all operators A satisfying*

(i) *A is essentially normal;*

(ii) *$\sigma(A) = \overline{\Omega}$, $\rho_F(A) \cap \sigma(A) = \Omega$;*

(iii) *$\operatorname{ind}(\lambda - A) = -n$ for all $\lambda \in \Omega$.*

It follows from the Riemann mapping theorem that there is an analytic function ϕ satisfying

(i) $\phi(0) = z_0$, where z_0 is a fixed point in Ω;

(ii) ϕ is injective;

(iii) $\phi(D) = \Omega$.

By the Schwarz reflection principle ϕ has an analytic continuation on $\overline{D}$ such that $\phi(\partial D) = \partial\Omega$. Let T_ϕ be the Toeplitz operator with symbol ϕ. Then $T_\phi^* \in \mathcal{B}_1(\Omega^*)$ and $\sigma(T_\phi^*) = \overline{\Omega}^*$.

In order to prove Theorems 5.4 and 5.5, we need several lemmas.

Lemma 5.6. *Let $\mathcal{M} = (\ker(T_\phi - \lambda)^*)^\perp$. Then $T_\phi|_\mathcal{M}$ is unitarily equivalent to T_ϕ, where $\lambda \in \Omega$.*

Proof. Since $\mathcal{M}$ is a hyperinvariant subspace of T_ϕ, $\mathcal{M}$ is an invariant subspace of T_z. By the Beurling theorem, $\mathcal{M} = T_g H^2$, where g is an inner function. Since $T_\phi^* \in \mathcal{B}_1(\Omega^*)$, $g = (z-a)/(1-a^*z)$, $(0 \le |a| < 1)$. Thus T_g can be considered as a unitary operator from H^2 to $\mathcal{M}$ and $T_g^*(T_\phi|_\mathcal{M})T_g f = T_\phi f$ for $f \in H^2$, i.e., $T_g^*(T_\phi|_\mathcal{M})T_g = T_\phi$. □

Lemma 5.7. *Let $\mathcal{H} = \mathbb{C} \oplus H^2$ and $\lambda \in \Omega$ and let $T = \begin{bmatrix} \lambda & 0 \\ E & T_\phi \end{bmatrix} \in \mathcal{L}(\mathcal{H})$, where $E = \alpha e_0 \otimes 1$, $e_0 \in \ker(\lambda - T_\phi)^*$ and $\|e_0\| = 1$, $\alpha \in \mathbb{C}$ and $\alpha \neq 0$. Then $T \sim_{\mathcal{U}+\mathcal{K}} T_\phi$.*

Proof. Note that $e_0 \in \operatorname{ran}(T_\phi - \lambda)^\perp$. By Lemma 5.6,

$$T_\phi = \begin{bmatrix} \lambda & 0 \\ g_1 & T_\phi|_\mathcal{M} \end{bmatrix} \begin{matrix} e_0 \\ \mathcal{M} = [\ker(T_\phi - \lambda)^*]^\perp \end{matrix} \simeq \begin{bmatrix} \lambda & 0 \\ g_2 & T_\phi \end{bmatrix} \begin{matrix} \mathbb{C} \\ H^2 \end{matrix},$$

where g_1, g_2 are rank-1 operators and assume that $g_2 = f \otimes 1$, $f \in H^2$. Since $e_0 \in \operatorname{ran}(T_\phi - \lambda)^\perp$, $f = \beta e_0 + g_3$, where $\beta \in \mathbb{C}$ and $g_3 \in \operatorname{ran}(T_\phi - \lambda)$. If $\beta = 0$, let $g_3 = (T_\phi - \lambda)x$. Then computation shows that

$$\begin{bmatrix} 1 & 0 \\ x \otimes 1 & 1 \end{bmatrix} \begin{bmatrix} \lambda & 0 \\ g_2 & T_\phi \end{bmatrix} \begin{bmatrix} 1 & 0 \\ -x \otimes 1 & 1 \end{bmatrix} = \begin{bmatrix} \lambda & 0 \\ 0 & T_\phi \end{bmatrix}.$$

This is contradictory to $T_\phi \in (SI)$. Thus $\beta \neq 0$, and $X \begin{bmatrix} \lambda & 0 \\ g_2 & T_\phi \end{bmatrix} X^{-1} = \begin{bmatrix} \lambda & 0 \\ \beta e_0 \otimes 1 & T_\phi \end{bmatrix}$, where $X = \begin{bmatrix} 1 & 0 \\ x \otimes 1 & 1 \end{bmatrix} \in (\mathcal{U} + \mathcal{K})(\mathcal{H})$. Since $\begin{bmatrix} \lambda & 0 \\ \beta e_0 \otimes 1 & T_\phi \end{bmatrix} \sim_{\mathcal{U}+\mathcal{K}} \begin{bmatrix} \lambda & 0 \\ E & T_\phi \end{bmatrix}$, the proof of Lemma 5.7 is complete. □

By the same argument as Lemma 5.7, we can prove

Lemma 5.8. *Let $\mathcal{H} = \mathbb{C} \oplus H^2$, $g \notin \operatorname{ran}(T_\phi - \lambda)$, $\lambda \in \Omega$ and let $T = \begin{bmatrix} \lambda & 0 \\ g \otimes 1 & T_\phi \end{bmatrix} \in \mathcal{L}(\mathcal{H})$. Then $T \sim_{\mathcal{U}+\mathcal{K}} T_\phi$.*

Lemma 5.9. *Let $T \in \mathcal{L}(\mathbb{C}^n \oplus H^2)$ and $T = \begin{bmatrix} F & 0 \\ C_n & T_\phi \end{bmatrix} \begin{matrix} \mathbb{C}^n \\ H^2 \end{matrix}$, where $F \in \mathcal{L}(\mathbb{C}^n)$ with $\sigma(F) \subset \Omega$, and $C_n \in \mathcal{L}(\mathbb{C}^n, H^2)$ is a non-zero operator. Then for each $\epsilon > 0$, there exists a compact operator K, $\|K\| < \epsilon$, such that $T + K \sim_{\mathcal{U}+\mathcal{K}} T_\phi$.*

Proof. We will prove the lemma by induction on n.

If $n = 1$, $T = \begin{bmatrix} \lambda & 0 \\ C_1 & T_\phi \end{bmatrix}$, where $C_1 = f \otimes 1$ is a rank-1 operator. Choose $g \in H^2$, $\|g\| < \epsilon$, such that $f + g \notin \operatorname{ran}(T_\phi - \lambda)$. Set $K = \begin{bmatrix} 0 & 0 \\ g \otimes 1 & 0 \end{bmatrix}$; then $\|K\| < \epsilon$ and by Lemma 5.8, $T + K \sim_{\mathcal{U}+\mathcal{K}} T_\phi$.

Assume that the conclusion of the lemma is true when $n \leq k - 1$, and let $n = k$. It is obvious that there is a unitary operator U such that

$$UTU^* = \begin{bmatrix} \lambda & 0 & 0 \\ * & F_{n-1} & 0 \\ * & C_{n-1} & T_\phi \end{bmatrix} \begin{matrix} \mathbb{C} \\ \mathbb{C}^{n-1} \\ H^2 \end{matrix},$$

where $F_{n-1} \in \mathcal{L}(\mathbb{C}^{n-1})$, $\lambda \in \sigma(F)$. By the inductive assumption, we can find a compact operator K_1 with $\|K_1\| < \epsilon/2$ and $X_1 \in (\mathcal{U} + \mathcal{K})(\mathcal{H})$ such that

$$X_1(UTU^* + K_1)X_1^{-1} = \begin{bmatrix} \lambda & 0 \\ C_1 & T_\phi \end{bmatrix}.$$

By the assumption again, we can find a compact operator K_2 such that $\|K_2\| < \epsilon/(2\|X_1\|\|X_1^{-1}\|)$ and

$$X_2\left(\begin{bmatrix} \lambda & 0 \\ C_1 & T_\phi \end{bmatrix} + K_2\right)X_2^{-1} = T_\phi \quad \text{for some } X_2 \in (\mathcal{U}+\mathcal{K})(\mathcal{H}).$$

Thus $K = U^*K_1U + U^*X_1^{-1}K_2X_1U$ satisfies the requirements of the lemma. $\square$

Lemma 5.10. *Given $T \in \mathcal{A}(\Omega, n)$ and $\epsilon > 0$, then there exists a compact operator K with $\|K\| < \epsilon$ such that $T + K \sim_{\mathcal{U}+\mathcal{K}} T_\phi^{(n)}$.*

Proof. Note that T_ϕ admits a lower triangular matrix representation with respect to the ONB $\{e^{ik\theta}\}_{k=0}^{\infty}$ of H^2. Set $\mathcal{M}_k = \bigvee\{e^{ij\theta} : j = 0, 1, \cdots, k\}$ and denote by P_k the orthogonal projection onto $\bigoplus_{k=1}^{n} \mathcal{M}_k$. By the Brown–Douglas–Fillmore theorem $UTU^* = T_\phi^{(n)} + K$, where U is a unitary operator and K is compact. Set $K_1 = P_mKP_m - K$; m will be determined later. By Lemma 5.6, we can find a unitary operator U_1 such that

$$U_1(UTU^* + K_1)U_1^* = \begin{bmatrix} F & 0 & \dots & 0 \\ C_1 & T_\phi & & 0 \\ \vdots & & \ddots & \\ C_n & & 0 & T_\phi \end{bmatrix},$$

where $F \in \mathcal{L}(\mathbb{C}^{nm})$, and C_k is a finite rank operator $(k = 1, \cdots, n)$. Fix m so that $\|K_1\| < \epsilon/8$ and $\sigma(F) \subset \Omega_{\epsilon/8}$. Thus we can find an operator $C \in \mathcal{L}(\mathbb{C}^{nm})$ such that $\|C\| < \epsilon/4$ and $\sigma(F') \subset \Omega$, where $F' = F + C$. Therefore there exists a compact operator K_2 with $\|K_2\| < \epsilon/4$ such that

$$U_1(UTU^* + K_1)U_1^* + K_2 = \begin{bmatrix} F' & 0 & \dots & 0 \\ C_1 & T_\phi & & 0 \\ \vdots & & \ddots & \\ C_n & & 0 & T_\phi \end{bmatrix} = A_1.$$

By Lemma 5.9, there is an $X_1' \in (\mathcal{U}+\mathcal{K})(\mathbb{C}^{nm} \oplus H^2)$ such that $X_1'\begin{bmatrix} F' & 0 \\ C_1 & T_\phi \end{bmatrix}X_1'^{-1} = T_\phi$. Thus we can find $X_1 \in (\mathcal{U}+\mathcal{K})(\mathbb{C}^{nm} \oplus (H^2)^{(n)})$ such that

$$X_1A_1X_1^{-1} = \begin{bmatrix} T_\phi & 0 & \dots & 0 \\ C_1' & T_\phi & & 0 \\ \vdots & & \ddots & \\ C_{n-1}' & & 0 & T_\phi \end{bmatrix} = A_2,$$

where C'_k $(k = 1, 2, \cdots, n-1)$ is a finite rank operator. Since $\mathcal{A}'(T_\phi)$ does not contain any compact operators, $\ker \tau_{T_\phi T_\phi}|_{\mathcal{K}(H^2)} = \{0\}$. By Lemma 1.10 $[\operatorname{ran} \tau_{T_\phi T_\phi}|_{C^1(H^2)}]^- = C^1(H^2)$. Thus for each k $(1 \leq k \leq n-1)$ we can find compact operators D_k and E_k satisfying $T_\phi E_k - E_k T_\phi = -C'_k - D_k$ and $\|D_k\| < \epsilon/(\|X_1\|\|X_1^{-1}\|8^k)$.

Set

$$K_3 = \begin{bmatrix} 0 & 0 & \ldots & 0 \\ D_1 & 0 & \ldots & 0 \\ \vdots & & \ddots & \vdots \\ D_{n-1} & 0 & \ldots & 0 \end{bmatrix}, \quad X_2 = \begin{bmatrix} 1 & 0 & \ldots & 0 \\ E_1 & 1 & \ldots & 0 \\ \vdots & & \ddots & \vdots \\ E_{n-1} & 0 & \ldots & 1 \end{bmatrix}.$$

Then K_3 is compact, $\|K_3\| < \epsilon/(8\|X_1\|\|X_1^{-1}\|)$ and $X_2 \in (\mathcal{U}+\mathcal{K})((H^2)^{(n)})$. Calculation indicates that $X_2(A_2 + K_3)X_2^{-1} = T_\phi^{(n)}$. Thus, $X(T+K)X^{-1} = T_\phi^{(n)}$, where $K = U^*K_1U + U^*U_1^*K_2U_1U + X_1^{-1}K_3X_1$ is a compact operator with $\|K\| < \epsilon$, $X = X_2X_1U_1U$ is invertible and of the form unitary plus compact. □

Lemma 5.10 implies that $\mathcal{A}(\Omega, n) \subset (\mathcal{U}+\mathcal{K})(T_\phi^{(n)})^-$.

Lemma 5.11. *Let $T \in \mathcal{A}(D, n)$, $\epsilon > 0$. Then there exist an operator $T' \in \mathcal{A}(D, n)$ and a compact operator K such that $\|K\| < \epsilon$, T' admits a lower triangular matrix representation with respect to some ONB of $\mathcal{H}$ and $T' + K \sim_{\mathcal{U}+\mathcal{K}} T$.*

Proof. By Apostol's triangular representation theorem $T = \begin{bmatrix} T_0 & B \\ 0 & T_l \end{bmatrix} \begin{matrix} \mathcal{H}_0(T) \\ \mathcal{H}_l(T) \end{matrix}$ and

$$\mathcal{H}_0 \text{ is a hyperinvariant subspace of } T, \tag{5.1}$$

$$T_l \text{ admits a lower triangular matrix representation,} \tag{5.2}$$

$$\rho_{s-F}(T) \subset \rho(T_0). \tag{5.3}$$

By the Brown–Douglas–Fillmore theorem $T \simeq T_z^{(n)} + K$, K is compact, and $\pi(T)$ is a unitary element in the Calkin algebra $\mathcal{A}(\mathcal{H}) = \mathcal{L}(\mathcal{H})/\mathcal{K}(\mathcal{H})$, i.e., $\pi(T^*)\pi(T) - \pi(1) = 0$. Thus

$$\begin{bmatrix} \pi(T_0^*T_0) - \pi(1) & \pi(T_0^*B) \\ \pi(B^*T_0) & \pi(B^*B + T_l^*T_l) - \pi(1) \end{bmatrix} = \begin{bmatrix} 0 & 0 \\ 0 & 0 \end{bmatrix}.$$

Since $D \subset \rho(T_0) \subset \rho(\pi(T_0))$, $\pi(T_0)$ is invertible and $\pi(T_0)^{-1} = \pi(T_0)^*$. Thus $\pi(T_0)$ is a unitary element in $\mathcal{A}(\mathcal{H}_0(T))$. Since T_0 is invertible, by the Brown–Douglas–Fillmore theorem $T_0 \simeq U_0 + K_0$, where U_0 is unitary and K_0 is compact. By (5.3) $\sigma(U_0 + K_0) \subset \partial D$. Since $\pi(T)$ is a unitary element in $\mathcal{A}(\mathcal{H})$, $TT^* - 1 \in \mathcal{K}(\mathcal{H})$, i.e., $\begin{bmatrix} T_0T_0^* + BB^* - 1 & BT_l^* \\ T_lB^* & T_lT_l^* - 1 \end{bmatrix}$ is compact. Since $\pi(T_0)$ is a unitary element, $T_0T_0^* - 1$

is compact, thus BB^* and B are compact. Since $T_l^* \in \mathcal{B}_n(D)$ and $D \cap \sigma(T_0^*) = \varnothing$, by Proposition 1.14 $\ker \tau_{T_0^* T_l^*} = \{0\}$ and therefore $\ker \tau_{T_l T_0} = \{0\}$. By Lemma 1.10, $\operatorname{ran} \tau_{T_0 T_l}$ is dense in $\mathcal{K}(\mathcal{H}_l, \mathcal{H}_0)$. Thus for $\delta > 0$, there exist compact operators E, $G \in \mathcal{L}(\mathcal{H}_l(T), \mathcal{H}_0(T))$ such that $T_0 G - G T_l = B + E$ and $\|E\| < \delta$.

Set

$$X_1 = \begin{bmatrix} 1 & -G \\ 0 & 1 \end{bmatrix} \begin{matrix} \mathcal{H}_0(T) \\ \mathcal{H}_l(T) \end{matrix} \quad \text{and} \quad K_1 = \begin{bmatrix} 0 & -E \\ 0 & 0 \end{bmatrix} \begin{matrix} \mathcal{H}_0(T) \\ \mathcal{H}_l(T). \end{matrix}$$

Then $X_1((T_0 \oplus T_l) + K_1) X_1^{-1} = T$, where $X_1 \in (\mathcal{U} + \mathcal{K})(\mathcal{H})$, K_1 is compact and $\|K_1\| < \delta/2$. Since $\sigma(T_0) \subset \partial D$, Lemma 5.3 indicates that there are compact operator K_2 and $X_2 \in (\mathcal{U} + \mathcal{K})(\mathcal{H})$ such that $\|K_2\| < \delta$ and $X_2((N_0 \oplus T_l) + K_2) X_2^{-1} = T_0 \oplus T_l$, where N_0 is a diagonal normal operator with $\sigma(N_0) = \sigma(T_0)$. By Theorem 1.29 we can find a compact operator K_3 and a unitary operator U such that $\|K_3\| < \delta$ and $U((N_0 \oplus T_l) + K_3) U^* = T_l$, i.e., $U^*(T_l - U K_3 U^*) U = N_0 \oplus T_l$. Thus letting $\delta = \epsilon / (4 \|X_2\| \|X_2^{-1}\|)$, $K = U K_2 U^* - U^2 K_3 U^{*2} + U X_2^{-1} K_1 X_2 U^*$, $T' = T_l$ satisfy the requirements of the lemma. □

Lemma 5.12. *Let $T \in \mathcal{A}(D, n)$, $0 < \epsilon < 1/10$ and $\|T^*T - 1\| < \epsilon$. Then there exists a unitary operator W such that $W^*TW - T_z^{(n)}$ is compact and $\|W^*TW - T_z^{(n)}\| < \epsilon$.*

Proof. Consider the polar decomposition $T = U|T|$ of T, where U is a partial isometry, $\operatorname{ran} U = \operatorname{ran} T$. By the Wold decomposition theorem (see A.L. Shilds [1]), $U \simeq T_z^{(n)} \oplus V$, where V is a unitary operator. By the assumption $\||T|^2 - 1\| < \epsilon$, $|z^2 - 1| < \epsilon$ for all $z \in \sigma(|T|)$. Since $\epsilon < 1/10$, $z > 1/2$ and $\epsilon > |z^2 - 1| = |z+1||z-1| > 3|z-1|/2$. Thus $\sigma(|T|) \subset \{x \in \mathbb{R} : |x - 1| < 2\epsilon/3\}$. Since $|T|$ is self-adjoint, $\||T| - 1\| < 2\epsilon/3$ and therefore $\|T - U\| = \|U|T| - U\| = \|U(|T| - 1)\| \le \|U\| \||T| - 1\| < 2\epsilon/3$. Note that $T \simeq T_z^{(n)} + K$, where K is compact, thus $|T|^2 - 1 = T^*T - 1 \simeq (T_z^{(n)} + K)^*(T_z^{(n)} + K) - 1$, and $|T|^2 = 1 + K_1$ and $|T| = 1 + K_2$ for some K_1, $K_2 \in \mathcal{K}(\mathcal{H})$.

Since V is unitary and $U \simeq T_z^{(n)} \oplus V$, by Theorem 1.29 there is a unitary operator W such that

$$\|W^*UW - T_z^{(n)}\| < \epsilon/3, \tag{5.4}$$

$$K_3 := W^*UW - T_z^{(n)} \text{ is compact.} \tag{5.5}$$

Thus $W^*TW - T_z^{(n)} = W^*(T - U)W + W^*UW - T_z^{(n)} = W^*[U(|T| - 1)]W + K_3 = W^*(UK_2)W + K_3$. It is compact, and

$$\|W^*TW - T_z^{(n)}\| \le \|W^*(T - U)W\| + \|K_3\| \le \|T - U\| + \frac{\epsilon}{3} < \epsilon. \qquad \square$$

Proposition 5.13. *Let $T \in \mathcal{A}(D,n)$ admit a lower triangular matrix representation* $T = \begin{bmatrix} F & 0 \\ X & S \end{bmatrix} \begin{matrix} \mathcal{H}_1 \\ \mathcal{H}_2 \end{matrix}$, *where F is a diagonal operator on a finite dimensional Hilbert space $\mathcal{H}_1$, $S \in \mathcal{A}(D,n)$ and $\|S^*S - 1\| < \delta < 1/10$. Then there exists $Q \in (U+K)(T)^-$ satisfying* $\operatorname{dist}(Q, \mathcal{U}(T_z^{(n)})) < 4\delta$ *and $Q - W^*(T_z^{(n)})W \in \mathcal{K}(\mathcal{H})$, for some unitary operator W.*

Proof. Case 1. Assume that $\sigma(F) \subset \{\lambda : |\lambda| < 1/(1+3\delta)\}$. We will proceed by induction on the dimension m of $\mathcal{H}_1$ to prove that there exists $Q \in (\mathcal{U}+\mathcal{K})(T)$ such that $\operatorname{dist}(Q, \mathcal{U}(T_z^{(n)})) < \delta$ and $Q - W^*T_z^{(n)}W$ is compact for some unitary W.

When $m = 1$,

$$T = \begin{bmatrix} \lambda & 0 \\ x \otimes e & S \end{bmatrix} \begin{matrix} \bigvee\{e\} = \mathcal{H}_1 \\ \mathcal{H}_2, \end{matrix}$$

where $x \in \mathcal{H}_2$ and $x \neq 0$ ($x = 0$ is contradictory to $T \in \mathcal{A}(D,n)$). If $x \in \operatorname{ran}(S - \lambda)$, computation shows that $T \sim \lambda \oplus S$. This is also contradictory to $T \in \mathcal{A}(D,n)$. Thus $x \notin \operatorname{ran}(S - \lambda)$. For $z \otimes e \in \mathcal{L}(\mathcal{H}_1, \mathcal{H}_2)$,

$$T \sim_{\mathcal{U}+\mathcal{K}} \begin{bmatrix} 1 & 0 \\ z \otimes e & 1 \end{bmatrix} \begin{bmatrix} \lambda & 0 \\ x \otimes e & S \end{bmatrix} \begin{bmatrix} 1 & 0 \\ -z \otimes e & 1 \end{bmatrix} = \begin{bmatrix} \lambda & 0 \\ y \otimes e & S \end{bmatrix},$$

where $y = x + (\lambda - S)z$. Since $\lambda - S$ is a Fredholm operator, we can always choose z so that $y \in \ker(\lambda - S)^*$ and $y \neq 0$. For $\alpha \in \mathbb{C}$ ($\alpha \neq 0$) and $\omega \in \mathcal{H}_2$, we have

$$T \sim_{\mathcal{U}+\mathcal{K}} \begin{bmatrix} 1 & 0 \\ \omega \otimes e & 1 \end{bmatrix} \begin{bmatrix} \alpha^{-1} & 0 \\ 0 & 1 \end{bmatrix} \begin{bmatrix} \lambda & 0 \\ y \otimes e & S \end{bmatrix} \begin{bmatrix} \alpha & 0 \\ 0 & 1 \end{bmatrix} \begin{bmatrix} 1 & 0 \\ -\omega \otimes e & 1 \end{bmatrix} = \begin{bmatrix} \lambda & 0 \\ v \otimes e & S \end{bmatrix},$$

where $v = \alpha y + (\lambda - S)\omega$.

Set $Q_1 = \begin{bmatrix} \lambda & 0 \\ v \otimes e & S \end{bmatrix}$. We will choose an adequate α and ω so that $\|Q_1^*Q_1 - 1\| < \delta$. Note that

$$Q_1^*Q_1 = \begin{bmatrix} |\lambda|^2 + (e \otimes v)(v \otimes e) & e \otimes (S^*v) \\ (S^*v) \otimes e & S^*S \end{bmatrix}.$$

We intend to choose α and ω so that

$$|\lambda|^2 + \|v\|^2 - 1 = 0 \quad \text{or} \quad \|v\| = \sqrt{1 - |\lambda|^2} \tag{5.6}$$

$$S^*v = 0 \tag{5.7}$$

and $\alpha \neq 0$.

Since S^* is a Fredholm operator and $\operatorname{nul} S = 0$, S^*S is invertible. Set $S^*S = A$. Thus

$$S^*v = S^*[\alpha y + (\lambda - S)\omega] = \alpha S^*y + (\lambda S^* - A)\omega = \alpha S^*y + A(\lambda A^{-1}S^* - 1)\omega.$$

Since $A > 0$, $\sigma(A) \subset (1-\delta, 1+\delta)$. Therefore $\|A^{-1}\| < 1/(1-\delta)$. It follows from $\|S^*\| \leq 1+\delta$ that $\|A^{-1}S^*\| \leq \|A^{-1}\|\|S^*\| < (1+\delta)/(1-\delta) < 1+3\delta$. By assumption, $\|\lambda A^{-1}S^*\| < 1$. Thus $(\lambda A^{-1}S^* - 1)$ and $(\lambda S^* - A)$ are invertible. Choosing $\omega = (-\alpha\lambda^*)(\lambda S^* - A)^{-1}y$, we have

$$S^*v = \alpha S^*y + (\lambda S^* - A)(-\alpha\lambda^*)(\lambda S^* - A)^{-1}y = \alpha\lambda^*y - \alpha\lambda^*y = 0.$$

Since

$$\begin{aligned}
\|v\| &= \|\alpha y + (\lambda - S)\omega\| \\
&= \left[\|\alpha y\|^2 + \|(\lambda - S)(-\alpha\lambda^*)(\lambda S^* - A)^{-1}y\|^2\right]^{1/2} \\
&= |\alpha|\left[\|y\|^2 + \|\lambda^*(\lambda - S)(\lambda S^* - A)^{-1}y\|^2\right]^{1/2}
\end{aligned}$$

and since λ and y are fixed, we can choose $\alpha \neq 0$ so that $\|v\| = (1-|\lambda|^2)^{1/2}$. Thus for the chosen v,

$$Q_1^*Q_1 = \begin{bmatrix} 1 & 0 \\ 0 & S^*S \end{bmatrix} \quad \text{and} \quad \|Q_1^*Q_1 - 1\| < \delta.$$

Since $Q_1 \in (\mathcal{U}+\mathcal{K})(T)$, $Q_1 \in \mathcal{A}(D,n)$. By Lemma 5.12 $\operatorname{dist}(Q_1, \mathcal{U}(T_z^{(n)})) < \delta$ and $Q_1 - W^*T_z^{(n)}W$ is compact for some unitary operator W.

To complete the induction step, we now assume that the result is true when F is an $(m-1)\times(m-1)$ diagonal matrix.

Let $T = \begin{bmatrix} F & 0 \\ x & S \end{bmatrix}$ satisfy the condition of the lemma, where F is an $m \times m$ matrix. Then

$$T \simeq \begin{bmatrix} F' & 0 & 0 \\ 0 & \lambda & 0 \\ X_1 & x_2 & S \end{bmatrix} = \begin{bmatrix} F' & 0 \\ X_1' & T_1 \end{bmatrix},$$

where F' is an $(m-1)\times(m-1)$ diagonal matrix, $X_1' = \begin{bmatrix} 0 \\ X_1 \end{bmatrix}$ and $T_1 = \begin{bmatrix} \lambda & 0 \\ x_2 & S \end{bmatrix}$. But T_1 is precisely of the form handled when $m = 1$, and so we can find $R \in (\mathcal{U}+\mathcal{K})(\mathcal{H}_2')$ such that if $R^{-1}T_1R := Q_1$, then $Q_1 \in \mathcal{A}(D,n)$, $\|Q_1^*Q_1 - 1\| < \delta$,

$\|R^{-1}T_1R - T_z^{(n)}\| < \delta$ and $R^{-1}T_1R - T_z^{(n)}$ is compact, where T_1 is acting on $\mathcal{H}_2'$. Then

$$T \sim_{\mathcal{U}+\mathcal{K}} \begin{bmatrix} 1 & 0 \\ 0 & R^{-1} \end{bmatrix} \begin{bmatrix} F' & 0 \\ X_1' & T_1 \end{bmatrix} \begin{bmatrix} 1 & 0 \\ 0 & R \end{bmatrix} = \begin{bmatrix} F' & 0 \\ R^{-1}X_1' & R^{-1}T_1R \end{bmatrix} := T'.$$

Thus $T' \in (\mathcal{U}+\mathcal{K})(T)$ and therefore $T' \in \mathcal{A}(D,n)$ and satisfies the same condition as T does. By the inductive assumption, we can find $Q \in (\mathcal{U}+\mathcal{K})(T') = (\mathcal{U}+\mathcal{K})(T)$ such that $\operatorname{dist}(Q, u(T_z^{(n)})) < \delta$ and $Q - W^*T_z^{(n)}W \in \mathcal{K}(\mathcal{H})$ for some unitary operator W. This completes the proof of case 1. Note that our distance estimate is actually δ as opposed to 4δ in this case.

Case 2. $\sigma(F) \cap \{\lambda : |\lambda| \geq 1/(1+3\delta)\} \neq \varnothing$.

By an appropriate choice of basis, we can assume that the eigenvalues $\{\lambda_1, \cdots, \lambda_m\}$ of F are listed in non-increasing order of absolute value. Thus

$$T \simeq \begin{bmatrix} \lambda_1 & & & & & & \\ & \ddots & & & 0 & & \\ & & \lambda_r & & & & \\ & & & \lambda_{r+1} & & & 0 \\ & 0 & & & \ddots & & \\ & & & & & \lambda_m & \\ x_1 & \cdots & x_r & x_{r+1} & \cdots & x_m & S \end{bmatrix} = \begin{bmatrix} \bigoplus_1^r \lambda_i & 0 \\ G & T_0 \end{bmatrix} \begin{matrix} \mathcal{H}_1' \\ \mathcal{H}_2' \end{matrix}$$

where $|\lambda_i| \in \big[1/(1+3\delta), 1\big)$ if and only if $1 \leq i \leq r$, $G = (x_1, \cdots, x_r) \in \mathcal{L}(\mathcal{H}_1', \mathcal{H}_2')$,

$$T_0 = \begin{bmatrix} \lambda_{r+1} & & & \\ & \ddots & 0 & 0 \\ & 0 & \lambda_m & \\ x_{r+1} & \cdots & x_m & S \end{bmatrix} \in \mathcal{L}(\mathcal{H}_2').$$

Set $Y_k = \begin{bmatrix} k & 0 \\ 0 & 1 \end{bmatrix} \begin{matrix} \mathcal{H}_1' \\ \mathcal{H}_2' \end{matrix} \in (\mathcal{U}+\mathcal{K})(\mathcal{H})$. Then

$$Y_kTY_k^{-1} = \begin{bmatrix} \bigoplus_1^r \lambda_i & 0 \\ \frac{1}{k}G & T_0 \end{bmatrix} \to \Big(\bigoplus_{i=1}^r \lambda_i\Big) \bigoplus T_0 \qquad (k \to \infty),$$

i.e., $T \to_{\mathcal{U}+\mathcal{K}} \big(\bigoplus_{i=1}^r \lambda_i\big) \bigoplus T_0$. Moreover, it is not difficult to check that T_0 satisfies all the conditions of case 1. Because of this, we can conclude that

$$\Big(\bigoplus_{i=1}^r \lambda_i\Big) \bigoplus T_0 \sim_{\mathcal{U}+\mathcal{K}} \Big(\bigoplus_{i=1}^r \lambda_i\Big) \bigoplus Q_1,$$

where $Q_1 \in (\mathcal{U}+\mathcal{K})(T_0)$, $\|Q_1 - T_z^{(n)}\| < \delta$ and $Q_1 - T_z^{(n)}$ is compact. Let $Q = (\bigoplus_{i=1}^r \lambda_i)\bigoplus Q_1$, thus $Q \in (\mathcal{U}+\mathcal{K})(T)^-$. Since $||\lambda_i| - 1| < |1/(1+3\delta) - 1| < 3\delta$ $(1 \le i \le r)$, we may easily find $\lambda_i' \in \partial D$ such that $|\lambda_i - \lambda_i'| < 3\delta$ $(1 \le i \le r)$. Since $T_z^{(n)}$ is essentially normal and $\partial D \subset \sigma_e(T_z^{(n)})$, by Proposition 1.28 we have $T_z^{(n)} = W\left[(\bigoplus_{i=1}^r \lambda_i')\bigoplus T_z^{(n)}\right]W^* + K$, where K is a compact operator with $\|K\| < \delta$, and W is a unitary operator. Thus

$$\begin{aligned}
\|Q - W^*T_z^{(n)}W\| &= \left\|\Big(\bigoplus\nolimits_{i=1}^r \lambda_i\Big)\bigoplus Q_1 - \Big(\bigoplus\nolimits_{i=1}^r \lambda_i'\Big)\bigoplus T_z^{(n)} - W^*KW\right\| \\
&\le \left\|\Big(\bigoplus_{i=1}^r (\lambda_i - \lambda_i')\Big)\bigoplus (Q_1 - T_z^{(n)})\right\| + \|K\| \\
&\le 3\delta + \delta = 4\delta.
\end{aligned}$$

Moreover, $Q - W^*T_z^{(n)}W$ is compact, since $Q_1 - T_z^{(n)}$ is compact. □

Proposition 5.14. *Suppose that $T \in \mathcal{A}(D,n)$. Then given δ, $0 < \delta < 1/10$, we can find a compact operator K with $\|K\| < \delta$ such that $T_z^{(n)} + K \sim_{\mathcal{U}+\mathcal{K}} T$. In particular, $T_z^{(n)} \in (\mathcal{U}+\mathcal{K})(T)^-$, and so $(\mathcal{U}+\mathcal{K})(T_z^{(n)})^- = (\mathcal{U}+\mathcal{K})(T)^-$.*

Proof. By Lemma 5.11, $T \to_{\mathcal{U}+\mathcal{K}} T'$, where $T' \in \mathcal{A}(D,n)$ and is lower triangular. Thus $T'^* \in \mathcal{B}_n(D)$. From the definition of the $\mathcal{B}_n(D)$ operators we may assume that the diagonal entries $\{\lambda_k\}_{k=1}^\infty$ of T' are pairwise distinct and form a dense subset of D. Thus, for each $k \ge 1$, we have

$$T' = \begin{bmatrix} \lambda_1 & & & 0 \\ & \ddots & & \vdots \\ & * & \lambda_k & 0 \\ Z_{11} & \dots & Z_{1k} & T_k' \end{bmatrix} := \begin{bmatrix} G_k & 0 \\ Z_k & T_k' \end{bmatrix}.$$

Since $T'^*T' - 1 =: K \in \mathcal{K}(\mathcal{H})$, we obtain

$$\begin{bmatrix} G_k^*G_k + Z_k^*Z_k & Z_k^*T_k' \\ T_k'Z_k & T_k'^*T_k' \end{bmatrix} = \begin{bmatrix} 1 & 0 \\ 0 & 1 \end{bmatrix} + \begin{bmatrix} K_{1k} & K_{2k} \\ K_{3k} & K_{4k} \end{bmatrix}$$

with respect to these decompositions of the spaces. Since K is compact, we can choose N large enough so that if $k \ge N$, $\|K_{4k}\| < \delta/4$. Let $S = T_N'$; then $S \in \mathcal{A}(D,n)$ and $\|S^*S - 1\| < \delta/4$. By Lemma 5.12, $d(S, \mathcal{U}(T_z^{(n)})) < \delta/4$. Since the eigenvalues of G_N are all distinct, we can find an invertible matrix R such that $F = R^{-1}G_NR$ is diagonal. Let $X = Z_NR$; then

$$T' \sim_{\mathcal{U}+\mathcal{K}} \begin{bmatrix} R^{-1} & 0 \\ 0 & 1 \end{bmatrix}\begin{bmatrix} G_N & 0 \\ Z_N & S \end{bmatrix}\begin{bmatrix} R & 0 \\ 0 & 1 \end{bmatrix} = \begin{bmatrix} F & 0 \\ X & S \end{bmatrix} = T_0.$$

By Proposition 5.13 we can find an operator $Q \in (\mathcal{U}+\mathcal{K})(T_0)^- = (\mathcal{U}+\mathcal{K})(T)^-$ and a unitary operator W such that

(i) $Q - W^*T_z^{(n)}W$ is compact,

(ii) $\|Q - W^*T_z^{(n)}W\| < \delta$. Thus, setting $K = WQW^* - T_z^{(n)}$, we have $\|K\| < \delta$ and $T_z^{(n)} + K \sim_{\mathcal{U}+\mathcal{K}} T$. Therefore $T_z^{(n)} \in (\mathcal{U}+\mathcal{K})(T)^-$. By Lemma 5.10, $(\mathcal{U}+\mathcal{K})(T_z^{(n)})^- = (\mathcal{U}+\mathcal{K})(T)^-$. □

Now we are in a position to prove Theorems 5.4 and 5.5.

Proof of Theorem 5.4. By Lemma 5.10, there is a compact operator K_1 with $\|K_1\| < \epsilon/2$ and an operator $Y \in (\mathcal{U}+\mathcal{K})(\mathcal{H})$ such that $T_1 + K_1 = YT_\phi^{(n)}Y^{-1}$. Let ϕ be the analytic function in Lemma 5.6; then ϕ is analytic in $\overline{D}$ and $\phi^{-1} = \psi$ is analytic in $\overline{\Omega}$. We can find δ_1, $\delta_2 > 0$ such that $\psi(\Omega_{\delta_1}) \supset D_{\delta_2}$. Set $A = \psi(T_2) \in \mathcal{A}(D, n)$, $B = T_z^{(n)}$. Let $m_1 = \max\{|\phi(z)| : |z| = 1 + \delta_2\}$ and $m_2 = \max\{\|(z-B)^{-1}\| : |z| = 1 + \delta_2\}$. By Proposition 5.14, there is a compact operator K_2 and an operator $X \in (\mathcal{U}+\mathcal{K})(\mathcal{H})$ such that $X(B+K_2)X^{-1} = A$, and $\|K_2\| < \min\{1/[(m+1)m_2], \delta_2\}$, where $m \in \mathbb{N}$ satisfies $[m_1m_2(1+\delta_2)]/m < \epsilon/(2\|Y\|\|Y^{-1}\|)$. This implies that $\sigma(B + K_1) = \overline{D}$. Therefore

$$\phi(X(B+K_1)X^{-1}) = X\,\phi(B+K_1)X^{-1} = \phi(A) = \phi \circ \psi(T) = T_2.$$

Set

$$\begin{aligned} K_3 &= \phi(B+K_2) - T_\phi^{(n)} = \phi(B+K_2) - \phi(B) \\ &= \frac{1}{2\pi i}\int_{|z|=1+\delta_2} \phi(z)\left[(z-B-K_2)^{-1} - (z-B)^{-1}\right] dz \\ &= \frac{1}{2\pi i}\int_{|z|=1+\delta_2} \phi(z)(z-B)^{-1}\sum_{n=1}^{\infty}\left[(z-B)^{-1}K_2\right]^n dz. \end{aligned}$$

Thus

$$\begin{aligned} \|K_3\| &\le \frac{2\pi m_1(1+\delta_2)m_2}{2\pi}\sum_{n=1}^{\infty}\|(z-B)^{-1}\|^n\|K_2\|^n \\ &< m_1m_2(1+\delta_2)\sum_{n=1}^{\infty}\frac{1}{(m+1)^n} \\ &= \frac{m_1m_2(1+\delta_2)}{m} < \frac{\epsilon}{2\|Y\|\|Y^{-1}\|}. \end{aligned}$$

Since K_2 is compact, K_3 is compact. Thus $T_\phi^{(n)} + K_3 \sim_{\mathcal{U}+\mathcal{K}} T_2$, i.e., there is a $Z \in (\mathcal{U}+\mathcal{K})(\mathcal{H})$ such that $T_\phi^{(n)} + K_2 = ZT_2Z^{-1}$. Thus $T_1 + K_1 + YK_3Y^{-1} = Y(T_\phi^{(n)} + K_3)Y^{-1} = YZT_2Z^{-1}$, and $K = K_1 + YK_3Y^{-1}$ satisfies the requirements of the theorem. □

Proof of Theorem 5.5. If A satisfies (i), (ii) and (iii), then for each $\epsilon > 0$ by Theorem 1.23, there exists a compact operator K, $\|K\| < \epsilon$, such that $A + K \in \mathcal{A}(\Omega, n)$. By Theorem 5.4, $A + K \in (\mathcal{U}+\mathcal{K})(T)^-$. Thus $A \in (\mathcal{U}+\mathcal{K})(T)^-$. If $A \in (\mathcal{U}+\mathcal{K})(T)^-$, then $X_nTX_n^{-1} \to A$ $(n \to \infty)$ for a sequence of invertible operators $\{X_n\} \subset (\mathcal{U}+\mathcal{K})(\mathcal{H})$. Since $\pi(X_n)\pi(T)\pi(X_n^{-1}) \to \pi(A)$ $(n \to \infty)$ and since each $\pi(X_n)$ is a unitary element, $\sigma_e(T) = \sigma_e(A) = \partial\Omega$ and $\operatorname{nul}(\lambda - T) = \operatorname{nul}(\lambda - A)$ for all $\lambda \in [\sigma(T) \cup \sigma(A)] \backslash \sigma_e(T)$. Thus A is essentially normal, $\overline{\Omega} \subset \sigma(A)$ and for $\lambda \in \Omega$, $\operatorname{ind}(\lambda - A) = -n$. Conversely, assume that $\lambda \in \sigma(A) \cap \rho(A)$ but $\lambda \notin \overline{\Omega}$. Since $\lambda \in \rho(X_nTX_n^{-1})$, $\operatorname{ind}(\lambda - A) = 0 = \operatorname{nul}(\lambda - A)$. Thus $\lambda \in \rho(A)$, a contradiction. Thus $\sigma(A) = \overline{\Omega}$. □

Using Theorem 5.4 and some results of Chapter 3, we can get

Proposition 5.15. *Let $T \in \mathcal{A}(\Omega, n)$ and $\epsilon > 0$. Then there exists a compact operator K with $\|K\| < \epsilon$ such that $T + K \in (SI)$.*

Proof. By Theorem 5.4, it suffices to prove that there exists a compact operator K, $\|K\| < \epsilon$, such that $T_\phi^{(n)} + K \in (SI)$, or $(T_\phi^*)^{(n)} + K \in (SI)$, where ϕ is an analytic homeomorphism from D onto Ω. Since $T_\phi^* \in \mathcal{B}_1(\Omega)$, by Proposition 3.18 we can find compact operators $K_1, K_2, \cdots, K_n$ such that $\|K_i\| < \epsilon/2$, $A_i := T_\phi^* + K_i \in \mathcal{B}_1(\Omega)$ $(i = 1, 2, \cdots, n)$ and $\ker \tau_{A_iA_j} = \{0\}$ $(i \neq j)$. Since $\sigma_r(A_1) \cap \sigma_l(A_i) \neq \varnothing$ $(i = 2, 3, \cdots, n)$, by Theorem 1.11 there are compact operators $C_2, C_3, \cdots, C_n$ such that $C_i \notin \operatorname{ran} \tau_{A_1A_i}$ and $\|C_i\| < \epsilon/2^i$ $(i = 2, 3, \cdots, n)$.

Set

$$K = \begin{bmatrix} K_1 & C_2 & \cdots & C_n \\ & K_2 & & \\ & & \ddots & \\ & 0 & & K_n \end{bmatrix}.$$

Then K is compact and $\|K\| < \epsilon$. By Corollary 3.14 $(T_\phi^*)^{(n)} + K \in (SI)$. □

5.2 Essentially normal + compact = strongly irreducible

Since the operators in $\mathcal{A}(\Omega, n)$ are essentially normal, after we get Proposition 5.15 it is very natural to ask the following question.

Question 5.16. *If T is an essentially normal operator with connected spectrum, does there exist a compact operator K such that $T+K \in (SI)$? If there does, can the norm $\|K\|$ of K be arbitrarily small?*

Here we only consider the operators with connected spectrum, because if $\sigma(T)$ is disconnected, then there exists a positive number δ such that $T+B$ has disconnected spectrum for all operators B provided $\|B\| < \delta$. Thus by the Riesz decomposition theorem $T+B \notin (SI)$. The main results of this section partially answer Question 5.16.

Theorem 5.17. *Let $T \in \mathcal{L}(\mathcal{H})$ be essentially normal with connected $\sigma(T)$ and $\operatorname{ind}(\lambda - T) > 0$ for all $\lambda \in \rho_F(T) \cap \sigma(T)$. Then for each $\epsilon > 0$, there exists a compact operator K, $\|K\| < \epsilon$, such that $T+K \in (SI)(\mathcal{H})$.*

In order to prove Theorem 5.17, we need following lemmas.

Lemma 5.18. *Let $B \in \mathcal{L}(\mathcal{H})$, $\lambda \in \rho_l(B) \cap \sigma(B)$ and $\mathcal{M} = [\ker(\lambda - B)^*]^\perp$. Then $B|_\mathcal{M} \sim B$ and $P_\mathcal{M} B^*|_\mathcal{M} \sim B^*$.*

Proof. Note that $\mathcal{M} = \operatorname{ran}(\lambda - B)$ and $\ker(\lambda - B) = \{0\}$. Set $A_1 \in \mathcal{L}(\mathcal{H}, \mathcal{M})$ by $A_1 x = (B-\lambda)x$ for $x \in \mathcal{H}$. Then $A_1^{-1} \in \mathcal{L}(\mathcal{M}, \mathcal{H})$. For $x \in \mathcal{M}$,

$$A_1^{-1}(B-\lambda)A_1 x = A_1^{-1}(B-\lambda)(B-\lambda)x = A_1^{-1}A_1(B-\lambda)x = (B-\lambda)x.$$

Thus $B|_\mathcal{M} \sim B$ and $P_\mathcal{M} B^*|_\mathcal{M} \sim B^*$. □

Lemma 5.19. *Let $B_n \in \mathcal{B}_1(\Omega_n)$, $\lambda_n \in \Omega_n$ $(n = 1, 2, \cdots, l;\ 1 \le l \le \infty)$ and let $\{\Omega_n\}$ be a sequence of uniformly bounded connected open subsets of $\mathbb{C}$ and $\{B_n\}$ be uniformly bounded. Let $T = \bigoplus_{n=1}^l B_n \in \mathcal{L}(\mathcal{H})$ and $P_\mathcal{M}$ be the orthogonal projection onto $\mathcal{M} = \bigoplus_{n=1}^m \ker(\lambda_n - B_n)^m$ $(1 \le m \le l)$. Then $P_\mathcal{M}^\perp T|_{P_\mathcal{M}^\perp \mathcal{H}} \sim T$, where $P_\mathcal{M}^\perp = 1 - P_\mathcal{M}$ and $\mathcal{H} = \bigoplus_{n=1}^l \mathcal{H}_n$.*

Proof. Denote $\mathcal{M}_n = \ker(\lambda_n - B_n)^m$ $(n = 1, 2, \cdots, m)$. Then we have

$$P_\mathcal{M}^\perp T|_{P_\mathcal{M}^\perp \mathcal{H}} = \Big(\bigoplus_{n=1}^m P_{\mathcal{M}_n}^\perp B_n|_{P_{\mathcal{M}_n}^\perp \mathcal{H}_n}\Big) \oplus \Big(\bigoplus_{n=m+1}^l B_n\Big).$$

By Lemma 5.18 there exists an invertible operator X_n such that

$$X_n(P_{\mathcal{M}_n}^{\perp} B_n|_{P_{\mathcal{M}_n^{\perp}}\mathcal{H}_n})X_n^{-1} = B_n \quad (n = 1, 2, \cdots, m).$$

Set $X = (\bigoplus_{n=1}^{m} X_n) \oplus (\bigoplus_{n=m+1}^{l} 1)$. Then $X(P_{\mathcal{M}}^{\perp} T|_{P_{\mathcal{M}}^{\perp}\mathcal{H}})X^{-1} = T$. □

By the argument of Lemma 5.9, we can get

Lemma 5.20. *Let* $T = \begin{bmatrix} F & C \\ 0 & A \end{bmatrix} \in \mathcal{L}(\mathbb{C}^n \oplus \mathcal{H})$, *where* $F \in \mathcal{L}(\mathbb{C}^n)$ *with* $\sigma(F) \subset \Omega$, $A \in \mathcal{B}_1(\Omega)$. *Then for each* $\epsilon > 0$ *there exists a compact operator* K, $\|K\| < \epsilon$, *such that* $T + K \sim A$.

Let $\{\Omega_i\}_{i=1}^{l}$ $(1 \le l \le \infty)$ be a sequence of connected uniformly bounded open subsets of $\mathbb{C}$, $\Omega_i \cap \Omega_j = \varnothing$ $(i \ne j)$ and for each i let $B_i \in \mathcal{B}_1(\Omega_i)$ be an operator on $\mathcal{H}_i$, $\sigma(B_i) = \overline{\Omega}_i$. Set $T_k = B_k^{(n_k)} \in \mathcal{L}(\mathcal{H}^{(n_k)})$ $(1 \le n_k \le \infty,\ k = 1, 2, \cdots)$ and $A = \bigoplus_{k=1}^{l} T_k \in \mathcal{L}(\bigoplus_{k=1}^{l} \mathcal{H}_k^{(n_k)})$. Given $F \in \mathcal{L}(\mathbb{C}^m)$ satisfies $\sigma(F) \subset \sigma(A) \cap \rho_F(A)$, then we have

Lemma 5.21. *For each* $\epsilon > 0$, *there exists a compact operator* K *with* $\|K\| < \epsilon$ *such that* $\begin{bmatrix} F & G \\ 0 & A \end{bmatrix} + K \sim A$.

Proof. From the construction $A = \bigoplus_{k=1}^{l_1} A_k$, where each $A_k \in \{B_i\}_{i=1}^{l}$. It is easily seen that $\sigma(A) \cap \rho_F(A) \subset \bigcup_k [\sigma(A_k) \cap \rho_F(A_k)]$. We may assume that $\sigma(F) \subset \bigcup_{k=1}^{r} [\sigma(A_k) \cap \rho_F(A_k)]$ for some $r < \infty$ and F is similar to a block diagonal: $F \sim \operatorname{diag}(F_1, F_2, \cdots, F_r)$, where $F_i \in \mathcal{L}(\mathbb{C}^{m_i})$ and $\sigma(F_i) \subset \sigma(A_i) \cap \rho_F(A_i)$ $(i = 1, 2, \cdots, r)$. Therefore

$$\begin{bmatrix} F & G \\ 0 & A \end{bmatrix} \sim \begin{bmatrix} F_1 & & & & & & & * \\ & A_1 & & & & & & \\ & & F_2 & & & & & \\ & & & A_2 & & & & \\ & & & & \ddots & & & \\ & & & & & F_r & & \\ & 0 & & & & & A_r & \\ & & & & & & & \bigoplus_{i=r+1}^{l_1} A_i \end{bmatrix}.$$

By Lemma 5.20, there exists a compact operator K_1, $\|K_1\| < \epsilon/2$, such that

$$\begin{bmatrix} F & G \\ 0 & A \end{bmatrix} + K_1 \sim \begin{bmatrix} A_1 & G_{12} & \cdots & G_{1r} & G_{1\,r+1} \\ & A_2 & \cdots & G_{2r} & G_{2\,r+1} \\ & & \ddots & & \vdots \\ & & & A_r & G_{r\,r+1} \\ & 0 & & & \bigoplus_{i>r} A_i \end{bmatrix},$$

where each G_{ij} is a finite rank operator ($i < j$).

Claim. $\ker \tau_{A_i A_j}$ does not contain any non-zero compact operators. In fact, if $A_i = A_j$ and $A_i K = K A_i$ for some compact operator K, then $K\, e(\lambda) = t(\lambda)\, e(\lambda)$, where $\lambda \in \Omega_i$, $e(\lambda) \in \ker(A_i - \lambda)$, $\|e(\lambda)\| = 1$ and $t(\lambda) \in \mathbb{C}$, since $\operatorname{nul}(A_i - \lambda) = 1$. Note that $e(\lambda)$ ($\lambda \in \Omega_i$) is a continuous vector valued function, so is $K\, e(\lambda)$. Thus $t(\lambda)$ ($\lambda \in \Omega_i$) is a continuous function. If $t(\lambda_0) \neq 0$ for some $\lambda_0 \in \Omega_i$, then we can find a $\Phi = \{z \in \mathbb{C} : |z - \lambda_0| < \delta\} \subset \overline{\Phi} \subset \Omega_i$ such that $|t(\lambda)| > \gamma > 0$ for all $\lambda \in \overline{\Phi}$ and for a γ. This is contradictory to the compactness of K. Thus $t(\lambda) \equiv 0$ and $K = 0$.

If $A_i \neq A_j$, then $\Omega_i \cap \Omega_j = \varnothing$. By Proposition 1.14 $\ker \tau_{A_i A_j} = \{0\}$.

Since $\ker \tau_{A_i A_j}$ does not contain any non-zero compact operators, $\ker \tau_{\bigoplus_{i>r} A_i, A_k}$ does not contain any non-zero compact operators. In fact if there is a compact operator $Q = \begin{bmatrix} Q_1 \\ Q_2 \\ \vdots \end{bmatrix}$ such that $A_{r+j} Q_j = Q_j A_k$ ($j = 1, 2, \cdots$) then by the claim, $Q = 0$. Thus it follows from Lemma 1.10 that $\operatorname{ran} \tau_{A_r \bigoplus_{i>r} A_i} | \mathcal{K}(\bigoplus_{i>r} \mathcal{H}_i, \mathcal{H}_r)$ is dense in $\mathcal{K}(\bigoplus_{i>r} \mathcal{H}_i, \mathcal{H}_r)$. So we can find G_r and Y_r in $\mathcal{K}(\bigoplus_{i>r} \mathcal{H}_i, \mathcal{H}_r)$ such that

$$A_r Y_r - Y_r\Big(\bigoplus_{i>r} A_i\Big) = G_{r\,r+1} + G_r \quad \text{and} \quad \|G_r\| < \frac{\epsilon}{2^{r+1}}.$$

Thus

$$\begin{bmatrix} 1 & Y_r \\ 0 & \bigoplus_{i>r} 1 \end{bmatrix} \begin{bmatrix} A_r & G_{r\,r+1} + G_r \\ 0 & \bigoplus_{i>r} A_i \end{bmatrix} \begin{bmatrix} 1 & -Y_r \\ 0 & \bigoplus_{i>r} 1 \end{bmatrix} = \begin{bmatrix} A_r & 0 \\ 0 & \bigoplus_{i>r} A_i \end{bmatrix},$$

and we can find a compact operator K_2, $\|K_2\| < \epsilon/2^{r+1}$, such that

$$\begin{bmatrix} F & G \\ 0 & A \end{bmatrix} + K_1 + K_2 \sim \begin{bmatrix} A_1 & & & & * \\ & A_2 & & & \\ & & \ddots & & \\ & & & A_{r-1} & \\ & 0 & & & \bigoplus_{i\geq r} A_i \end{bmatrix}.$$

By the same argument we can find compact operators $K_3, \cdots, K_{r+1}$ such that

$$\begin{bmatrix} F & G \\ 0 & A \end{bmatrix} + \sum_{i=1}^{r+1} K_i \sim \bigoplus_{i=1}^{l_1} A_i \quad \text{and} \quad \|K_i\| < \frac{\epsilon}{2^{r+i}}.$$

Thus $K = \sum_{i=1}^{r+1} K_i$ satisfies the requirements of the lemma. □

Let $T \in \mathcal{L}(\mathcal{H})$ be the essentially normal operator given in Theorem 5.17, and let $\{\Omega_i\}_{i=1}^l$ $(1 \le l \le \infty)$ be the components of $\rho_F(T) \cap \sigma(T)$ and $n_i = \operatorname{ind}(\lambda - T)$ $(\lambda \in \Omega_i)$. Denote $\sigma = \sigma(T) \backslash \bigcup_{i=1}^l \overline{\Omega}_i$ and $D = \operatorname{diag}\{\lambda_1, \lambda_2, \cdots\}$, the diagonal operator with diagonal $\{\lambda_i\}_{i=1}^\infty$, where $\{\lambda_i\}_{i=1}^\infty \subset \sigma^\circ$, $\{\lambda_i\}^- = \sigma$ and for each k, $\operatorname{card}\{n : \lambda_n = \lambda_k\} = \infty$. By Lemma 3.21, there is an essentially normal operator $B_i \in \mathcal{B}_1(\Omega_i)$ such that $\sigma(B_i) = \overline{\Omega}_i$ and $\Omega = \rho_F(B_i) \cap \sigma(B_i)$.

Set $T_k = B_k^{(n_k)}$ $(k = 1, 2, \cdots, l)$ and $G = \big(\bigoplus_{k=1}^l T_k\big) \bigoplus D$. It is obvious that G is essentially normal.

Lemma 5.22. *Let $T \in \mathcal{L}(\mathcal{H})$ be given as in Theorem 5.17. Then*

(i) $\Lambda(G) = \Lambda(T)$;

(ii) *For each $\epsilon > 0$, there exists a compact operator K with $\|K\| < \epsilon$ such that $T + K \sim G$.*

Proof. It is obvious that $\Lambda(G) = \Lambda(T)$. By the Brown–Douglas–Fillmore theorem there exist a unitary operator U and a compact operator K_0 such that $UTU^* = G + K_0$.

Suppose that $D \in \mathcal{L}(\mathcal{K})$ and $\{f_k\}_{k=1}^\infty$ is an ONB of $\mathcal{K}$ such that $Df_k = \lambda_k f_k$ $(k = 1, 2, \cdots)$. Also, let $\mu_k \in \Omega_k$ $(k = 1, 2, \cdots, l)$ be a fixed point. Set $\mathcal{M}_1(m) = \bigoplus_{k=1}^m \ker(T_k - \mu_k)^m$, $\mathcal{M}_2(m) = \bigvee\{f_k : k = 1, 2, \cdots, m\}$ and denote by P_{1m} and P_{2m} the orthogonal projections onto $\mathcal{M}_1(m)$ and $\mathcal{M}_2(m)$ respectively. Set $P_m = P_{1m} + P_{2m}$. Then $P_m \to 1$ $(m \to \infty)$ in the strong operator topology. Therefore there exists an $N \in \mathbb{N}$ such that $\|K_1\| < \epsilon/8$, where $K_1 = P_N K_0 P_N - K_0$. Thus

$$
\begin{aligned}
G + K_0 + K_1 &= G + P_N K_0 P_N \\
&= \begin{bmatrix} P_{1N}(\bigoplus T_k)P_{1N} & P_{1N}(\bigoplus T_k)P_{1N}^\perp & 0 & 0 \\ 0 & P_{1N}^\perp(\bigoplus T_k)P_{1N}^\perp & 0 & 0 \\ 0 & 0 & P_{2N}DP_{2N} & 0 \\ 0 & 0 & 0 & P_{2N}^\perp D P_{2N}^\perp \end{bmatrix} \begin{matrix} \mathcal{M}_1(N) \\ \\ \mathcal{M}_2(N) \\ \mathcal{K} \ominus \mathcal{M}_2(N) \end{matrix} \\
&\quad + \begin{bmatrix} P_{1N}K_0P_{1N} & 0 & P_{1N}K_0P_{2N} & 0 \\ 0 & 0 & 0 & 0 \\ P_{2N}K_0P_{2N} & 0 & P_{2N}K_0P_{2N} & 0 \\ 0 & 0 & 0 & 0 \end{bmatrix} \begin{matrix} \mathcal{M}_1(N) \\ \\ \mathcal{M}_2(N) \\ \mathcal{K} \ominus \mathcal{M}_2(N) \end{matrix}
\end{aligned}
$$

$$= \begin{bmatrix} K_{11} & K_{13} & K_{12} & 0 \\ 0 & P_{1N}^{\perp}(\bigoplus T_k)P_{1N}^{\perp} & 0 & 0 \\ K_{21} & 0 & K_{22} & 0 \\ 0 & 0 & 0 & P_{2N}^{\perp}DP_{2N}^{\perp} \end{bmatrix} \simeq \begin{bmatrix} K_{11} & K_{12} & K_{13} & 0 \\ K_{21} & K_{22} & 0 & 0 \\ 0 & 0 & P_{1N}^{\perp}(\bigoplus T_k)P_{1N}^{\perp} & 0 \\ 0 & 0 & 0 & D \end{bmatrix},$$

where $K_{11} = P_{1N}(\bigoplus T_k)P_{1N} + P_{1N}K_0P_{1N}$, $K_{12} = P_{1N}K_0P_{1N}$, $K_{13} = P_{1N}(\bigoplus T_k)P_{1N}^{\perp}$, $K_{21} = P_{2N}K_0P_{2N}$, $K_{22} = P_{2N}DP_{2N} + P_{2N}K_0P_{2N}$. They are compact operators. Note that $P_{2N}^{\perp}DP_{2N}^{\perp} \simeq D$. Thus, we have

$$U_1(G + K_0 + K_1)U_1^* = \begin{bmatrix} L_{11} & L_{12} & 0 \\ 0 & P_{1N}^{\perp}(\bigoplus T_k)P_{1N}^{\perp} & 0 \\ 0 & 0 & D \end{bmatrix}$$

for some unitary operator U_1, where $L_{11} = \begin{bmatrix} K_{11} & K_{12} \\ K_{21} & K_{22} \end{bmatrix}$ is acting on a space of dimension $d = (N+1)\sum_{k=1}^{N} n_k$ and $L_{12} = \begin{bmatrix} K_{13} \\ 0 \end{bmatrix}$. We can choose $\|K_1\|$ so small that $\sigma(L_{11}) \subset \sigma(T)_{\epsilon/8}$. Thus there is a compact operator L and an invertible operator X such that $\sigma(L_{11} + L) \subset \sigma(T)$, $X(L_{11} + L)X^{-1} = \operatorname{diag}\{\gamma_1, \cdots, \gamma_d\} := D_1$ and each $\gamma_i \in \sigma(\bigoplus T_k) \cap \rho_F(\bigoplus T_k)$ or $\gamma_i = \lambda_k$ for some k. Without loss of generality we can assume that $\{\gamma_1, \cdots, \gamma_p\} \subset \sigma_p(D)$ and $\{\gamma_{p+1}, \cdots, \gamma_d\} \subset \sigma(\bigoplus T_k) \cap \rho_F(\bigoplus T_k)$. Then $D_1 = D_2 + D_3$, where $D_2 = \operatorname{diag}\{\gamma_1, \cdots, \gamma_p\}$, $D_3 = \operatorname{diag}\{\gamma_{p+1}, \cdots, \gamma_d\}$. By Lemma 5.19, $P_{1N}^{\perp}(\bigoplus T_k)|_{\operatorname{ran} P_{1N}^{\perp}} \sim \bigoplus T_k$. Thus, there is a compact operator K_2 with $\|K_2\| < \epsilon/4$ and an invertible operator X_1 such that

$$\begin{aligned} & X_1U_1(G + K_0 + K_1 + U_1^*K_2U_1)U_1^*X_1^{-1} \\ &= X_1\left(\begin{bmatrix} L_{11} & L_{12} & 0 \\ 0 & P_{1N}^{\perp}(\bigoplus T_k)|_{\operatorname{ran} P_{1N}^{\perp}(\bigoplus T_k)} & 0 \\ 0 & 0 & D \end{bmatrix} + K_2\right)X_1^{-1} \\ &= \begin{bmatrix} D_2 & 0 & L_1 & 0 \\ 0 & D_3 & L_2 & 0 \\ 0 & 0 & \bigoplus T_k & 0 \\ 0 & 0 & 0 & D \end{bmatrix} := G_1, \end{aligned}$$

where L_1, L_2 are compact. Since $\sigma(D_2) \cap \sigma(\bigoplus T_k) = \varnothing$, by Theorem 1.11 we can find an operator Y so that $D_2Y - Y(\bigoplus T_k) = L_1$. Set $X_2 = \begin{bmatrix} 1 & 0 & Y & 0 \\ 0 & 1 & 0 & 0 \\ 0 & 0 & 1 & 0 \\ 0 & 0 & 0 & 1 \end{bmatrix}$. Then

$$X_2G_1X_2^{-1} = \begin{bmatrix} D_2 & 0 & 0 & 0 \\ 0 & D_3 & L_2 & 0 \\ 0 & 0 & \bigoplus T_k & 0 \\ 0 & 0 & 0 & D \end{bmatrix} := G_2.$$ By Lemma 5.21, there exist a compact operator K_3' with $\|K_3'\| < \epsilon(8\|X_1\|\|X_1^{-1}\|\|X_2\|\|X_2^{-1}\|)^{-1}$ and an invertible operator X' such that

$$X'\left(\begin{bmatrix} D_3 & L_2 \\ 0 & \bigoplus T_k \end{bmatrix} + K_3'\right)X'^{-1} = \bigoplus T_k.$$

Therefore we can find a compact operator K_3, $\|K_3\| < \epsilon(8\|X_1\|\|X_1^{-1}\|\|X_2\|\|X_2^{-1}\|)^{-1}$, and an invertible operator X_3 such that

$$X_3(G_2+K_3)X_3^{-1} = \begin{bmatrix} D_2 & 0 & 0 \\ 0 & \bigoplus T_k & 0 \\ 0 & 0 & D \end{bmatrix} \simeq (\bigoplus T_k) \oplus D_2 \oplus D \simeq (\bigoplus T_k) \oplus D = G.$$

Note that since $\sigma(D_2) \subset d(D)$ (the diagonal of D), we have the last unitary equivalence. □

Now we are in a position to prove Theorem 5.17.

Proof of Theorem 5.17. By Lemma 5.22, we have an operator G and a compact operator K_1, $\|K_1\| < \epsilon/2$, such that $T + K_1 \sim G$. Thus it suffices to prove that for $\delta > 0$ there is a compact operator K_2 with $\|K_2\| < \delta$ such that $G + K_2 \in (SI)$.

Observe that $G = G_1 \oplus B_1^{(n_1-1)}$, where $G_1 = B_1 \oplus (\bigoplus_{k=2}^{l} T_k) \oplus D$. It is obvious that $\sigma(G_1) = \sigma_W(G_1)$ is connected and $\operatorname{ind}(\lambda - G_1) = 1$ for $\lambda \in \Omega_1$. By Theorem 1.44 there exists a compact operator K, $\|K\| < \delta/2$, such that $A_1 := G_1 + K \in \mathcal{B}_1(\Omega_1)$. Pick a $\lambda_0 \in \Omega_1$; then

$$A_1 \simeq \begin{bmatrix} \lambda_0 & a_{12} & & * \\ & \lambda_0 & a_{23} & \\ & & \lambda_0 & \ddots \\ & 0 & & \ddots \end{bmatrix} \quad \text{and} \quad B_1 \simeq \begin{bmatrix} \lambda_0 & b_{12} & & * \\ & \lambda_0 & b_{23} & \\ & & \lambda_0 & \ddots \\ & 0 & & \ddots \end{bmatrix}.$$

Consider $r(A_1, B_1) = \overline{\lim}_{n\to\infty}[\prod_{k=1}^{n}(|a_{k\,k+1}|/|b_{k\,k+1}|)]^{1/n}$. If $r(A_1, B_1) = 1$, by the argument of Proposition 3.17, there are compact operators $F_2, \cdots, F_{n_1}$ with $\max\{\|F_i\|\} < \delta/4$ such that $E_i \in \mathcal{B}_1(\Omega_1)$ $(i=2,\cdots,n_1)$, $\ker \tau_{A_1E_i} = \ker \tau_{E_iA_1} = 0$ $(i = 2, 3, \cdots, n_1)$ and $\ker \tau_{E_iE_j} = \{0\}$ $(i \neq j)$, where $E_i = B_1 + F_i$ $(i = 2, \cdots, n_1)$. Since $\sigma_r(A_1) \cap \sigma_l(E_j) \neq \varnothing$, by Theorem 1.11 we can find compact operators $C_2, C_3, \cdots, C_{n_1}$ with $\max\{\|C_j\|\} < \delta/4$ such that $C_j \notin \operatorname{ran} \tau_{A_1E_j}$ $(j = 2, 3, \cdots, n_1)$.

Set $L = \begin{bmatrix} 0 & C_2 & \cdots & C_{n_1} \\ & F_2 & & \\ & & \ddots & \\ 0 & & & F_{n_1} \end{bmatrix}$. Then L is compact, $\|L\| < \delta/4$ and $A_1 \oplus B_1^{(n_1-1)} + L = \begin{bmatrix} A_1 & C_2 & \cdots & C_{n_1} \\ & E_2 & & \\ & & \ddots & \\ 0 & & & E_{n_1} \end{bmatrix}$. By Corollary 3.14 $A_1 \oplus B_1^{(n_1-1)} + L \in (SI)$. Thus, setting $K_2 = K \oplus 0 + L$, we have $G + K_2 \in (SI)$.

If $r(A_1, B_1) > 1$, by Proposition 3.18 we can find compact operators F_2, F_3, $\cdots$, F_{n_1} with $\max\{\|F_i\|\} < \delta/4$ such that

$$E_i = B_1 + F_i \in \mathcal{B}_1(\Omega_1) \qquad (i = 2, 3, \cdots, n_i)$$

and $\ker \tau_{E_i E_j} = \{0\}$ ($i \neq j$ and $i, j \geq 2$). By Proposition 3.17, $r(A_1, E_i) > 1$ ($i = 2, 3, \cdots, n_1$), and $\ker \tau_{E_i A_1} = \{0\}$. By the same argument as in the case of $r(A_1, B_1) = 1$, we can find a compact operator K_2 with $\|K_2\| < \delta$ such that $G + K_2 \in (SI)$.

If $r(A_1, B_1) < 1$,

$$A_1 \oplus B_1^{(n_1-1)} + L \simeq \begin{bmatrix} E_{n_1} & & & C_{n_1} \\ & \ddots & & \vdots \\ & & E_2 & C_2 \\ 0 & & & A_1 \end{bmatrix},$$

where $\ker \tau_{A_1 E_i} = \{0\}$ ($i = 2, 3, \cdots, n_1$) and $C_i \notin \operatorname{ran} \tau_{E_i A_1}$ and $\ker \tau_{E_i E_j} = \{0\}$ ($i \neq j$; $i, j \geq 2$). Then by the same arguments as in the cases $r(A_1, B_1) = 1$ and $r(A_1, B_1) > 1$, we can find a compact operator K_2 with $\|K_2\| < \delta$ such that $G + K_2 \in (SI)$. □

If the norm of the compact perturbation is not required to be arbitrarily small, we have

Theorem 5.23. *Let $T \in \mathcal{L}(\mathcal{H})$ be an essentially normal operator with connected spectrum; then there exists a compact operator K such that $T + K \in (SI)$.*

In order to construct an essentially normal operator $A \in (SI)$ with $\Lambda(A) = \Lambda(T)$, we need the following two lemmas.

Lemma 5.24. *Given a bounded and connected open subset Ω of $\mathbb{C}$, there exists an essentially normal and co-subnormal operator N such that for each $\epsilon > 0$, there is a compact operator K with $\|K\| < \epsilon$ satisfying*

(i) $A(\Omega) = N + K \in \mathcal{B}_1(\Omega)$;

(ii) $\sigma(A(\Omega)) = \overline{\Omega}$ *and* $\sigma_e(A(\Omega)) = \partial\Omega$;

(iii) $\sigma_p(A(\Omega)) \cap \partial\overline{\Omega} = \varnothing$.

Proof. Let $N_+(\Omega_1^*)$ be the Bergman operator on $\mathcal{H}_0 = A^2(\Omega_1^*)$ and $S_0 = N_+(\Omega_1^*)^*$, where $\Omega_1 = (\overline{\Omega})^\circ$. Denote $\sigma = \Omega_1 \backslash \Omega$. Let $\{\mu_n\}_{n=1}^l$ $(0 \le l \le \infty)$ be a dense subset of σ and let $\{\mathcal{H}_n\}_{n=1}^\infty$ be a sequence of Hilbert spaces and $\{e_k^n\}_{k=1}^\infty$ be an ONB of $\mathcal{H}_n$ $(n = 0, 1, 2, \cdots)$. Define $A_n e_k^n = \epsilon/(k!)^n e_{k+1}^n = a_k^n e_{k+1}^n$ $(k = 1, 2, \cdots)$ and $S_n = \mu_n + A_n$. By Proposition 1.51, there is an analytic function $f : \Omega_1 \to \mathcal{H}_0$ such that $f(\lambda) \neq 0$ and $(S_0 - \lambda) f(\lambda) \equiv 0$ $(\lambda \in \Omega_1)$.

Define

$$A(\Omega) = \begin{bmatrix} S_0 & & & \\ C_1 & S_1 & & \\ C_2 & & S_2 & \\ \vdots & & & \ddots \end{bmatrix} \begin{matrix} \mathcal{H}_0 \\ \mathcal{H}_1 \\ \mathcal{H}_2 \\ \vdots \end{matrix}$$

with respect to $\mathcal{H} = \bigoplus_{i=0}^l \mathcal{H}_i$, where $C_n = \epsilon/(2^n \|f(\mu_n)\|)\, e_1^n \otimes f(\mu_n)$ $(n = 1, 2, \cdots)$. Then $A(\Omega) = N + K$, where $N = S_0 \bigoplus \left(\bigoplus_{n=1}^l \mu_n\right)$ is essentially normal and co-subnormal and

$$K = \begin{bmatrix} 0 & & & \\ C_1 & A_1 & & \\ C_2 & & A_2 & \\ \vdots & & & \ddots \end{bmatrix}$$

is compact with $\|K\| < \epsilon$.

Thus $\sigma(A(\Omega)) \supset \Omega_1$, $\operatorname{ind}(\lambda - A(\Omega)) = 1$ $(\lambda \in \Omega_1)$ and $\sigma_e(A(\Omega)) = \sigma_e(N) = \partial\Omega$. Since $\sigma_p(S_n^*) \cap \Omega = \varnothing$, $\ker(\lambda - A(\Omega))^* = 0$ $(\lambda \in \Omega_1)$. Set

$$\begin{aligned}\chi(\lambda) = f(\lambda) &\oplus \Big[- \frac{\epsilon(f(\lambda), f(\mu_1))}{2\|f(\mu_1)\|} \sum_{k=1}^\infty \frac{(-1)^{k+1} a_1^1 \cdots a_{k-1}^1}{(\mu_1 - \lambda)^k} e_k^1 \Big] \oplus \cdots \\ &\oplus \Big[- \frac{\epsilon(f(\lambda), f(\mu_n))}{2^n \|f(\mu_n)\|} \sum_{k=1}^\infty \frac{(-1)^{k+1} a_1^n \cdots a_{k-1}^n}{(\mu_n - \lambda)^k} e_k^n \Big] \oplus \cdots .\end{aligned}$$

Then computation shows that $\chi(\lambda) \in \ker(A(\Omega) - \lambda)$ $(\lambda \in \Omega_1)$.

Denote $\bigvee\{\chi(\lambda) : \lambda \in \Omega\} = \mathcal{H}'$ and assume that $\bigoplus_{n=0}^{\infty} y_n \in \mathcal{H}'^{\perp}$, where $y_n \in \mathcal{H}_n$ and $y_n = \sum_{k=1}^{\infty}(b_k^n)^* e_k^n$ $(n = 0, 1, 2, \cdots)$; then

$$(f(\lambda), y_0) = \frac{\epsilon(f(\lambda), f(\mu_1))}{2\|f(\mu_1)\|} \sum_{k=1}^{\infty} \frac{(-1)^{k+1} a_1^1 \cdots a_{k-1}^1 b_k^1}{(\mu_1 - \lambda)^k} + \cdots$$
$$+ \frac{\epsilon(f(\lambda), f(\mu_n))}{2^n \|f(\mu_n)\|} \sum_{k=1}^{\infty} \frac{a_1^n \cdots a_{k-1}^n b_k^n}{(\mu_n - \lambda)^k} + \cdots .$$

Since $(f(\lambda), y_0)$ is analytic in Ω_1 $(\Omega_1 \supset \{\mu_k\}_{k=1}^{\infty})$ and $f(\mu_k) \neq 0$, $b_k^n = 0$ $(k = 1, 2, \cdots; n = 1, 2, \cdots)$. Therefore $\bigvee\{\chi(\lambda) : \lambda \in \Omega\} = \mathcal{H}$ and $A(\Omega) \in \mathcal{B}_1(\Omega)$.

If $\lambda \notin \overline{\Omega}$, then $\lambda \in \rho(N)$. Since $\lambda \notin \sigma_p(A(\Omega)^*)^*$, $\lambda \in \rho(A(\Omega))$ and $\sigma(A(\Omega)) = \overline{\Omega}$. Since $\sigma_p(S_0) \cap \partial\Omega_1 = \varnothing$ and since $\partial\Omega_1 \subset \rho(S_n)$ $(n = 1, 2, \cdots)$, $\sigma_p(A(\Omega)) \cap \partial\Omega_1 = \varnothing$. In fact, if there are $\lambda \in \partial\Omega_1$ and $\chi = \bigoplus_{k=0}^{\infty} \chi_k \neq 0$ where $\chi_k \in \mathcal{H}_k$ such that $(A(\Omega) - \lambda)\chi = 0$, then we have

$$(S_0 - \lambda)\chi_0 = 0,\ C_1\chi_0 = (\lambda - S_1)\chi_1,\ \cdots,\ C_k\chi_0 = (\lambda - S_k)\chi_k \cdots .$$

Since $\sigma_p(S_0) \cap \partial\Omega_1 = \varnothing$, $\chi_0 = 0$. Since $\partial\Omega_1 \subset \rho(S_n)$, $\chi_n = 0$, thus $\chi = 0$. □

Lemma 5.25. *Let Ω be a bounded, connected open subset of $\mathbb{C}$, $n \in \mathbb{N}$, $n \geq 2$ and $\epsilon > 0$. Then there exist operators A, $\overline{A}(\Omega) \in (SI)$ and compact operators $Q(\Omega)$ and K with $\|Q(\Omega)\| < \epsilon$ and $\|K\| < \epsilon$ satisfying*

(i) *$A \in \mathcal{B}_1(\Omega)$, $\sigma_p(A) \cap \partial\overline{\Omega} = \varnothing$, $\rho_F(A) \cap \sigma(A) = \Omega$ and $\sigma(A) = \overline{\Omega}$;*

(ii) *$\overline{A}(\Omega) = A^{(n-1)} + K \in \mathcal{B}_{n-1}(\Omega)$, $Q(\Omega) \notin \operatorname{ran} \tau_{\overline{A}(\Omega)\,A}$ and $\ker \tau_{A_1\,\overline{A}(\Omega)} = \{0\}$.*

Proof. Let A be the operator $A(\Omega)$ obtained in Lemma 5.24. Then A satisfies (i).

Using A, Lemma 3.12 and Proposition 3.13 we can get $\overline{A}(\Omega) = A^{(n-1)} + K \in \mathcal{B}_{n-1}(\Omega)$, where K is compact with $\|K\| < \epsilon$. By the construction it is easily seen that $\ker \tau_{A\,\overline{A}(\Omega)} = \{0\}$. Note that $\sigma_r(\overline{A}(\Omega)) \cap \sigma_l(A) \neq \varnothing$, and by Theorem 1.11 we can find a compact operator $Q(\Omega)$ with $\|Q(\Omega)\| < \epsilon$, such that $Q(\Omega) \notin \operatorname{ran} \tau_{\overline{A}(\Omega)A}$.

□

Let $\sigma \subset (a, b)^2 = G$ be a compact subset of $\mathbb{C}$. In the proof of Theorem 3.7 we know that for each $\mu \in \sigma$, there is a vector $k_\mu \in \ker(\mu - M_\lambda^\circ(\sigma))^*$ corresponding to the multiplicative functional: $f \mapsto (f, k_\mu) = f(\mu)$ $(f \in W_0^{22}(G) \ominus m_0(\sigma))$. Thus if P is idempotent in $\mathcal{A}'(M_\lambda^\circ(\sigma)^*)$, then $Pk_\mu \in \ker(\mu - M_\lambda^\circ(\sigma))^*$. Since $\dim\ker(\mu - M_\lambda^\circ(\sigma))^* = 1$, $Pk_\mu = a(\mu)k_\mu$ for $a(\mu) \in \mathbb{C}$. Since $P^2 = P$, $a(\mu) = 0$ or 1 $(\mu \in \sigma)$.

Lemma 5.26. *Let Ω be a bounded, connected open subset of $\mathbb{C}$, $A \in \mathcal{L}(\mathcal{H})$ is in $\mathcal{B}_1(\Omega)$ such that $\sigma(A) = \overline{\Omega}$ and $\sigma_p(A) \cap \partial\overline{\Omega} = \varnothing$. Also let σ be a compact subset of $\mathbb{C}$, $\sigma \cap \partial\overline{\Omega} \neq \varnothing$ and $\{\sigma_n\}_{n\in\Lambda}$ be a sequence of pairwise disjoint closed maximal connected components of σ, each σ_n consisting of more than one point and $\sigma = [\bigcup \sigma_n]^-$. Let $\sigma' = \bigcup\{\sigma_n : n \in \Lambda,\ \sigma_n \cap \partial\overline{\Omega} \neq \varnothing\}$ and G be an open square containing σ^*, and set $B = M_\lambda^\circ(\sigma^*)^*$ on $\mathcal{H}_1 = W_0^{22}(G) \ominus m_0(\sigma^*)$. Then for each $\epsilon > 0$, there is a compact operator $K \in \mathcal{L}(\mathcal{H}_1, \mathcal{H})$ with $\|K\| < \epsilon$ such that $K \notin \operatorname{ran} \tau_{AB}$ and having the property: if there is an idempotent P in $\mathcal{A}'(B)$ and an operator $X \in \mathcal{L}(\mathcal{H}_1, \mathcal{H})$ such that $AXP - XPB = KP$, then $a(\lambda)|_{(\sigma')^*} = 0$ and $Pk_\mu = a(\mu)k_\mu$ for $\mu \in \sigma^*$.*

Similarly, there is a compact operator $K' \in \mathcal{L}(\mathcal{H}, \mathcal{H}_1)$ with $\|K'\| < \epsilon$ such that $K' \notin \operatorname{ran} \tau_{BA}$ and having the property: if there is an idempotent P in $\mathcal{A}'(B)$ and an operator $X \in \mathcal{L}(\mathcal{H}, \mathcal{H}_1)$ such that $BPX - PXA = PK'$, then $a(\lambda)|_{(\sigma')^*} = 0$ and $Pk_\mu = a(\mu)k_\mu$ for $\mu \in \sigma^*$.

Proof. It is obvious that k_μ is a weakly continuous function on σ^*. Choose two sets $\{\lambda_i'\}_{i=1}^\infty$ and $\{\mu_i'\}_{i=1}^\infty$ of pairwise distinct complex numbers satisfying

(i) $\{\lambda_i'\} \subset \Omega$ and $\{\mu_i'\} \subset \sigma'$;
(ii) $\operatorname{card}\{i : \mu_i' \in \sigma_n\} = \infty$ for each σ_n, $\sigma_n \cap \partial\overline{\Omega} \neq \varnothing$;
(iii) $|\mu_i' - \lambda_i'| < 1/2^i$ $(i = 1, 2, \cdots)$.

Choose dense subsets $\{\lambda_k\}_{k=1}^\infty$ of Ω and $\{\mu_k\}_{k=1}^\infty$ of σ respectively, such that $\lambda_i' = \lambda_{2i}$, $\mu_i' = \mu_{2i}$ $(i = 1, 2, \cdots)$ and $\lambda_i \neq \lambda_j$, $\mu_i \neq \mu_j$ $(i \neq j)$. Since $\bigvee\{\ker(\lambda_i - A) : i = 1, 2, \cdots\} = \mathcal{H}$ and $\bigvee\{\ker(\mu_i - B) : i = 1, 2, \cdots\} = \mathcal{H}_1$, it is completely apparent that A and B have upper triangular matrix representations with respect to the ONB $\{e_n\}_{n=1}^\infty$ of $\mathcal{H}$ and, respectively, $\{f_n\}_{n=1}^\infty$ of $\mathcal{H}_1$ obtained by the Gram–Schmidt orthonormalization of unit vectors in $\bigcup_{i=1}^\infty\{\ker(\lambda_i - A)\}$ and $\bigcup_{i=1}^\infty\{\ker(\mu_i - B)\}$ respectively, i.e.,

$$A = \begin{bmatrix} \lambda_1 & & * \\ & \lambda_2 & \\ 0 & & \ddots \end{bmatrix} \begin{matrix} e_1 \\ e_2 \\ \vdots \end{matrix}, \quad B = \begin{bmatrix} \mu_1 & & * \\ & \mu_2 & \\ 0 & & \ddots \end{bmatrix} \begin{matrix} f_1 \\ f_2 \\ \vdots \end{matrix}$$

and

$$UBU^* = B' = \begin{bmatrix} \mu_1 & & * \\ & \mu_2 & \\ 0 & & \ddots \end{bmatrix} \begin{matrix} e_1 \\ e_2 \\ \vdots \end{matrix}$$

for the unitary operator U: $Uf_i = e_i$ $(i = 1, 2, \cdots)$.

Define $De_k = a_k e_k$, where

$$a_k = \begin{cases} \epsilon/2i, & \text{if } k = 2i, \\ 0, & \text{otherwise;} \end{cases}$$

then D is compact with $\|D\| < \epsilon$.

If $AX - XB' = D$ for some X, then calculation shows that X admits an upper triangular matrix representation with respect to $\{e_n\}$, i.e., $X = (x_{ij})$, $x_{ij} = 0\,(i > j)$. Thus $|x_{2i\,2i}| = a_{2i}/|\lambda_{2i} - \mu_{2i}| \geq 2^{i-1}/i \to \infty\ (i \to \infty)$. This contradiction implies that $D \notin \operatorname{ran} \tau_{A,B'}$ or $K = DU \notin \operatorname{ran} \tau_{AB}$.

Suppose that $P \in \mathcal{A}'(B)$ is an idempotent and $Pk_\mu = a(\mu)k_\mu$ $(\mu \in \sigma^*)$. If $a(\mu)$ takes both values 0 and 1 on some σ_n^*, then we can find $\{h_k\}_{k=0}^{\infty} \subset \sigma_n^*$ such that $h_k \to h_0$ $(k \to \infty)$, $a(h_k) = 0$ $(k \geq 1)$ and $a(h_0) = 1$ (if $a(h_k) = 1$ $(k \geq 1)$ and $a(h_0) = 0$, consider the idempotent $1 - P$, where 1 is the identity on $\mathcal{H}_1$).

Set

$$\mathcal{M} = \bigvee\{k_\mu : a(\mu) = 1,\ \mu \in \sigma^*\}, \quad \mathcal{N} = \bigvee\{k_\mu : a(\mu) = 0,\ \mu \in \sigma^*\}.$$

Thus $k_{h_k} \in \mathcal{N}$ $(k \geq 1)$ and $k_{h_0} \in \mathcal{M}$. Note that $\mathcal{M} \subset \operatorname{ran} P$ and $\mathcal{N} \subset \ker P$, therefore $\mathcal{M} \cap \mathcal{N} = \{0\}$. Since $k_{h_k} \to k_{h_0}$ weakly, $k_{h_0} \in \mathcal{M} \cap \mathcal{N}$. This contradiction implies that $a(\mu)$ takes same value (0 or 1) on each σ_n^*. Also, since $\mu \mapsto k_\mu$ is weakly continuous on σ^*, $a(\mu)$ is continuous on σ^*.

Assume that P satisfies $AXP - XPB = KP$ for some $X \in \mathcal{L}(\mathcal{H}_1, \mathcal{H})$. Note that P has an upper triangular matrix representation

$$P = \begin{bmatrix} p_1 & & * \\ & p_2 & \\ & 0 & \ddots \end{bmatrix}$$

with respect to the $ONB\,\{f_i\}$. Thus XP admits an upper triangular matrix representation

$$XP = \begin{bmatrix} y_1 & & * \\ & y_2 & \\ & 0 & \ddots \end{bmatrix}$$

with respect to the ONB's $\{e_i\}$ and $\{f_i\}$.

Suppose that $\sigma_n \cap \partial\overline{\Omega} \neq \varnothing$ and $a(\mu) = 1$ $(\mu \in \sigma_n^*)$. Also suppose that $\mu'_{i_j} \in \sigma_n$ $(j = 1, 2, \cdots)$. Thus $AXP - XPB = KP$ indicates that $\lambda_{2i_j} y_{2i_j} - y_{2i_j}\mu_{2i_j} = a_{2i_j} p_{2i_j}$.

Since $a(\mu^*_{2i_j}) = 1, p_{2i_j} = 1$. Thus $y_{2i_j} = 2^{i_j-1}\epsilon/i_j \to \infty\ (j \to \infty)$. This contradiction implies that $a(\mu)|_{(\sigma')^*} = 0$.

By the same argument we can prove the second part of the lemma. □

Proof of Theorem 5.23. The spectral picture $\Lambda(T)$ of T is determined by the following sets.

$$\rho^\circ_F(T) \cap \sigma(T) = \{\lambda \in \sigma(T) : \operatorname{ind}(\lambda - T) = 0\} = \bigcup_{i\in\Lambda_1} \Omega_{1i},$$

$$\rho^+_F(T) = \{\lambda \in \sigma(T) : \operatorname{ind}(\lambda - T) > 0\} = \bigcup_{i\in\Lambda_2} \Omega_{2i},$$

$$\rho^-_F(T) = \{\lambda \in \sigma(T) : \operatorname{ind}(\lambda - T) < 0\} = \bigcup_{i\in\Lambda_3} \Omega_{3i},$$

$$\sigma = \sigma(T)\backslash[\rho_F(T)^-]^\circ,$$

where Ω_{1i}, Ω_{2i}, Ω_{3i} are connected components.

Choose countably many closed maximal connected components $\{\sigma_l\}_{l\in\Lambda_4}$ of σ satisfying

(i) $\sigma_l \cap \sigma_{l'} = \varnothing\ (l \neq l')$ and $[\bigcup_{l\in\Lambda_4} \sigma_l]^- = \sigma$;

(ii) for each component Ω of $\rho_F(T)\cap\sigma(T)$, there is an $l \in \Lambda_4$ such that $\partial\overline{\Omega} \subset \sigma_l$;

(iii) each σ_l consists of more than one point and $\sigma_l \cap \partial\Omega \neq 0$ for some component Ω of $\rho_F(T)\cap\sigma(T)$ or $\sigma_l \cap (\bigcup_{l'\in\Lambda'} \sigma_{l'})^- \neq \varnothing$, where $\Lambda' = \{l \in \Lambda_4 : \sigma_l \cap \partial\Omega \neq \varnothing$ for some component Ω of $\rho_F(T)\cap\sigma(T)\}$.

Construct operators $A(\Omega_{1i}) \in \mathcal{B}_1(\Omega_{1i})$ and $A(\Omega^*_{1i}) \in \mathcal{B}_1(\Omega^*_{1i})$ with respect to Ω_{1i} and, respectively, Ω^*_{1i} $(i \in \Lambda_1)$ (Lemma 5.24) so that $A(\Omega_{1i})$ and $A(\Omega^*_{1i})$ are essentially normal and co-subnormal plus compact and let the norms of compact operators in Lemma 5.24 be so small that $\bigoplus_{i\in\Lambda_1} A(\Omega_{1i})$ and $\bigoplus_{i\in\Lambda_1} A(\Omega^*_{1i})$ are essentially normal and co-subnormal plus compact. Set $C(\Omega_{1i}) = A(\Omega^*_{1i})^*$ $(i \in \Lambda_1)$.

Construct an operator $A(\Omega_{2j}) \in \mathcal{B}_1(\Omega_{2j})$ $(j \in \Lambda_2)$ (Lemma 5.24), $\overline{A}(\Omega_{2j}) \in \mathcal{B}_{n_j-1}(\Omega_{2j})\cap(SI)$ and a compact operator $Q(\Omega_{2j})$ if $n_j > 1$ (Lemma 5.25) such that

$$Q(\Omega_{2j}) \notin \operatorname{ran} \tau_{\overline{A}(\Omega_{2j})A(\Omega_{2j})}, \quad \ker \tau_{A(\Omega_{2j})\overline{A}(\Omega_{2j})} = \{0\}$$

and let the norms of $Q(\Omega_{2j})$ and the compact operator K in Lemma 5.25 be so small that $\bigoplus Q(\Omega_{2j})$ is compact, $\bigoplus_{j\in\Lambda_2} A(\Omega_{2j})$ and $\bigoplus \overline{A}(\Omega_{2j})$ are essentially normal and co-subnormal plus compact.

Construct $A(\Omega_{3k}) \in \mathcal{B}_1(\Omega_{3k})$ and $A(\Omega^*_{3k}) \in \mathcal{B}_1(\Omega^*_{3k})$ $(k \in \Lambda_3)$ (Lemma 5.24), $\overline{A}(\Omega^*_{3k}) \in \mathcal{B}_{|n_k|}(\Omega^*_{3k})$ and the compact operator $Q(\Omega^*_{3k})$ if $|n_k| > 1$ (Lemma 5.25) such that

$$Q(\Omega^*_{3k}) \notin \operatorname{ran} \tau_{\overline{A}(\Omega^*_{3k})A(\Omega^*_{3k})}, \quad \ker \tau_{A(\Omega^*_{3k})\overline{A}(\Omega^*_{3k})} = \{0\}$$

and let the norms of $Q(\Omega^*_{3k})$ and the compact operator K in Lemma 5.25 be so small that $\bigoplus Q(\Omega^*_{3k})$ is compact, $\bigoplus_{k\in\Lambda_3} A(\Omega_{3k})$, $\bigoplus_{k\in\Lambda_3} A(\Omega^*_{3k})$ and $\bigoplus \overline{A}(\Omega^*_{3k})$ are essentially normal and co-subnormal plus compact.

Set $C(\Omega_{3k}) = A(\Omega^*_{3k})^*$ and $\overline{C}_k = \overline{A}(\Omega^*_{3k})^*$.

Construct $B = [M^\circ_\lambda(\sigma^*)]^*$ on the Hilbert space $W^{22}_0(\Delta) \ominus m_0(\sigma^*)$, where $\Delta \subset (a,b)^2 \supset \sigma^*$.

Our proof will be divided into four steps.

First step. Rearrange $\{A(\Omega_{2j}) : j \in \Lambda_2\} = \{A_k\}_{k=1}^{l_1}$ $(0 \le l_1 \le \infty)$, and $\{A(\Omega_{1j}) : j \in \Lambda_1,\ A(\Omega_{3k}) : k \in \Lambda_3\} = \{A^a_k\}_{k=1}^{l_2}$ $(0 \le l_2 \le \infty)$. Thus from the construction $\bigoplus_{k=1}^{l_1} A_k$ and $\bigoplus_{k=1}^{l_2} A^a_k$ are essentially normal and co-subnormal plus compact. Therefore,

$$\sigma\Big(\bigoplus_{k=1}^{l_1} A_k\Big) = \Big(\bigcup_{k=1}^{l_1} \sigma(A_k)\Big)^-, \quad \sigma\Big(\bigoplus_{k=1}^{l_1} A^a_k\Big) = \Big(\bigcup_{k=1}^{l_1} \sigma(A^a_k)\Big)^-.$$

Define

$$G = \begin{bmatrix} \bigoplus A_i & D_1 & 0 \\ 0 & B & D_2 \\ 0 & 0 & \bigoplus A^a_k \end{bmatrix}$$

on $\mathcal{H}_1 \oplus \mathcal{H}_2 \oplus \mathcal{H}_3$ as follows, where $\mathcal{H}_1 = \bigoplus_{i=1}^{l_1} \mathcal{H}^1_i$, $\mathcal{H}_2 = W^{22}_0(\Delta) \ominus m_0(\sigma^*)$, $\mathcal{H}_3 = \bigoplus_{k=1}^{l_2} \mathcal{H}^3_k$ and for each k, $A_k \in \mathcal{L}(\mathcal{H}^1_k)$, $A^a_k \in \mathcal{L}(\mathcal{H}^3_k)$.

Assume that $D_1 = \begin{bmatrix} D^1_1 \\ D^1_2 \\ \vdots \end{bmatrix}$ and $D_2 = [\,D^2_1 \quad D^2_2 \quad \ldots]$. Choose $D^1_k \notin \operatorname{ran} \tau_{A_k B}$ and $D^2_k \notin \operatorname{ran} \tau_{BA^a_k}$, D^1_k, D^2_k are compact and their norms are so small that D_1, D_2 are compact (Lemma 5.26).

Suppose that $P \in \mathcal{A}'(G)$ is idempotent and

$$P = \begin{bmatrix} P_{11} & P_{12} & P_{13} \\ P_{21} & P_{22} & P_{23} \\ P_{31} & P_{32} & P_{33} \end{bmatrix} \begin{matrix} \mathcal{H}_1 \\ \mathcal{H}_2 \\ \mathcal{H}_3. \end{matrix}$$

Observe that A_k and A_k^a are Cowen–Douglas operators of index 1 and Lemma 5.25 indicates that $\sigma_p(A_k^a) \cap \partial\sigma(A_k^a) = \varnothing$ and therefore $\sigma_p(B) \cap \sigma_p(\bigoplus A_k^a) = \varnothing$. Comparing both sides of $PG = GP$ and applying Proposition 1.14, we get $P_{31} = P_{21} = P_{32} = 0$. Using Proposition 1.14 again and noting that $A_k, A_k^a \in (SI)$, we get $P_{11} = \bigoplus_{i=1}^{l_1} \delta_i^1$, $P_{33} = \bigoplus_{i=1}^{l_2} \delta_i^3$, where δ_i^1 (δ_i^3) is 0 or 1. Since $P_{22} \in \mathcal{A}'(B)$ is idempotent, $P_{22}k_{\mu^*} = a(\mu^*)k_{\mu^*}$ $(\mu \in \sigma)$ and $a(\mu^*)$ takes constant values (1 or 0) on each σ_l and is continuous on σ (Lemma 5.26).

Set $\Delta_1 = \{\sigma_l : a(\mu^*)|_{\sigma_l^*} \equiv 0\}$, $\Delta_2 = \{\Delta_l : a(\mu^*)|_{\sigma_l^*} \equiv 1\}$ and denote $\Gamma_1 = \bigcup_{\Delta_1} \sigma_l$, $\Gamma_2 = \bigcup_{\Delta_2} \sigma_l$. Thus $\Gamma_1 \cup \Gamma_2 = \sigma$ and $\Gamma_1 \cap \Gamma_2 = \varnothing$. Since $a(\mu^*)$ is continuous on σ^*, $\overline{\Gamma}_1 \cap \overline{\Gamma}_2 = \varnothing$. If $\overline{\Gamma}_1$ and $\overline{\Gamma}_2$ are not empty, σ is a disconnected compact subset of $\mathbb{C}$. It follows from the connectedness of $\sigma(G) = \sigma(T)$ that there exists an A_k (or A_k^a) such that $\partial\sigma(A_k) \cap \overline{\Gamma}_1 \neq \varnothing$ and $\partial\sigma(A_k) \cap \overline{\Gamma}_2 \neq \varnothing$ (or $\partial\sigma(A_k^a) \cap \overline{\Gamma}_1 \neq \varnothing$ and $\partial\sigma(A_k^a) \cap \overline{\Gamma}_2 \neq \varnothing$). Denote $[\sigma(A_k)]^\circ = \Omega$; then Ω is one of the components of $\sigma(T) \cap \rho_F(T)$. It follows from $\overline{\Gamma}_1 \cap \overline{\Gamma}_2 = \varnothing$ and property (ii) of $\{\sigma_l\}_{l \in \Lambda_4}$ that there are two components Φ_1 and Φ_2 of $\mathbb{C} \backslash \Omega$ and $\sigma_{l_1} \subset \Delta_1$, $\sigma_{l_2} \subset \Delta_2$ such that $\partial\Phi_1 \cap \sigma_{l_1} \neq \varnothing$ and $\partial\Phi_2 \cap \sigma_{l_2} \neq \varnothing$. Let $P_{12} = \begin{bmatrix} P_1^1 \\ P_2^1 \\ \vdots \end{bmatrix}$, where $P_i^1 \in \mathcal{L}(\mathcal{H}_2, \mathcal{H}_i^1)$. Since $PG = GP$, $A_k P_k^1 + D_k^1 P_{22} = \delta_k^1 D_k^1 + P_k^1 B$. If $\delta_k^1 = 0$ we have $A_k P_k^1 - P_k^1 B = -D_k^1 P_{22}$. By Lemma 5.26, $a(\mu^*) = 0$ when $\mu \in \sigma \cap \partial\overline{\Omega} \neq \varnothing$; therefore $a(\mu^*)|_{\sigma_{l_1}} = a(\mu^*)|_{\sigma_{l_2}} = 0$, a contradiction. If $\delta_k^1 = 1$, we have $A_k P_k^1 - P_k^1 B = D_k^1(1 - P_{22})$, and therefore $a(\mu^*) = 1$ for $\mu \in \sigma \cap \partial\overline{\Omega} \neq \varnothing$. Thus $a(\mu^*)|_{\sigma_{l_1}} = a(\mu^*)|_{\sigma_{l_2}} = 1$, a contradiction. The contradictions indicate that one of $\overline{\Gamma}_1$ and $\overline{\Gamma}_2$ must be empty. Thus $P_{22} = \delta_2$, $\delta_2 = 0$ or 1.

Let $P_{23} = [\, P_1^2 \quad p_2^2 \quad \ldots]$, where $P_i^2 \in \mathcal{L}(\mathcal{H}_i^3, \mathcal{H}^2)$. Note that

$$\big(\bigoplus \delta_i^1\big) D_1 + P_{12} B = \big(\bigoplus A_i\big) P_{12} + D_1 P_{22}$$

and

$$P_{22} D_2 + P_{23}\big(\bigoplus A_k^a\big) = B P_{23} + D_2\big(\bigoplus \delta_k^3\big).$$

Thus

$$A_i P_i^1 - P_i^1 B = D_i^1(\delta_i^1 - \delta_2) \quad (i = 1, 2, \cdots, l_1)$$

and

$$B P_k^2 - P_k^2 A_k^a = (\delta_2 - \delta_k^3) D_k^2 \quad (k = 1, 2, \cdots, l_2).$$

Using Lemma 5.26 with respect to the idemponents $(\delta_i^1 - \delta_2)$ and $(\delta_2 - \delta_k^3)$, we get $\delta_i^1 = \delta_k^2 = \delta_2$ for all i and k, i.e., $P = 0$ or 1. Thus $G \in (SI)$.

Second step: recovery of the positive indices.

Rearrange and reexpress $\{\overline{A}(\Omega_{2j})\} = \{\overline{A}_k\}$. Then $\sigma(\bigoplus \overline{A}_k) = \left[\bigcup \sigma(\overline{A}_k)\right]^-$.

Define

$$M = \begin{bmatrix} \bigoplus \overline{A}_j & D_3 & & 0 \\ & \bigoplus A_i & D_1 & \\ & & B & D \\ 0 & & & \bigoplus A_k^a \end{bmatrix}$$

on $\mathcal{H}_4 \oplus \mathcal{H}_1 \oplus \mathcal{H}_2 \oplus \mathcal{H}_3$, where $\mathcal{H}_4 = \bigoplus \mathcal{H}_j^4$, $\overline{A}_j \in \mathcal{L}(\overline{\mathcal{H}}_j)$ $(j = 1, 2, \cdots)$, and

$$D_3 = \begin{bmatrix} D_{11}^3 & D_{12}^3 & \cdots \\ D_{21}^3 & D_{22}^3 & \cdots \\ \cdots & \cdots & \end{bmatrix}, \quad D_{ji}^3 = \begin{cases} Q(\Omega_{2k}), & \text{if } \sigma(\overline{A}_j) = \sigma(A_i) = \overline{\Omega}_{2k}, \\ 0, & \text{otherwise.} \end{cases}$$

Thus D_3 is compact.

Suppose that $E \in \mathcal{A}'(M)$ is idempotent and

$$E = \begin{bmatrix} E_{11} & E_{12} & E_{13} & E_{14} \\ E_{21} & E_{22} & E_{23} & E_{24} \\ E_{31} & E_{32} & E_{33} & E_{34} \\ E_{41} & E_{42} & E_{43} & E_{44} \end{bmatrix} \begin{matrix} \mathcal{H}_4 \\ \mathcal{H}_1 \\ \mathcal{H}_2 \\ \mathcal{H}_3 \end{matrix} = \begin{bmatrix} E_{11} & E'_{12} \\ E'_{21} & E'_{22} \end{bmatrix} \begin{matrix} \mathcal{H}_4 \\ \mathcal{H}_1 \oplus \mathcal{H}_2 \oplus \mathcal{H}_3 \end{matrix}.$$

Then by Proposition 1.14 $E_{41} = E_{31} = 0$. By Lemma 5.25 and Proposition 1.14 $E_{21} = 0$. Since $\left[\sigma(\overline{A}_k)\right]^\circ \cap \left[\sigma(\overline{A}_j)\right]^\circ = \varnothing$ $(i \neq j)$ and $\overline{A}_j \in (SI)$, $E_{11} = \bigoplus \delta_j^4$, $\delta_j^4 = 0$ or 1 $(j = 1, 2, \cdots)$. Since $G \in (SI)$, $E'_{22} = 0$ or 1. Assume that $E'_{22} = 0$ (if $E'_{22} = 1$, consider $1 - E$). Thus we have $E_{11}D_3 + E_{12}(\bigoplus A_k) = (\bigoplus \overline{A}_j)E_{12}$. Suppose that

$$E_{12} = \begin{bmatrix} L_{11} & L_{12} & \cdots \\ L_{21} & L_{22} & \cdots \\ \cdots & \cdots & \end{bmatrix}.$$

For each k, find i such that $\sigma(\overline{A}_k) = \sigma(A_i)$; then $\overline{A}_k L_{ki} - L_{ki} A_i = \delta_k^4 D_{ki}^3$ implies that $\delta_k^4 = 0$. Therefore $E = 0$ and $M \in (SI)$.

Third step: recovery of the zero indices.

Rearrange $\{C(\Omega_{1i})(i \in \Lambda_1),\ A(\Omega_{3k})(k \in \Lambda_3)\} = \{C_k\}_{k=1}^{l_2}$, and define $C = \bigoplus_{k=1}^{l_2} C_k$. Then $\sigma(C) = \left[\bigcup_{k=1}^{l_2} \sigma(C_k)\right]^-$.

Define

$$R = \begin{bmatrix} \bigoplus \overline{A}_j & D_3 & & & 0 \\ & \bigoplus A_i & D_1 & & \\ & & B & D_2 & \\ & & & \bigoplus A_k^a & D_4 \\ & 0 & & & C \end{bmatrix} \begin{matrix} \mathcal{H}_4 \\ \mathcal{H}_1 \\ \mathcal{H}_2 \\ \mathcal{H}_3 \\ \mathcal{H}_5 \end{matrix},$$

where $D_4 = (D_{ij}^4)$,

$$D_{ij}^4 \begin{cases} = 0, & \text{if } \sigma(A_i^a) \cap \sigma(C_j) = \varnothing, \\ \notin \operatorname{ran} \tau_{A_i^a C_j} \text{ and } D_{ij}^4 \text{ is compact}, & \text{if } \sigma(A_i^a) \cap \sigma(C_j) \neq \varnothing. \end{cases}$$

By Theorem 1.11 we can choose $\|D_{ij}^4\|$ so small that D_4 is a compact operator. Then by the same argument used in the first and second steps, we can prove that $R \in (SI)$.

Fourth step: recovery of the negative indices.

Set $\overline{C} = \bigoplus \overline{C}_k \in \mathcal{L}(\mathcal{H}_6)$; then $\sigma(\overline{C}) = \left[\bigcup \sigma(\overline{C}_k)\right]^-$.

Define

$$A = \begin{bmatrix} \bigoplus \overline{A}_j & D_3 & & & & 0 \\ & \bigoplus A_i & D_1 & & & \\ & & B & D_2 & & \\ & & & \bigoplus A_k^a & D_4 & \\ & 0 & & & C & D_5 \\ & & & & & \overline{C} \end{bmatrix} \begin{matrix} \mathcal{H}_4 \\ \mathcal{H}_1 \\ \mathcal{H}_2 \\ \mathcal{H}_3 \\ \mathcal{H}_5 \\ \mathcal{H}_6, \end{matrix}$$

where

$$D_5 = \begin{bmatrix} D_{11}^5 & D_{12}^5 & \cdots \\ D_{21}^5 & D_{22}^5 & \cdots \\ \cdots & \cdots & \end{bmatrix} \quad \text{and} \quad D_{ik}^5 = \begin{cases} Q(\Omega_{3k}^*)^*, & \text{if } \sigma(C_i) = \sigma(\overline{C}_k), \\ 0, & \text{if } \sigma(C_i) \neq \sigma(\overline{C}_k), \end{cases}$$

note that $\sigma(C_i) = \sigma(\overline{C}_k)$ when and only when $C_i = A(\Omega_{3k}^*)^*$, $\overline{C}_k = \overline{A}(\Omega_{3k}^*)^*$. Thus D_5 is compact and we can prove that $A \in (SI)$. From the whole construction of A it is easily seen that A is essentially normal and $\Lambda(A) = \Lambda(T)$. Thus by the Brown–Douglas–Fillmore theorem $A \simeq T + K$ for some compact operator K; thus the proof is now complete. □

Although there exists a normal operator N such that N cannot be expressed as a direct sum of (SI) operators, we have the following result.

Theorem 5.27. *Let $T \in \mathcal{L}(\mathcal{H})$ be essentially normal. Then there exists a compact operator K such that $T+K$ has a unique (SI) decomposition.*

Proof. Choose countably many compact connected subsets $\{\sigma_k\}_{k=1}^{\infty}$ of $\sigma(T)$ satisfying

(i) $\{\sigma_k\}$ is pairwise disjoint;

(ii) each component Ω of $\rho_F \cap \sigma(T)$ is contained in some σ_k, i.e., $\Omega \subset \sigma_k$ for some k;

(iii) $\left[\bigcup \sigma_k\right]^- = \sigma(T)$.

For each k, by Lemmas 5.24, 3.12, and Corollary 3.14 there is an essentially normal operator $A_k \in (SI)$ such that $\operatorname{ind}(\lambda - A_k) = \operatorname{ind}(\lambda - T)$, $\lambda \in \rho_F(T) \cap \sigma_k$. In fact in the proofs of Lemmas 5.24, 3.12 and Corollary 3.14 we can choose the compact perturbations to be so small that $[A_k^*, A_k] \le m(\sigma_k) + 1/2^k$. Set $G = \bigoplus_{k=1}^{\infty} A_k$; then G is essentially normal. Since $\sigma(A_k) \cap \sigma(A_j) = \varnothing$ $(k \ne j)$, Proposition 4.16 asserts that G has a unique (SI) decomposition. Note that $\Lambda(G) = \Lambda(T)$; thus by the Brown–Douglas–Fillmore theorem $G = T + K$ for some compact operator K. $\square$

Theorem 5.28. *Let $T \in \mathcal{L}(\mathcal{H})$ be essentially normal with* $\operatorname{ind}(\lambda - T) \ge 0$ *for all $\lambda \in \rho_F(T)$. Then for each $\epsilon > 0$, there exists a compact operator K, $\|K\| < \epsilon$, such that $T + K$ has a unique (SI) decomposition.*

Proof. By Proposition 1.22 and the proof of Theorem 1.23, there is a compact operator K, $\|K\| < \epsilon/4$, such that $\min\operatorname{ind}(T + K - \lambda) = 0$ for $\lambda \in \rho_F(T)\backslash\sigma_0(T)$. Suppose that $\rho_F^+(T) \cap \sigma(T) = \{\Omega_i\}_{i=1}^{l_1}$ $(0 \le l_1 \le \infty)$, $\sigma_0(T) = \{\lambda_i\}_{i=1}^{l_2}$ $(0 \le l_2 \le \infty)$. Let $\{\mu_i\}_{i=1}^{\infty}$ be a dense subset of $\sigma(T)\backslash[\rho_F(T)^-]^\circ$ and $\mu_i \ne \mu_j$ $(i \ne j)$. For each Ω_i, Lemma 3.21 gives an operator $A(\Omega_i) \in \mathcal{B}_1(\Omega_i) \cap (SI)$ such that $\sigma(A(\Omega_i)) = \overline{\Omega}_i$ and $\rho_F(A(\Omega_i)) \cap \sigma(A(\Omega_i)) = \Omega_i$.

Set

$$G_1 = \bigoplus_{i=1}^{l_1} [A(\Omega_i)]^{(n_i)}, \qquad \text{where } n_i = \operatorname{ind}(\lambda - T),\ \lambda \in \Omega_i;$$

$$G_2 = \bigoplus_{i=1}^{l_2} \lambda_i q_{m_i}, \qquad \text{where } m_i = \dim E(\lambda_i, T)\mathcal{H};$$

$$G_3 = \bigoplus_{i=1}^{\infty} \mu_i \in \mathcal{L}\left(\bigoplus_{i=1}^{\infty} \mathcal{H}_i\right), \qquad \dim \mathcal{H}_i = \infty\ (i = 1, 2, \cdots),$$

and $G = G_1 \oplus G_2 \oplus G_3$. By the same argument as Lemma 5.22, we can find a compact operator K', $\|K'\| < \epsilon/4$, such that $T + K + K' \sim G$. Thus it suffices to prove that for each $\delta > 0$, there exists a compact operator F with $\|F\| < \delta$ such that $G + F$ has a unique (SI) decomposition.

By Lemmas 3.21, 3.12 and Proposition 3.13, there exists a compact operator K_i, $\|K_i\| < \delta/4^i$ such that $[A(\Omega_i)]^{(n_i)} + K_i \in (SI)$ $(i = 1, 2, \cdots, l_1)$.

Set $B_i = \mu_i + (\delta/4^i)B^*$ $(i = 1, 2, \cdots)$, where B is the unilateral weighted shift with weight $\{1/k\}_{k=1}^{\infty}$.

Define $\overline{G}_1 = \bigoplus_{i=1}^{l_1} A_i$, $\overline{G}_2 = G_2$, $\overline{G}_3 = \bigoplus_{i=1}^{\infty} B_i$, and $F = \left(\bigoplus_{i=1}^{l_1} K_i\right) \oplus 0 \oplus \left(\bigoplus_{i=1}^{\infty} B_i^*/4^i\right)$. Then F is compact with $\|F\| < \delta$ and $G + F = \overline{G}_1 \oplus \overline{G}_2 \oplus \overline{G}_3$. By Proposition 4.16, $G + F$ has a unique (SI) decomposition. □

5.3 Biquasitriangular + small compact = strongly irreducible

By Theorem 5.17 and its proof, we have

Theorem 5.29. *Let $T \in \mathcal{L}(\mathcal{H})$ be an essentially normal quasitriangular operator with connected $\sigma(T)$ and $\sigma_W(T)$. Then for each $\epsilon > 0$, there is a compact operator K, $\|K\| < \epsilon$, such that $T + K \in (SI)$.*

It follows from Theorem 5.29 that there is a small compact perturbation of a normal or essentially normal operator with connected spectrum which is an (SI) operator. This result can be generalized to the biquasitriangular operator case.

Theorem 5.30. *Let $T \in \mathcal{L}(\mathcal{H})$ be a biquasitriangular operator with connected $\sigma(T)$. Then for each $\epsilon > 0$, there exists a compact operator K, $\|K\| < \epsilon$, such that $T + K \in (SI)(\mathcal{H})$.*

Observe that T is a (BQT) operator with connected spectrum $\sigma(T)$ if and only if T satisfies at least one of the following conditions:

(i) $\sigma(T) = \sigma_{lre}(T)$ is a connected perfect set;

(ii) $\sigma(T) = \{\lambda_0\}$ for some $\lambda_0 \in \mathbb{C}$;

(iii) $T \in (BQT)$ with connected $\sigma(T)$ and $\rho_{s-F}(T) \cap \sigma(T) \neq \emptyset$.

We will prove Theorem 5.30 when T satisfies the above three conditions.

Case 1. *$\sigma(T) = \sigma_{lre}(T)$ is a connected perfect set.*

Lemma 5.31. *Let $N \in \mathcal{L}(\mathcal{H})$ be a normal operator with connected perfect spectrum $\sigma(N) = \sigma$ and let $\{\mu_m\}_{m=1}^{\infty} \subset \sigma$. Then for each $\epsilon > 0$, there exists an $A_1 \in \mathcal{L}(\mathcal{H})$ satisfying*

(i) A_1 *is an almost normal* (SI) *operator;*

(ii) $\sigma(A_1) = \sigma_{lre}(A_1) = \sigma$, $\sigma_p(A_1^*) = \varnothing$ *and* $\sigma_p(A_1) \cap \{\mu_m\} = \varnothing$;

(iii) *there exists a subset* $\{\lambda_j\}_{j=1}^{\infty}$ *of* $\sigma_p(A_1)$ *such that* $\bigvee\{\ker(A_1 - \lambda_j)^k : j, k = 1, 2, \cdots\} = \mathcal{H}$;

(iv) $\|N - A_1\| < \epsilon$ *and* $N - A_1 \in \mathcal{K}(\mathcal{H})$.

Proof. By Proposition 3.27, there is an almost normal operator A' and a normal operator N' such that A' satisfies (i) and (ii), $\|N' - A'\| < \epsilon/2$, $K_1 := A' - N' \in \mathcal{K}(\mathcal{H})$, and $\sigma(N') = \sigma$. Since $\sigma(N) = \sigma(N') = \sigma$, by the Voiculescu theorem $N = UN'U^* + K_2$ for some unitary operator U and compact operator K_2 with $\|K_2\| < \epsilon/2$. Thus

$$N = U(A' - K_1)U^* + K_2 = UA'U^* - UK_1U^* + K_2 = A_1 + K,$$

where $A_1 = UA'U^*$ and $K = K_2 - UK_1U^*$. It is easily seen that A_1 satisfies all the requirements of the lemma. □

If $T = \bigoplus_{k=1}^{\infty} D_k \in \mathcal{L}(\mathcal{H})$, where $\mathcal{H} = \bigoplus_{k=1}^{\infty} \mathcal{H}_k$, $D_k \in \mathcal{L}(\mathcal{H}_k)$ and $\dim \mathcal{H}_k < \infty$ $(k = 1, 2, \cdots)$ then T is called a *block diagonal operator*.

Lemma 5.32. *Given a block diagonal operator* T *and a normal operator* N *with connected perfect spectrum* $\sigma(N) = \sigma_e(N) \supset \sigma(T)$, *then for each* $\epsilon > 0$, *there exists a compact operator* K, $\|K\| < \epsilon$, *such that* $(N \oplus T) + K \in (SI)(\mathcal{H})$.

Proof. Since T is block diagonal, T admits an upper triangular matrix representation

$$T = \begin{bmatrix} \lambda_1 & & * \\ & \lambda_2 & \\ 0 & & \ddots \end{bmatrix}$$

with respect to some ONB $\{e_n\}_{n=1}^{\infty}$; assume that $\{e_{kj}\}_{j=1}^{m_k}$ is the ONB of $\ker(T - \lambda_k)$ $(k = 1, 2, \cdots)$. Denote $\Gamma = \{\gamma = (k, j) \in \mathbb{N}^2 : k = 1, 2, \cdots; j = 1, 2, \cdots, m_k\}$. By Voiculescu's theorem there exists a compact operator K_0, $\|K_0\| < \epsilon/4$, such that $N + K_0 \simeq N \oplus (\bigoplus_{\Gamma} N_\gamma)$, where N_γ is normal and $N_\gamma \simeq N$. Rearrange $\{N_\gamma\}_{\gamma \in \Gamma}$ as $\{N_i\}_{i=2}^{\infty}$. By Lemma 5.31 there exists a compact operator K_{11}, $\|K_{11}\| < \epsilon/4$, such that $A_1 := N + K_{11} \in (SI)$, $\sigma_p(A_1) \cap \sigma_p(T) = \varnothing$ (note that since T is block diagonal, $\sigma_p(T)$ is an at most countable set), and A_1 admits an upper triangular matrix representation with the diagonal elements $\{\lambda_i^1\}_{i=1}^{\infty}$ with respect to some ONB of the space.

By Lemma 5.31 again, there exists a compact operator K_{12}, $\|K_{12}\| < \epsilon/4^2$, such that $A_2 := N_2 + K_{12} \in (SI)$, $\sigma_p(A_2) \cap [\sigma_p(T) \cup \{\lambda_i^1\}_{i=1}^{\infty}] = \varnothing$ and A_2 admits an upper matrix representation with diagonal elements $\{\lambda_i^2\}_{i=1}^{\infty}$.

In general, there exists a compact operator K_{1n}, $\|K_{1n}\| < \epsilon/4^n$, such that $A_n := N_n + K_{1n} \in (SI)$, $\sigma_p(A_n) \cap [\sigma_p(T) \cup \{\lambda_i^j : j = 1, 2, \cdots, n-1;\ i = 1, 2, \cdots\}] = \varnothing$ and A_n has an upper triangular matrix representation with diagonal elements $\{\lambda_i^n\}_{i=1}^{\infty}$.

Since $\sigma_r(A_1) \cap \sigma_l(A_{n+1}) \neq \varnothing$, by Theorem 1.11 there exists a compact operator C_n with $\|C_n\| < \epsilon/4^n$ such that $C_n \notin \operatorname{ran} \tau_{A_1 A_{n+1}}$ $(n = 1, 2, \cdots)$.

Set

$$A = \begin{bmatrix} A_1 & C_1 & C_2 & \cdots \\ & A_2 & & \\ & & A_3 & \\ & 0 & & \ddots \end{bmatrix}.$$

Then by Proposition 1.14 and Corollary 3.14 $A \in (SI)$.

For each $\gamma = (k, j) \in \Gamma$, $\lambda_k \in \sigma(T) \subset \sigma_e(A_\gamma)$. Thus there is a vector $y_\gamma \notin \operatorname{ran}(A_\gamma - \lambda_k)$.

Define $E_\gamma = [(\epsilon y_\gamma / (2^{|\gamma|+2} \|y_\gamma\|)] \otimes e_{kj}$, where $|\gamma| = k + j$. E_γ is a rank-1 operator.

Define $G = \begin{bmatrix} A & E \\ 0 & T \end{bmatrix}$, where $E = \begin{bmatrix} 0 \\ E_2 \\ E_3 \\ \vdots \end{bmatrix}$ is compact and $\|E\| < \epsilon/8$. If F is an idempotent commuting with G, assume that $F = \begin{bmatrix} F_{11} & F_{12} \\ F_{21} & F_{22} \end{bmatrix}$ and $F_{21} = [L_1\ L_2\ \cdots]$. Since $F_{21}A = TF_{21}$, $L_1 A_1 = TL_1$ or $L_1 \in \ker \tau_{TA_1}$. It follows from $\sigma_p(A_1) \cap \sigma_p(T) = \varnothing$ and Proposition 1.14 that $L_1 = 0$. For the same reason we can conclude that each $L_n = 0$ and therefore $F_{21} = 0$.

Since $F_{11}A = AF_{11}$ and F_{11} is an idempotent, and since $A \in (SI)$, $F_{11} = 0$ or 1. Without loss of generality, we can assume that $F_{11} = 0$.

Set $T_1 = T|_{\operatorname{ran} F_{22}}$ and $T_1 = T|_{\operatorname{ran}(1-F_{22})}$. Thus $T = T_1 \dot{+} T_2$. It is easy to see that $\sigma_p(T) = \sigma_p(T_1) \cup \sigma_p(T_2)$ and $\ker(T - \lambda_k)^n = \ker(T_1 - \lambda_k)^n \dot{+} \ker(T_2 - \lambda_k)^n$ $(n, k \geq 1)$.

If there is some $\lambda_k \in \sigma_p(T_1)\ (\subset \sigma_p(T))$, choose $f \in \ker(T_1 - \lambda_k) \subset \ker(T - \lambda_k)$. Assume that $f = \sum_{j=1}^{m_k} \alpha_j e_{kj}$. Noting that $F_{12}T = AF_{12} + EF_{22}$ and $f \in \operatorname{ran} F_{22}$, we have $F_{12}Tf - AF_{12}f = Ef$ or $(A - \lambda_k)F_{12}f = -Ef$. Assume that $F_{12} = \begin{bmatrix} F_1 \\ F_2 \\ \vdots \end{bmatrix}$.

Then

$$\begin{bmatrix} A_1-\lambda_k & C_1 & C_2 & \cdots & C_{n-1} & \cdots \\ & A_2-\lambda_k & & & & \\ & & A_3-\lambda_k & & & \\ & & & \ddots & & \\ & 0 & & & A_n-\lambda_k & \\ & & & & & \ddots \end{bmatrix} \begin{bmatrix} F_1 f \\ F_2 f \\ F_3 f \\ \vdots \\ F_n f \\ \vdots \end{bmatrix} = - \begin{bmatrix} 0 \\ E_2 f \\ E_3 f \\ \vdots \\ E_n f \\ \vdots \end{bmatrix}.$$

If $\alpha_1 \neq 0$, consider the index n which corresponds to $(k,1)\in\Gamma$. Since $(A_n-\lambda_k)F_n f = -E_n f = -(\epsilon y_{(k1)}\alpha_1)/(2^{k+1+2}\|y_{(k1)}\|)$ and since $y_{(k1)} \notin \operatorname{ran}(A_n-\lambda_k)$, $\alpha_1 = 0$. By the same argument, $\alpha_j = 0$ $(j = 2,3,\cdots,m_k)$, i.e., $f=0$ and $\sigma_p(T_1)=\varnothing$.

Since each $\lambda_k \notin \sigma_p(T_1)$, $(T_1-\lambda_k)$ is injective and $\ker(T_1-\lambda_k)^n = 0$ $(k,n\geq 1)$. Thus $\bigvee\{\ker(T-\lambda_k)^n : n,k\geq 1\} = \mathcal{H}$ implies that $\bigvee\{\ker(T_2-\lambda_k)^n : n,k\geq 1\} = \mathcal{H}$, i.e., $\operatorname{ran}(1-F_{22}) = \mathcal{H}$ or $F_{22}=0$. Thus $F=0$ and $G = \begin{bmatrix} N & 0 \\ 0 & T \end{bmatrix} + K \in (SI)$, where $K = \begin{bmatrix} K_0 & 0 \\ 0 & 0 \end{bmatrix} + U \begin{bmatrix} \bigoplus_{n=1}^{\infty} K_{1n} & E \\ 0 & 0 \end{bmatrix} U^*$ is compact, $\|K\|<\epsilon$, and U is a unitary operator. □

Proposition 5.33. *Given $T \in (BQT)$ with connected perfect spectrum $\sigma(T) = \sigma_{lre}(T)$, for each $\epsilon > 0$, there exists a compact operator K, $\|K\| < \epsilon$, such that $T+K \in (SI)$.*

Proof. By a result of D. Voiculescu [2], there exists a compact operator K_1 with $\|K_1\| < \epsilon/3$ such that

$$T+K_1 \simeq \begin{bmatrix} N_1\oplus D_1 & & & * \\ & N_2\oplus D_2 & & \\ & & N_3\oplus D_3 & \\ & 0 & & N_4\oplus D_4 \end{bmatrix},$$

where N_i is normal, D_i is block diagonal and $\sigma(D_i) \subset \sigma(N_i) = \sigma_e(N_i) = \sigma(T)$ $(i=1,2,3,4)$. For each i choose a compact operator K_{2i}, i.e., slightly change the diagonal elements of D_i, with $\|K_{2i}\| < \epsilon/6$ such that

$$\sigma_p(D_i+K_{2i}) \cap \sigma_p(D_j+K_{2j}) = \varnothing \qquad (i\neq j).$$

Denote $T_i = D_i + K_{2i}$ $(i=1,2,3,4)$. Each T_i is still block diagonal. Using Lemma 5.32 find a compact operator K_{3i} $(i = 1,2,3,4)$, $\|K_{3i}\| < \epsilon/6$, such that

$$G_i := N_i \oplus T_i + K_{3i} = \begin{bmatrix} A_1^i & C_1^i & C_2^i & \dots & 0 \\ & A_2^i & & & E_2^i \\ & & A_3^i & & E_3^i \\ & & & \ddots & \vdots \\ & & & & T_i \end{bmatrix} = \begin{bmatrix} A^i & E^i \\ 0 & T_i \end{bmatrix} \in (SI),$$

where $A^i = \begin{bmatrix} A_1^i & C_1^i & \dots \\ & A_2^i & \\ 0 & & \ddots \end{bmatrix}$, $E^i = \begin{bmatrix} 0 \\ E_2^i \\ \vdots \end{bmatrix}$, and $\{A_j^i : i = 1,2,3,4;\ j = 1,2,3,\cdots\}$ satisfy

(i) $\sigma_p(A_j^i) \cap \sigma_p(T_k) = \varnothing$ $(i,k = 1,2,3,4;\ j = 1,2,\cdots)$;

(ii) $\sigma_p(A_j^i)$ does not intersect the diagonal elements of A_k^i $(1 \le k < j;\ i = 1,2,3,4)$;

(iii) $\sigma_p(A_j^i)$ does not intersect the diagonal elements of A_n^k $(k < i \le 4;\ n = 1,2,\cdots)$.

Thus $T + K_1 + K_2 \simeq \begin{bmatrix} G_1 & T_{12} & & * \\ & G_2 & T_{23} & \\ & & G_3 & T_{34} \\ 0 & & & G_4 \end{bmatrix}$, where

$$K_2 \simeq \begin{bmatrix} K_{21}+K_{31} & & & 0 \\ & K_{22}+K_{32} & & \\ & & K_{23}+K_{33} & \\ 0 & & & K_{24}+K_{34} \end{bmatrix}$$

is compact and $\|K_2\| < \epsilon/3$.

If $X \in \ker \tau_{G_i G_j}$ $(i > j)$, assume that

$$X = \begin{bmatrix} X_{11} & X_{12} & \dots & X_1 \\ X_{21} & X_{22} & \dots & X_2 \\ & \dots & & \dots \\ Y_1 & Y_2 & \dots & Z \end{bmatrix} = \begin{bmatrix} X' & X'' \\ Y & Z \end{bmatrix}$$

with respect to the decompositions of the space, where

$$X' = \begin{bmatrix} X_{11} & X_{12} & \dots \\ X_{21} & X_{22} & \dots \\ \dots & \dots & \end{bmatrix}, \quad X'' = \begin{bmatrix} X_1 \\ X_2 \\ \vdots \end{bmatrix}, \quad Y = [Y_1 \quad Y_2 \quad \dots].$$

Since $T_iY = YA^j$, $T_i[Y_1 \quad Y_2 \quad \ldots] = [Y_1A_1^j \quad Y_1C_1^j + Y_2A_2^j \quad \ldots]$. Thus $T_iY_1 = Y_1A_1^j$. By Proposition 1.14 $Y_1 = 0$. By the same argument $Y_n = 0$ $(n = 2, 3, \cdots)$, i.e., $Y = 0$.

Since $A^iX' = X'A^j$, $X_{21} \in \ker \tau_{A_2^i A_1^j}$. By Proposition 1.14 $X_{21} = 0$. Similarly, $X_{nm} = 0$ $(n \geq 2;\ m \geq 1)$ and $X' = \begin{bmatrix} X_{11} & X_{12} & X_{13} & \cdots \\ & 0 & & \end{bmatrix}$.

Since $A_1^iX_{11} = X_{11}A_1^j$, Proposition 1.14 asserts that $X_{11} = 0$. Similarly, $X_{1n} = 0$ $(n = 2, 3, \cdots)$ and $X' = 0$.

Since $T_iZ = ZT_j$ and since $\sigma_p(T_i) \cap \sigma_p(T_j) = \varnothing$, $Z = 0$. By the same argument $X'' = 0$. Therefore $\ker \tau_{G_iG_j} = \{0\}$ $(i > j)$.

Since $\sigma_r(G_i) \cap \sigma_l(G_{i+1}) \neq \varnothing$, by Theorem 1.11 there exists a compact operator $G_{i\,i+1}$, $\|G_{i\,i+1}\| < \epsilon/3$, such that $T_{i\,i+1} + G_{i\,i+1} \notin \operatorname{ran} \tau_{G_iG_{i+1}}$ $(i = 1, 2, 3)$. Thus

$$A := T + K_1 + K_2 + K_3 \simeq \begin{bmatrix} G_1 & T_{12} + G_{12} & & * \\ & G_2 & T_{23} + G_{23} & \\ & & G_3 & T_{34} + G_{34} \\ & 0 & & G_4 \end{bmatrix},$$

where $K_3 \simeq \begin{bmatrix} 0 & G_{12} & & 0 \\ & 0 & G_{23} & \\ & & 0 & G_{34} \\ & 0 & & 0 \end{bmatrix}$ is compact and $\|K_3\| < \epsilon/3$. By Proposition 3.16 $A \in (SI)$. Thus $K = K_1 + K_2 + K_3$ satisfies the requirements of the proposition.

□

Case 2. *$\sigma(T)$ is a singleton.*

Lemma 5.34. *Given block diagonal operators D_1 and D_2 with $\sigma(D_1) = \sigma(D_2) = \{0\}$, for each $\epsilon > 0$ there exist compact operators K_1 and K_2, $\|K_1\| < \epsilon$ and $\|K_2\| < \epsilon$, such that $B_i = D_i + K_i \in (SI)$ $(i = 1, 2)$ and $\ker \tau_{B_2B_1} = \{0\}$.*

Proof. Without loss of generality, we may assume that $\|D_i\| < 1$ $(i = 1, 2)$. Suppose that

$$D_1 = \begin{bmatrix} M_1 & & 0 \\ & M_2 & \\ 0 & & \ddots \end{bmatrix} \begin{matrix} \mathcal{H}_1 \\ \mathcal{H}_2 \\ \vdots \end{matrix} \quad \text{and} \quad D_2 = \begin{bmatrix} L_1 & & 0 \\ & L_2 & \\ 0 & & \ddots \end{bmatrix} \begin{matrix} \mathcal{K}_1 \\ \mathcal{K}_2 \\ \vdots \end{matrix}$$

and $\dim \mathcal{H}_n = h_n$, $\dim \mathcal{K}_n = k_n$, $h_n, k_n < \infty$ $(n \geq 1)$. Choose an ONB $\{e_i^n\}_{i=1}^{h_n}$ of

$\mathcal{H}_n$ so that M_n admits an upper triangular matrix representation

$$M_n = \begin{bmatrix} 0 & & * \\ & \ddots & \\ 0 & & 0 \end{bmatrix}$$

with respect to $\{e_i^n\}_{i=1}^{h_n}$ $(n \geq 1)$. Choose a compact operator E_1 with $\|E_1\| < \epsilon/2$ such that

$$D_1 + E_1 = \begin{bmatrix} N_1 & & 0 \\ & N_2 & \\ 0 & & \ddots \end{bmatrix}, \qquad N_n = \begin{bmatrix} 0 & & * \\ & \ddots & \\ 0 & & 0 \end{bmatrix} \begin{matrix} e_1^n \\ \vdots \\ e_{h_n}^n \end{matrix}$$

and $(N_n e_j^n, e_{j-1}^n) \neq 0$ $(j = 2, \cdots, h_n;\ n \geq 1)$.

Set $C_n = (\epsilon/4^n) e_{h_n}^n \otimes e_1^{n+1}$ $(n \geq 1)$ and define

$$B_1 = D_1 + E_1 + E_2 = \begin{bmatrix} N_1 & C_1 & & 0 \\ & N_2 & C_2 & \\ & & N_3 & \ddots \\ & 0 & & \ddots \end{bmatrix} \begin{matrix} \mathcal{H}_1 \\ \mathcal{H}_2 \\ \mathcal{H}_3 \\ \vdots \end{matrix},$$

where $E_2 = \begin{bmatrix} 0 & C_1 & & 0 \\ & 0 & C_2 & \\ & & 0 & \ddots \\ 0 & & & \ddots \end{bmatrix}$ is compact and $\|E_2\| < \epsilon/4$.

Suppose that $B_1 x = 0$ for some vector $x = \sum_{n=1}^{\infty} x_n$, $x_n \in \mathcal{H}_n$ $(n \geq 1)$. Since $N_1 x_1 + C_1 x_2 = 0$ or $N_1 x_1 + \epsilon/4(e_1^2, x_2) e_{h_1}^1 = 0$, $(N_1 x_1, e_{h_1}^1) + \epsilon/4(e_1^2, x_2) = 0$, i.e., $(e_1^2, x_2) = 0$. Thus $N_1 x_1 = 0$ and $x_1 = \alpha e_1^1$ for some $\alpha \in \mathbb{C}$.

Since $N_2 x_2 + C_2 x_3 = 0$ or $N_2 x_2 + \epsilon/16(e_1^3, x_3) e_{h_2}^2 = 0$, $(N_2 x_2, e_{h_2}^2) + \epsilon/16(e_1^3, x_3) = 0$, i.e., $(e_1^3, x_3) = 0$. It follows from $N_2 x_2 = 0$ and $(e_1^2, x_2) = 0$ that $x_2 = 0$. Similarly, $x_n = 0$ $(n = 3, 4, \cdots)$, i.e., $x = \alpha e_1^1$. Thus $\ker B_1 = \bigvee\{e_1^1\}$.

Assume that $P \in \mathcal{A}'(B_1)$ is idempotent and

$$P = \begin{bmatrix} P_{11} & P_{12} & \cdots \\ P_{21} & P_{22} & \cdots \\ \cdots & \cdots & \end{bmatrix} \begin{matrix} \mathcal{H}_1 \\ \mathcal{H}_2 \\ \vdots \end{matrix} = \begin{bmatrix} p_{11} & p_{12} & \cdots \\ p_{21} & p_{22} & \cdots \\ \cdots & \cdots & \end{bmatrix} \begin{matrix} e_1 \\ e_2 \\ \vdots \end{matrix},$$

where $\{e_n\}_{n=1}^{\infty}$ is the rearrangement of $\{e_j^n : j = 1, 2, \cdots, h_n;\ n \geq 1\}$ preserving the order of the spaces $\{\mathcal{H}_n\}_{n=1}^{\infty}$ and the order of elements in each ONB $\{e_j^n\}_{j=1}^{h_n}$.

Since $B_1Pe_1 = PB_1e_1 = 0$, $Pe_1 = p_{11}e_1$ and $p_{n1} = 0\ (n > 1)$. Since $B_1^2Pe_2 = PB_1^2e_2 = 0$, $Pe_2 = p_{12}e_1 + p_{22}e_2$ and $p_{n2} = 0\ (n > 2)$. In general, $Pe_k = \sum_{i=1}^k p_{ik}e_i$ and $p_{nk} = 0\ (n > k)$. Thus P admits an upper triangular matrix representation with respect to the $ONB\ \{e_n\}_{n=1}^\infty$. Since $(N_ne_j^n, e_{j-1}^n) \neq 0\ (j = 2, 3, \cdots, h_n)$, $N_n \sim q_{h_n}$, the Jordan nilpotent of dimension h_n. Thus $N_n \in (SI)$. Since $P_{ij} = 0\ (i > j)$, P_{nn} commutes with the (SI) operator N_n. Thus $P_{nn} = 0$ or 1_n, where, 1_n is the identity on $\mathcal{H}_n\ (n \geq 1)$. Assume that $P_{11} = 0$ (if $P_{11} = 1_1$, consider $1 - P$). From $P_{12}N_2 = N_1P_{12} + C_1P_{22}$ we have $N_1P_{12} - P_{12}N_2 = -C_1$ if $P_{22} = 1_2$. Checking the $(h_1, 1)$ entries of both sides, we get the contradiction $0 = -\epsilon/4$. Thus $P_{22} = 0$. Similarly, $P_{nn} = 0\ (n \geq 3)$, i.e., $P = 0$ and $B_1 \in (SI)$. Denote $K_1 = E_1 + E_2$; thus $B_1 = D_1 + K_1$, K_1 is compact and $\|K_1\| < \epsilon$. Denote $d_k^1 = (B_1e_{k+1}, e_k)$ $(k = 1, 2, \cdots)$.

Similarly, we can find a compact operator F_1, $\|F_1\| < \epsilon$, such that

$$D_2 + F_1 = \begin{bmatrix} G_1 & & 0 \\ & G_2 & \\ 0 & & \ddots \end{bmatrix} \begin{matrix} \mathcal{K}_1 \\ \mathcal{K}_2 \\ \vdots \end{matrix},$$

where G_n admits an upper triangular matrix representation

$$G_n = \begin{bmatrix} 0 & & * \\ & \ddots & \\ 0 & & 0 \end{bmatrix}$$

with respect to some $ONB\ \{f_i^n\}_{i=1}^{k_n}$ of $\mathcal{K}_n$ and $(G_nf_j^n, f_{j-1}^n) \neq 0\ (j = 2, 3, \cdots, k_n;\ n \geq 1)$. Rearrange $\{f_j^n : j = 1, 2, \cdots, k_n;\ n \geq 1\}$ as $\{f_i\}_{i=1}^\infty$ and preserve the order of the spaces $\{\mathcal{K}_n\}_{n=1}^\infty$ and the order of the elements in each $ONB\ \{f_j^n\}_{j=1}^{k_n}$. Denote $n_i = \sum_{j=1}^i k_j\ (i \geq 1)$ and denote $d_k^2 = \big((D_2 + F_2)f_{k+1}, f_k\big)\ (k \neq n_i;\ i \geq 1)$.

Define

$$d_{n_1}^2 = \frac{\epsilon(\prod_{i=1}^{n_1} d_i^1)(\prod_{i=n_1+1}^{m_1} d_i^1)}{4(\prod_{i=1}^{n_1-1} d_i^2)m_1!},$$

where m_1 is so large that $(\prod_{i=1}^{n_1-1} d_i^2)m_1! > 1$;

$$d_{n_2}^2 = \frac{\epsilon(\prod_{i=1}^{n_2} d_i^1)(\prod_{i=n_2+1}^{m_2} d_i^1)}{4(\prod_{i=1}^{n_2-1} d_i^2)m_2!},$$

where m_2 is so large that $(\prod_{i=1}^{n_2-1} d_i^2)m_2! > 2$ and $m_2 > m_1$; inductively,

$$d_{n_k}^2 = \frac{\epsilon(\prod_{i=1}^{n_k} d_i^1)(\prod_{i=n_k+1}^{m_k} d_i^1)}{4(\prod_{i=1}^{n_k-1} d_i^2)m_k!},$$

where m_k is so large that $(\prod_{i=1}^{n_k-1} d_i^2)m_k! > k$ and $m_k > m_{k-1}$ $(k = 2, 3, \cdots)$. Since $|d_i^1| < 1$ $(i \geq 1)$, $d_{n_k}^2 \to 0$ $(k \to \infty)$.

Define $F_2 = \sum_{k=1}^{\infty} d_{n_k}^2 f_{n_k} \otimes f_{n_k+1}$ and

$$B_2 = D_2 + K_2,$$

where $K_2 = F_1 + F_2$ is compact and $\|K_2\| < \epsilon$. Then by the same argument used to prove $B_1 \in (SI)$, we can show that $B_2 \in (SI)$.

If B_2' is the operator whose matrix representation with respect to the ONB $\{e_n\}$ is the same as that of B_2 with respect to the ONB $\{f_n\}$, then $B_2' \simeq B_2$. It is easily seen that $X \in \ker \tau_{B_2' B_1}$ if and only if $U^* X \in \ker \tau_{B_2 B_1}$, where U is the unitary operator satisfying $U^* B_2' U = B_2$. Therefore without loss of generality we may assume that $f_n = e_n$ $(n \geq 1)$.

If $B_2 X = X B_1$ for some operator X, then $B_2^n X e_n = X B_1^n e_n = 0$ for all $n \geq 1$. Since $\ker B_1 = \bigvee\{e_1\}$, $X e_n = \sum_{i=1}^{n} x_{in} e_i$, $x_{in} \in \mathbb{C}$, i.e., X admits an upper triangular matrix representation with respect to $\{e_n\}$.

Computations show that

$$x_{nn} = \frac{\prod_{i=1}^{n-1} d_i^1}{\prod_{i=1}^{n-1} d_i^2} x_{11} \qquad (n \geq 1).$$

If $x_{11} \neq 0$,

$$\begin{aligned} |x_{n_k+1\, n_k+1}| &= \left| \frac{4\left(\prod_{i=1}^{n_k} d_i^1\right)\left(\prod_{i=1}^{n_k-1} d_i^2\right) m_k!\, x_{11}}{\epsilon\left(\prod_{i=1}^{n_k-1} d_i^2\right)\left(\prod_{i=1}^{n_k} d_i^1\right)\left(\prod_{i=n_k+1}^{m_k} d_i^1\right)} \right| \\ &= \frac{4}{\epsilon} \frac{m_k!}{\left|\prod_{i=n_k+1}^{m_k} d_i^1\right|} |x_{11}| \geq \frac{4}{\epsilon} m_k!\, |x_{11}| \to \infty \quad (k \to \infty). \end{aligned}$$

Thus $x_{11} = x_{nn} = 0$ $(n \geq 2)$.

Similarly, calculation indicates that

$$x_{n\, n+1} = \frac{(\prod_{i=2}^{n-1} d_i^1) d_n^1}{(\prod_{i=2}^{n-1} d_i^2) d_1^2} x_{12} \qquad (n \geq 1).$$

If $x_{12} \neq 0$,

$$\begin{aligned} |x_{n_k+1\, n_k+2}| &= \left| \frac{4\left(\prod_{i=2}^{n_k} d_i^1\right)\left(\prod_{i=1}^{n_k-1} d_i^2\right) d_{n_k+1}^1 m_k!}{\epsilon\left(\prod_{i=2}^{n_k-1} d_i^2\right)\left(\prod_{i=1}^{n_k} d_i^1\right)\left(\prod_{i=n_k+1}^{m_k} d_i^1\right) d_1^2} x_{12} \right| \\ &= \left| \frac{4 m_k!}{\epsilon d_1^1 \prod_{i=n_k+2}^{m_k} d_i^1} x_{12} \right| \geq \frac{4 m_k!}{\epsilon} |x_{12}| \to \infty \quad (k \to \infty). \end{aligned}$$

Therefore $x_{12} = x_{n\, n+1} = 0$ $(n \geq 2)$. In general, by the same argument we can prove that $x_{n\, n+k} = 0$ $(k \geq 0;\ n \geq 1)$, i.e., $X = 0$ and $\ker \tau_{B_2 B_1} = \{0\}$. $\square$

Corollary 5.35. *Given a class* $\{D_i\}_{i=1}^n$ $(n < \infty)$ *of finitely many block diagonal operators with* $\sigma(D_i) = \{0\}$ $(i = 1, 2, \cdots, n)$, *for each* $\epsilon > 0$ *there exist compact operators* $\{K_i\}_{i=1}^n$, $\|K_i\| < \epsilon\,(1 \leq i \leq n)$, *such that* $B_i = D_i + K_i \in (SI)$ $(1 \leq i \leq n)$ *and* $\ker \tau_{B_i B_j} = \{0\}$ $(i > j)$.

Proposition 5.36. *Given an operator* $T \in \mathcal{L}(\mathcal{H})$ *with* $\sigma(T) = \{\lambda\}$, *for each* $\epsilon > 0$ *there exists a compact operator* K, $\|K\| < \epsilon$, *such that* $T + K \in (SI)$.

Proof. Without loss of generality, we may assume that $\sigma(T) = \{0\}$. By Voiculescu's theorem, there exists a compact operator K_1 with $\|K_1\| < \epsilon/2$ such that

$$T + K_1 = \begin{bmatrix} D_1 & T_{12} & & * \\ & D_2 & T_{23} & \\ & & D_3 & T_{34} \\ & 0 & & D_4 \end{bmatrix},$$

where D_i is block diagonal and $\sigma(D_i) = \{0\}$ $(i = 1, 2, 3, 4)$.

By Corollary 5.35, there are compact operators K_{2i} $(1 \leq i \leq 4)$, such that $\|K_{2i}\| < \epsilon/16$, $B_i := D_i + K_{2i} \in (SI)$ and $\ker \tau_{B_i B_j} = \{0\}$ $(i > j)$. Since $\sigma_r(B_i) \cap \sigma_l(B_{i+1}) \neq \varnothing$, we can find a compact operator C_i, $\|C_i\| < \epsilon/16$, such that $C_i + T_{i\,i+1} \notin \operatorname{ran} \tau_{B_i B_{i+1}}$ $(1 \leq i \leq 3)$.

Set

$$K_2 = \begin{bmatrix} K_{21} & C_1 & & 0 \\ & K_{22} & C_2 & \\ & & K_{23} & C_3 \\ & 0 & & K_{24} \end{bmatrix}.$$

Then K_2 is compact and $\|K_2\| < \epsilon/2$. By Proposition 3.16 $T + K_1 + K_2 \in (SI)$. Denote $K = K_1 + K_2$; then the proof of the proposition is now complete. □

Case 3. $T \in (BQT)$, *$\sigma(T)$ is connected and* $\rho_{s-F}(T) \cap \sigma(T) \neq \varnothing$.

Proposition 5.37. *Given* $T \in (BQT)$ *with connected spectrum* $\sigma(T)$ *and* $\rho_{s-F}(T) \cap \sigma(T) \neq \varnothing$, *for each* $\epsilon > 0$, *there exists a compact operator* K, $\|K\| < \epsilon$, *such that* $T + K \in (SI)$.

Proof. If $\min\operatorname{ind}(\lambda - T) > 1$ for all $\lambda \in \rho_{s-F}(T) \cap \sigma(T)$, by Theorem 1.24 there exists a compact operator K_1 with $\|K_1\| < \epsilon/8$ such that $\sigma(T + K_1) = \sigma(T)$ and $\min\operatorname{ind}(T + K_1 - \lambda) \geq 1$ for $\lambda \in \rho_{s-F}(T + K_1) \cap \sigma(T + K_1)$ and $\min\operatorname{ind}(T + K_1 - \lambda) = 1$ for $\lambda \in \Omega_1$, where Ω_1 is a component of $\rho_{s-F}(T + K_1) \cap \sigma(T + K_1)$. If min ind$(\lambda - T) = 1$ for λ in some component of $\rho_{s-F}(T) \cap \sigma(T)$, then let $K_1 = 0$ and let Ω_1 be this component.

By Voiculescu's theorem, there is a compact operator K_2, $\|K_2\| < \epsilon/8$, such that

$$T + K_1 + K_2 \simeq A \oplus (T + K_1) \oplus A,$$

where $\sigma(A) = \sigma_e(A) = \sigma_e(T) = \sigma_{lre}(T)$. By Apostol's triangular representation

$$T + K_1 = \begin{bmatrix} T_r & B & C \\ 0 & T_0 & D \\ 0 & 0 & T_l \end{bmatrix} \begin{matrix} \mathcal{H}_r(T + K_1) \\ \mathcal{H}_0(T + K_1) \\ \mathcal{H}_l(T + K_1), \end{matrix}$$

where T_r is a triangular operator, $\rho_{s-F}(T_r) \supset \rho_{s-F}(T+K_1)$, min ind$(T_r - \lambda) = 0$ and ind$(T_r - \lambda) \geq 1$ for $\lambda \in \rho_{s-F}(T_r) \cap \sigma(T_r)$. In particular, ind$(T_r - \lambda) = 1$ $(\lambda \in \Omega_1)$; T_l^* is a triangular operator, $\rho_{s-F}(T_l) \supset \rho_{s-F}(T + K_1)$, min ind$(T_l - \lambda) = 0$ and $-$ind$(T_l - \lambda) \geq 1$ for $\lambda \in \rho_{s-F}(T_l) \cap \sigma(T_l)$. In particular, ind$(T_l - \lambda) = -1$ $(\lambda \in \Omega_1)$; $\sigma(T_0) \subset \sigma_e(T + K_1)$. Therefore,

$$T + K_1 + K_2 \simeq \begin{bmatrix} A & 0 & 0 & 0 & 0 \\ 0 & T_r & B & C & 0 \\ 0 & 0 & T_0 & D & 0 \\ 0 & 0 & 0 & T_l & 0 \\ 0 & 0 & 0 & 0 & A \end{bmatrix} = \begin{bmatrix} T_1 & Q \\ 0 & T_2 \end{bmatrix} \begin{matrix} \mathcal{H}_1 \\ \mathcal{H}_2, \end{matrix}$$

where

$$T_1 = \begin{bmatrix} A & 0 & 0 \\ 0 & T_r & B \\ 0 & 0 & T_0 \end{bmatrix} \quad \text{and} \quad T_2 = \begin{bmatrix} T_l & 0 \\ 0 & A \end{bmatrix}.$$

Computation shows that $T_1 \in (QT)$, $\sigma(T_1) = \sigma(T + K_1) = \sigma_W(T_1)$ is connected, $\rho_{s\text{-}F}(T_1) = \rho_{s\text{-}F}(T+K_1)$, ind$(T_1 - \lambda) \geq 1$ for $\lambda \in \rho_{s\text{-}F}(T_1) \cap \sigma(T_1)$ and ind$(T_1 - \lambda) = 1$ for $\lambda \in \Omega_1$. Similarly, $T_2^* \in (QT)$, $\sigma(T_2) = \sigma(T + K_1) = \sigma_W(T_2)$ is connected, $\rho_{s\text{-}F}(T_2) = \rho_{s\text{-}F}(T+K_1)$, $-$ind$(T_2 - \lambda) \geq 1$ for $\lambda \in \rho_{s\text{-}F}(T_2) \cap \sigma(T_2)$ and ind$(T_2 - \lambda) = -1$ for $\lambda \in \Omega_1$. Thus by Theorem 1.44 we can find compact operators K_{31} and K_{32} with $\|K_{31}\| < \epsilon/8$ and $\|K_{32}\| < \epsilon/8$ such that $T_1 + K_{31} = M_1 \in \mathcal{B}_1(\Omega_1)$ and $(T_2 + K_{32})^* = M_2^* \in \mathcal{B}_1(\Omega_1^*)$. By Proposition 1.14 $\ker \tau_{M_2 M_1} = \{0\}$. Since $\sigma_r(M_1) \cap \sigma_l(M_2) \neq \varnothing$, we can find a compact operator K_{33}, $\|K_{33}\| < \epsilon/8$, such that $Q + K_{33} \notin \operatorname{ran} \tau_{M_1 M_2}$.

Set $K_3 \simeq \begin{bmatrix} K_{31} & K_{33} \\ 0 & K_{32} \end{bmatrix}$. Then by Proposition 3.16

$$T + K \simeq \begin{bmatrix} M_1 & Q + K_{33} \\ 0 & M_2 \end{bmatrix} \in (SI),$$

where $K = K_1 + K_2 + K_3$ is compact and $\|K\| < \epsilon$. □

Summarizing Propositions 5.33, 5.36 and 5.37, we get Theorem 5.30.

Corollary 5.38. *Let $T \in \mathcal{L}(\mathcal{H})$ be a biquasitriangular operator and $\sigma(T)$ consist of finitely many components. Then for each $\epsilon > 0$, there exists a compact operator K with $\|K\| < \epsilon$ such that $T + K$ has unique (SI) decomposition.*

Proof. Assume that $\sigma(T) = \bigcup_{k=1}^{n} \sigma_k$ $(n < \infty)$. Then by the Riesz theorem $T = \sum_{k=1}^{n} \dot{+} T_k$. By Theorem 5.30 and Proposition 4.16 we can prove the corollary. □

5.4 Hyponormal + compact = strongly irreducible

In this section, we will prove the following theorem.

Theorem 5.39. *Let $T \in \mathcal{L}(\mathcal{H})$ be hyponormal with connected spectrum $\sigma(T)$ and* $\operatorname{ind}(T - \lambda) \leq 0$ *for all $\lambda \in \rho_{s-F}(T)$. Then there exists a compact operator K such that $T + K \in (SI)(\mathcal{H})$.*

Proof. Denote $A = T^*$. Then $A \in (QT)$ and A is cohyponormal with connected spectrum $\sigma(A)$. It suffices to show that $A + K \in (SI)(\mathcal{H})$ for some compact operator K. We will prove the conclusion of the theorem in six cases.

Case 1. $1 < \min\{n : n = \operatorname{ind}(\lambda - A),\ \lambda \in \rho_{s-F}(A) \cap \sigma(A)\} < \infty$.

Suppose that $\{\Omega_i\}_{i=1}^{l}$ $(1 \leq l \leq \infty)$ are components of $\rho_{s-F}(A) \cap \sigma(A)$, and $n_i = \operatorname{ind}(\lambda - A)$ $(\lambda \in \Omega_i)$. Without loss of generality, we may assume that $n_1 = \min\{n_i\}$.

By Proposition 1.28 and Theorem 1.29, there exists a compact operator K_1 such that

$$A + K_1 \simeq \begin{bmatrix} A & 0 & 0 \\ 0 & \bigoplus_{i=1}^{l_1} \lambda_i & 0 \\ 0 & 0 & \bigoplus M(\Gamma_i)^{(n_i - 1)} \end{bmatrix} \begin{matrix} \mathcal{H} \\ \mathcal{H}^{(l_1)} \\ \bigoplus_{i=1}^{l} (\mathcal{H}_i^+ \oplus \mathcal{H}_i^-)^{(n_i - 1)}, \end{matrix}$$

where $\{\lambda_i\}_{i=1}^{l_1}$ is a dense subset of $(\bigcup_{i=1}^{l} \overline{\Omega}_i) \backslash [\bigcup_{i=1}^{l} (\Omega_i \cup \partial \overline{\Omega}_i)]$, $\Gamma_i = \partial \overline{\Omega}_i$ and $M(\Gamma_i) = \begin{bmatrix} M_+(\Gamma_i) & Z_i \\ 0 & M_-(\Gamma_i) \end{bmatrix} \begin{matrix} \mathcal{H}_i^+ \\ \mathcal{H}_i^- \end{matrix}$ is given in Section 0.2. Note that $Z_1^{(n_1 - 1)}$ is compact.

Thus there exists a compact operator K_2 such that $A+K_1+K_2 \simeq A\oplus(\bigoplus_{i=1}^{l_1} \lambda_i)\oplus M_1$, where

$$M_1 = \begin{bmatrix} M_+(\Gamma_1)^{(n_1-1)} & 0 & 0 & 0 \\ 0 & \bigoplus\limits_{i=2}^{l} M_+(\Gamma_i)^{(n_i-1)} & 0 & E \\ 0 & 0 & M_-(\Gamma_1)^{(n_1-1)} & 0 \\ 0 & 0 & 0 & \bigoplus\limits_{i=2}^{l} M_-(\Gamma_i)^{(n_i-1)} \end{bmatrix}.$$

Denote $B = A \oplus (\bigoplus_{i=1}^{l_1} \lambda_i) \oplus M_+(\Gamma_1)^{(n_1-1)} \oplus (\bigoplus_{i=2}^{l} M_+(\Gamma_i)^{(n_i-1)})$. Then $B \in (QT)$, $\sigma(B) = \sigma_W(B) = \sigma(A)$ is connected and $\operatorname{ind}(\lambda - B) = 1$ for $\lambda \in \Omega_1$. By Theorem 1.44, there exists a compact operator K_3' such that $G := B + K_3' \in \mathcal{B}_1(\Omega_1)$ and $\sigma(G) = \sigma(A)$, i.e., there exists a compact operator K_3 such that

$$A + K_1 + K_2 + K_3 \simeq \begin{bmatrix} G & 0 & E_1 \\ 0 & M_-(\Gamma_1)^{(n_1-1)} & 0 \\ 0 & 0 & \bigoplus_{i=2}^{l} M_-(\Gamma_i)^{(n_i-1)} \end{bmatrix}.$$

By arguments similar to Theorem 5.17, we can find compact operators F_i, $\|F_i\| < 1/10^i$, such that

$$T_1 := (G \oplus M_-(\Gamma_1)^{(n_1-1)}) + F_1 \in \mathcal{B}_{n_1}(\Omega_1) \cap (SI),$$
$$\sigma(T_1) = \sigma(A)$$

and

$$T_i := M_-(\Gamma_i)^{(n_i-1)} + F_i \in \mathcal{B}_{(n_i-1)}(\Omega_i) \quad (i = 2, 3, \cdots, l).$$

Thus, we can find a compact operator K_4 such that

$$A + K_1 + K_2 + K_3 + K_4 \simeq \begin{bmatrix} T_1 & E_2 \\ 0 & \bigoplus_{i=2}^{l} T_i \end{bmatrix}.$$

Denote $E_2 = [E_{22}\, E_{23} \, \cdots \, E_{2l}]$.

Since $\Omega_i \cap \Omega_j = \varnothing$ $(i \neq j)$, by Proposition 1.14 $\ker \tau_{T_iT_j} = \{0\}$. Since $\sigma_r(T_1) \cap \sigma_l(T_i) \neq \varnothing$ $(i \geq 2)$, by Theorem 1.11 we can find a compact operator G_i, $\|G_i\| < 1/10^i$, such that $G_i + E_{2i} \notin \operatorname{ran} \tau_{T_1T_i}$ $(i = 2, 3, \cdots, l)$. Thus there is a compact operator K_5 such that

$$Q := A + \sum_{i=1}^{5} K_i \simeq \begin{bmatrix} T_1 & E_{22} + G_2 & E_{23} + G_3 & \cdots \\ & T_2 & & \\ & & T_3 & \\ & 0 & & \ddots \end{bmatrix}.$$

By Corollary 3.14, $Q \in (SI)$. Note that $K := \sum_{i=1}^{5} K_i$ is a compact operator.

Case 2. $\min\{n : n = \operatorname{ind}(\lambda - A),\ \lambda \in \rho_{s-F}(A) \cap \sigma(A)\} = 1$.

Without loss of generality, we can assume that $n_1 = 1$. By Theorem 1.44, there is a compact operator K such that $A + K \in \mathcal{B}_1(\Omega_1) \subset (SI)$.

Case 3. $\rho_{s-F}^0(A) \cap \sigma(A) \neq \varnothing$.

Assume that $\{\Omega_i\}_{i=1}^{l}$ are the components of $\rho_{s-F}^0(A) \cap \sigma(A)$, and $\{\lambda_i\}_{i=1}^{l_1}$ is a dense subset of $\left(\bigcup_{i=1}^{l} \overline{\Omega}_i\right)\backslash\left[\bigcup_{i=1}^{l}(\Omega_i \cup \partial\overline{\Omega}_i)\right]$.

Set

$$M(\Gamma_i) = \begin{bmatrix} M_+(\Gamma_i) & Z_i \\ 0 & M_-(\Gamma_i) \end{bmatrix} \qquad (i = 1, 2, \cdots, l),$$

where $\Gamma_i = \partial\overline{\Omega}_i$. Since Z_i is compact and $\|Z_i\| \leq \left(m(\Omega_i)/\pi\right)^{1/2}$, $\bigoplus_{i=1}^{l} Z_i$ is compact. By the argument of case 1,

$$A + K_1 \simeq \begin{bmatrix} A & 0 & 0 & 0 \\ 0 & \bigoplus\limits_{i=1}^{l} \lambda_i & 0 & 0 \\ 0 & 0 & \bigoplus\limits_{i=1}^{l} M_-(\Gamma_i) & 0 \\ 0 & 0 & 0 & \bigoplus\limits_{i=1}^{l} M_+(\Gamma_i) \end{bmatrix}$$

for some compact operator K_1.

Set $G = A \oplus \left(\bigoplus_{i=1}^{l} \lambda_i\right) \oplus \left(\bigoplus_{i=1}^{l} M_-(\Gamma_i)\right)$. Then $G \in (QT)$ and $\sigma(G) = \sigma_W(G) = \sigma(A)$ is connected, $\operatorname{ind}(\lambda - G) = 1$ for $\lambda \in \Omega_1$. By Theorem 1.44, we can find a compact operator F such that $G + F \in \mathcal{B}_1(\Omega_1) \subset (SI)$. By the argument of case 1, we can find a compact operator K such that $A + K \in (SI)$.

Case 4. $(\sigma(A))^\circ = \varnothing$.

When $(\sigma(A))^\circ = \varnothing$, $A \in (BQT)$. By Theorem 5.30, there is a compact operator K such that $A + K \in (SI)$.

Case 5. $\sigma(A) = \sigma_e(A)$ and there is a bounded component Ω of $\mathbb{C}\backslash(\sigma(A))^\circ$.

Let $\Gamma = \partial\Omega$; then $\Gamma \subset \sigma_{lre}(A)$. By the argument of case 1, $A + K_1 \simeq A \oplus M_-(\Gamma) \oplus M_+(\Gamma)$ for some compact operator K_1. Note that $G := A \oplus M_-(\Gamma) \in (QT)$, by Theorem 1.44, $G + F \in \mathcal{B}_1(\Omega)$ for some compact operator F. Since $\sigma_r(G + F) \cap \sigma_l(M_+(D)) \neq \varnothing$, we can find a compact operator L such that $L \notin \operatorname{ran} \tau_{G+F, M_+(\Gamma)}$.

It is apparent that $\ker \tau_{M_+(\Gamma),G+F} = \{0\}$. Thus $A + K \simeq \begin{bmatrix} G+F & L \\ 0 & M_+(\Gamma) \end{bmatrix}$ for some compact operator K. By Corollary 3.14, $A + K \in (SI)$.

Case 6. $\min\{n : n = \operatorname{ind}(A-\lambda),\ \lambda \in \rho_{s-F}(A) \cap \sigma(A)\} = \infty$ *and each component of* $(\sigma(A))^\circ$ *is simply connected..*

Let $\{\Omega_i\}_{i=1}^l$ be the components of $(\sigma(A))^\circ$. By a similar discussion to case 1, we can find a compact operator K_1 such that

$$A + K_1 \simeq \begin{bmatrix} A & 0 & 0 & 0 & 0 \\ 0 & \bigoplus_{i=1}^l M_+(\Gamma_i)^{(\infty)} & 0 & 0 & E \\ 0 & 0 & M_+(\Gamma_1) & Z_1 & 0 \\ 0 & 0 & 0 & M_-(\Gamma_1) & 0 \\ 0 & 0 & 0 & 0 & \bigoplus_{i=1}^l M_-(\Gamma_i)^{(\infty)} \end{bmatrix},$$

where $\Gamma_i = \partial\Omega_i$ $(i = 1, 2, \cdots, l)$. Since Z_1 is compact, we can find a compact operator K_2 such that

$$A + K_1 + K_2 \simeq \begin{bmatrix} G & E_1 \\ 0 & \bigoplus M_-(\Gamma_i)^{(\infty)} \end{bmatrix},$$

where $G = A \oplus \left(\bigoplus_{i=1}^l M_+(\Gamma_i)^{(\infty)}\right) \oplus M_+(\Gamma_1) \oplus M_-(\Gamma_1)$. A simple computation shows that $G \in (BQT)$ and $\sigma(G) = \sigma_{lre}(G) = \sigma(A)$. Since Ω_1 is simply connected, by 0.2 $M_-(\Gamma_1)$ is unitarily equivalent to T_ϕ^*, where ϕ is a conformal mapping from D onto Ω_1, and T_ϕ is the Toeplitz operator with symbol ϕ. By Theorem 1.50 there exists a compact operator F such that $T_1 = G + F$ is quasisimilar to $M_-(\Gamma_1) \in \mathcal{B}_1(\Omega_1) \subset (SI)$. Thus by Theorem 2.36 $T_1 \in (SI)$ and $A + K_1 + K_2 + K_3 \simeq \begin{bmatrix} T_1 & E_2 \\ 0 & \bigoplus_{i=1}^l M_-(\Gamma_i)^{(\infty)} \end{bmatrix}$ for some compact operator K_3. Let $\lambda_0 \in \Omega_1$, by Proposition 3.20 $T_1 - \lambda_0$ and $M_-(\Gamma_1) - \lambda_0$ admit upper triangular matrix representations (g_{ij}) and (h_{ij}) respectively, $g_{ij} = h_{ij} = 0$ $(i > j)$ and

$$\lim_{n\to\infty} \left[\prod_{k=1}^n \left|\frac{g_{k\,k+1}}{h_{k\,k+1}}\right|\right]^{1/n} = 1.$$

By the argument of Proposition 3.17, we can find compact operators K_{1i}, such that $\|K_{1i}\| < 1/2^{1+i}$, $B_{1i} := M_-(\Gamma_1) + K_{1i} \in \mathcal{B}_1(\Omega_1)$, $\ker \tau_{B_{1i}B_{1j}} = 0$ $(i \neq j)$ and $\ker \tau_{T_1 B_{1i}} = \ker \tau_{B_{1i} T_1} = \{0\}$ $(i = 1, 2, \cdots)$. Since $\sigma_r(B_{11}) \cap \sigma_l(B_{1j}) \neq \varnothing$

$(j = 2, 3, \cdots)$, by Theorem 1.11 and Proposition 3.13 we can find a compact operator F_2, $\|F_2\| < 1/2$, such that

$$T_2 := M_-(\Gamma_1)^{(\infty)} + F_2 = \begin{bmatrix} B_{11} & C_{12} & C_{13} & \cdots \\ & B_{12} & & \\ & & B_{13} & \\ & 0 & & \ddots \end{bmatrix} \in \mathcal{B}_\infty(\Omega_1) \cap (SI),$$

where $C_{1i} \notin \operatorname{ran} \tau_{B_{11} B_{1i}}$, C_{1i} is compact and $\|C_{1i}\| < 1/2^{1+i}$ $(i \geq 2)$. Since $\ker \tau_{B_{1i} T_1} = \{0\}$, computation indicates that $\ker \tau_{T_2 T_1} = \{0\}$. Similarly, we can find compact operators F_i, $\|F_i\| < 1/2^i$, such that $T_i = M_-(\Gamma_i)^{(\infty)} + F_i \in \mathcal{B}_\infty(\Omega_i) \cap (SI)$ $(i \geq 3)$. Since $\Omega_1 \cap \Omega_i = \varnothing$ $(i \geq 2)$, by Proposition 1.14 $\ker \tau_{T_i T_1} = \{0\}$ $(i \geq 3)$. Therefore there exists a compact operator K_4 such that

$$A + K_1 + K_2 + K_3 + K_4 \simeq \begin{bmatrix} T_1 & & & * \\ & T_2 & & \\ & & T_3 & \\ & 0 & & \ddots \end{bmatrix}.$$

Repeating the argument of case 1, we can prove that there is a compact operator K_5 such that

$$A + \sum_{i=1}^{5} K_i \in (SI).$$

Note that $A \in (QT)$ and A is cohyponormal with connected $\sigma(A)$ if and only if A is in at least one of the six cases. Thus the proof of Theorem 5.39 is now complete.

□

To end this section we introduce the following results about the application of Theorem 5.39 to the theory of analytic Toeplitz operators on Bergman spaces.

Let $\mathbb{B}_n$ denote the open unit ball of $\mathbb{C}^n$, $L^2_a(\mathbb{B}_n, dv)$ denote the Bergman space, i.e., the subspace of $L^2(\mathbb{B}_n, dv)$ spanned by analytic functions in $\mathbb{B}_n$, where dv is the normalized Lebesgue measure of $\mathbb{B}_n$. Let $H^\infty(\mathbb{B}_n)$ be the set of functions analytic in $\mathbb{B}_n$ and bounded in $\overline{\mathbb{B}}_n$. For $\phi \in H^\infty(\mathbb{B}_n)$, define the multiplication operator M_ϕ on $L^2_a(\mathbb{B}_n, dv)$ by $M_\phi h = \phi h$, $\forall h \in L^2_a(\mathbb{B}_n, dv)$. Let P denote the orthogonal projection from $L^2(\mathbb{B}_n, dv)$ onto $L^2_a(\mathbb{B}_n, dv)$. For $\phi \in L^2(\mathbb{B}_n, dv)$, define $T_\phi f = P\phi f$, $\forall f \in L^2_a(\mathbb{B}_n, dv)$. If T_ϕ is bounded on $L^2_a(\mathbb{B}_n, dv)$, T_ϕ is called a Toeplitz operator on $L^2_a(\mathbb{B}_n, dv)$.

In 3.4, we analysed the strong irreducibility of the analytic Toeplitz operators on Hardy space H^2. We will discuss the strong irreducibility of T_ϕ for $\phi \in H^\infty(\mathbb{B}_n)$.

Proposition 5.40. *Let $\phi \in H^\infty(\mathbb{B}_n)$ and let $\phi(z) = \sum_m F_{p_m}(z)$ be the homogeneous expansion of ϕ. If* g.c.d $\{p_1, p_2, \cdots\} = l > 1$, *then T_ϕ is strongly reducible.*

Proof. It is easily seen that $\{a_k z^k\}_{k=0}^{\infty}$ is an ONB of $L_a^2(\mathbb{B}_n, dv)$, where $k = (k_1, k_2, \cdots, k_n)$, k_i is a non-negative integer $(1 \leq i \leq n)$, $z^k = z_1^{k_1} z_2^{k_2} \cdots z_n^{k_n}$, and $a_k \in \mathbb{C}$.

Set $\mathcal{H}_0 = \bigvee\{a_k z^k : l \mid |k|\}$, $\mathcal{H}_1 = \bigvee\{a_k z^k : l \nmid |k|\}$, where $|k| = \sum_{i=1}^n k_i$, $l \mid |k|$ means that l is a divisor of $|k|$. Thus $L_a^2(\mathbb{B}_n, dv) = \mathcal{H}_0 \oplus \mathcal{H}_1$ and

$$\phi(z) a_k z^k = \sum_m a_k z^k F_{p_m}(z) \begin{cases} \in \mathcal{H}_0, & \text{if } l \mid |k|, \\ \in \mathcal{H}_1, & \text{if } l \nmid |k|. \end{cases}$$

This implies that $\mathcal{H}_0$ and $\mathcal{H}_1$ are invariant subspaces of T_ϕ, i.e., T_ϕ is strongly reducible. □

By $\mathcal{A}$ we shall denote the set $\{\phi \in L^2(\mathbb{B}_n, dv) : T_\phi$ is bounded on $L_a^2(\mathbb{B}_n, dv)\}$. For $\phi \in \mathcal{A}$, define $\|\phi\| := \|T_\phi\|$ and let $\mathrm{Sym}\,\mathbb{B}_n$ denote the closure of $\mathcal{A}$ with respect to the norm $\|\phi\|$. $\mathrm{Sym}\,\mathbb{B}_n$ is called the *symbol space* of Toeplitz operators. For $\phi \in \mathrm{Sym}\,\mathbb{B}_n$, the *generalized Toeplitz operator* T_ϕ with symbol ϕ is defined by

$$T_\phi = \lim_{i \to \infty} T_{\phi_i},$$

where $\phi_i \in \mathcal{A}$, $\phi_i \to \phi$ $(i \to \infty)$ and T_{ϕ_i} is the Toeplitz operator in the normal sense $(i = 1, 2, \cdots)$.

According to the definition, we have $\|T_\phi\| = \|\phi\|$ for each $\phi \in \mathrm{Sym}\,\mathbb{B}_n$. We can consider it as a generalization of the classical Toeplitz operators on the Hardy space H^2.

For $z \in \mathbb{B}_n$, let K_z denote a normal reproducing kernel on the Bergman space. For each bounded operator A on $L_a^2(\mathbb{B}_n, dv)$, its Berezin symbol is denoted by $\hat{A}(z) = (AK_z, K_z)$ $(z \in \mathbb{B}_n)$.

A result of M. Engliš [1] asserts that $\hat{A}(z)$ is a bounded smooth function on $\overline{\mathbb{B}}_n$, and $\hat{A} = \hat{B}$ if and only if $A = B$.

Let $C_0 = \{f : f$ is a bounded and smooth function on $\overline{\mathbb{B}}_n$ and $lim_{z \to \partial\mathbb{B}_n} f(z) = 0\}$. It is easily seen that if F is a compact operator, then since $K_z \to 0$ weakly when $z \to \partial\mathbb{B}_n$, $\hat{F}(z) \in C_0$.

For $z \in \mathrm{Sym}\,\mathbb{B}_n$, $\tilde{\phi} = \hat{T}_\phi$ is called the *Berezin Transform* of ϕ. The Toeplitz operator T_ϕ with $\phi \in \mathrm{Sym}\,\mathbb{B}_n$ is called *generalized analytic* if $\hat{T}_\phi(z) = \tilde{\phi}(z)$ is analytic in $\mathbb{B}_n$. If T_ϕ is analytic we have $\hat{T}_\phi(z) = \tilde{\phi}(z) = \hat{T}_{\tilde{\phi}}(z)$, i.e., $T_\phi = T_{\tilde{\phi}}$. Thus a generalized analytic Toeplitz operator is an analytic Toeplitz operator in the normal sense. $\phi \in \operatorname{Sym}\mathbb{B}_n$ is called *Berezin vanishing* if $\tilde{\phi} \in C_0$. ϕ_1 is called a *Berezin perturbation* of ϕ_2 if ϕ_1, $\phi_2 \in \operatorname{Sym}\mathbb{B}_n$ and $\widetilde{\phi_1 - \phi_2} \in C_0$.

Let F be a compact operator and $\phi(z) = (FK_z, K_z) = \hat{F}(z)$. Then from the above analysis, $\phi(z) \in C_0$. M. Engliš asserts that $\hat{T}_\phi(z) = \hat{F}(z)$. Therefore $F = T_\phi$. Note that each analytic Toeplitz operator T_ϕ is subnormal, and so it is hyponormal and $\sigma(T_\phi)$ is connected with $\operatorname{ind}(T_\phi - \lambda) \leq 0$ for $\lambda \in \rho_{s-F}(T_\phi)$. Thus by Theorem 5.39 we have

Theorem 5.41. *Let $\phi \in H^\infty(\mathbb{B}_n)$. Then there exists a Berezin perturbation ϕ' of ϕ such that $T_{\phi'} = T_\phi + T_{\phi'-\phi}$ is strongly irreducible.*

Theorem 5.41 implies that we can modify the boundary value of ϕ so as to get an (SI) analytic Toeplitz operator.

5.5 Some Cowen–Douglas operator + compact = strongly irreducible

Proposition 5.42. *Given $T \in \mathcal{L}(\mathcal{H})$ with connected $\sigma(T)$ and $\sigma_{lre}(T)$ satisfying $\rho_{s-F}^{\pm\infty}(T) = \varnothing$ and $\operatorname{ind}(T - \lambda) = 0$ for λ in some component Ω of $\rho_F(T) \cap \sigma(T)$, for each $\epsilon > 0$ there exists a compact operator K, $\|K\| < \epsilon$, such that $T + K \in (SI)(\mathcal{H})$.*

Proof. By Voiculescu's theorem $T + K_1 \simeq A \oplus T \oplus A$ for some compact operator K_1 with $\|K_1\| < \epsilon/3$, where $\sigma(A) = \sigma_e(A) = \sigma_{lre}(T)$. By the argument of Theorem 1.23 we can find a compact operator K_2', $\|K_2'\| < \epsilon/3$, such that $\min\operatorname{ind}(T + K_2' - \lambda) = 1$ for all $\lambda \in \Omega$ and $\sigma(T + K_2') = \sigma(T)$. Thus $T + K_1 + K_2 \simeq A \oplus B \oplus A$ for some compact operator K_2 with $\|K_2\| < \epsilon/3$, where $B = T + K_2'$. By Apostol's triangular representation

$$T + K_1 + K_2 \simeq \begin{bmatrix} A & 0 & 0 & 0 & 0 \\ 0 & B_r & B_{12} & B_{13} & 0 \\ 0 & 0 & B_0 & B_{23} & 0 \\ 0 & 0 & 0 & B_l & 0 \\ 0 & 0 & 0 & 0 & A \end{bmatrix}.$$

Set

$$B_1 = \begin{bmatrix} A & 0 & 0 \\ 0 & B_r & B_{12} \\ 0 & 0 & B_0 \end{bmatrix} \quad \text{and} \quad B_2 = \begin{bmatrix} B_l & 0 \\ 0 & A \end{bmatrix}.$$

Since $\sigma(A) = \sigma_{lre}(T)$ is connected, computation shows that $\sigma(B_1)$, $\sigma_W(B_1)$, $\sigma(B_2)$ and $\sigma_W(B_2)$ are connected, $\operatorname{ind}(\lambda - B_1) = 1$ and $\operatorname{ind}(\lambda - B_2) = -1$ for $\lambda \in \Omega$. It follows from Theorem 1.44 that there exist compact operators F_1 and F_2, $\|F_1\| < \epsilon/6$ and $\|F_2\| < \epsilon/6$, such that $A_1 := B_1 + F_1 \in \mathcal{B}_1(\Omega)$, $A_2^* := (B_2 + F_2)^* \in \mathcal{B}_1(\Omega^*)$. Thus $T+K_1+K_2+K_3 \simeq \begin{bmatrix} A_1 & E \\ 0 & A_2 \end{bmatrix}$ for some compact operator K_3 with $\|K_3\| < \epsilon/6$, where $E = \begin{bmatrix} 0 & 0 \\ B_{13} & 0 \\ B_{23} & 0 \end{bmatrix}$. Observe that $\sigma_r(A_1) \cap \sigma_l(A_2) \neq \varnothing$, applying Theorem 1.11 and Corollary 3.14 we can find a compact operator K_4, $\|K_4\| < \epsilon/6$, such that $T + \sum_{i=1}^4 K_i \in (SI)$. Thus $K = \sum_{i=1}^4 K_i$ satisfies all the requirements of the proposition. □

By the same argument as Proposition 5.42, we can prove Proposition 5.43.

Proposition 5.43. *Given $T \in \mathcal{L}(\mathcal{H})$ with connected $\sigma(T)$ and $\sigma_{lre}(T)$ satisfying $\rho_{s-F}^{\pm\infty}(T) = \varnothing$, $\rho_{s-F}^{1}(T) \neq 0$ and $\rho_{s-F}^{-1}(T) \neq \varnothing$, for each $\epsilon > 0$ there exists a compact operator K with $\|K\| < \epsilon$ such that $T + K \in (SI)(\mathcal{H})$.*

It is not certain that there always exists a compact perturbation of an operator, even a triangular operator with connected spectrum, which is strongly irreducible. But for a special class of Cowen–Douglas operators we have

Theorem 5.44. *Let Ω be an analytic, simply connected Cauchy domain and $T \in \mathcal{B}_n(\Omega)$ with $\sigma(T) = \overline{\Omega}$. Then for each $\epsilon > 0$, there is a compact operator K with $\|K\| < \epsilon$, such that $T + K \in (SI)$.*

In order to prove Theorem 5.44, we need the next lemma.

Lemma 5.45. *Let Ω be an analytic, simply connected Cauchy domain, $T_1, T_2 \in \mathcal{B}_1(\Omega)$ with $\sigma(T_1) = \sigma(T_2) = \overline{\Omega}$, and $\lambda_0 \in \Omega$. Assume that $T_1 - \lambda_0$ and $T_2 - \lambda_0$ admit upper triangular matrix representations*

$$T_1 - \lambda_0 = \begin{bmatrix} 0 & b_{12}^1 & & * \\ & 0 & b_{23}^1 & \\ & & 0 & \ddots \\ & 0 & & \ddots \end{bmatrix} \quad \text{and} \quad T_2 - \lambda_0 = \begin{bmatrix} 0 & b_{12}^2 & & * \\ & 0 & b_{23}^2 & \\ & & 0 & \ddots \\ & 0 & & \ddots \end{bmatrix}$$

with respect to the ONB $\{e_n\}_{n=1}^{\infty}$ and, respectively, $\{f_n\}_{n=1}^{\infty}$ of $\mathcal{H}$. Then

$$\lim_{n\to\infty} \left(\prod_{k=1}^{n} |b_{k\,k+1}^i| \right)^{1/n} = r > 0 \qquad (i = 1, 2).$$

Proof. Let f be the conformal mapping from Ω onto the unit disc D such that $f(\partial\Omega)=\partial D$ and $f(\lambda_0)=0$. Thus $A_i := f(T_i)\in\mathcal{B}_1(D)$ and $\sigma(A_i)=\overline{D}$ $(i=1,2)$, and A_1, A_2 admit upper triangular matrix representations (a^1_{ij}) and (a^2_{ij}) with respect to $\{e_n\}$ and, respectively, $\{f_n\}$. By Proposition 3.19

$$\lim_{n\to\infty}\left(\prod_{k=1}^{n}|a^i_{k\,k+1}|\right)^{1/n}=1 \qquad (i=1,2).$$

Denote $g=f^{-1}$; thus $g(A_i)=T_i$ $(i=1,2)$. Since $g(z)$ is analytic in $\overline{D}$,

$$g(z)=g(0)+g'(0)z+\cdots+\frac{1}{n!}f^{(n)}(0)z^n+\cdots.$$

Therefore

$$\begin{aligned} g(A_i) &= g(0)+g'(0)A_i+\frac{1}{2!}g''(0)A_i^2+\cdots \\ &= \begin{bmatrix} \lambda_0 & g'(0)a^i_{12} & & * \\ & \lambda_0 & g'(0)a^i_{23} & \\ & & \lambda_0 & \ddots \\ 0 & & & \ddots \end{bmatrix} \qquad (i=1,2), \end{aligned}$$

and $b^i_{k\,k+1}=g'(0)a^i_{k\,k+1}$ $(i=1,2;\ k\geq 1)$. Since g is also a conformal mapping, $|g'(0)|=r\neq 0$. Thus

$$\lim_{n\to\infty}\left(\prod_{k=1}^{n}|b^i_{k\,k+1}|\right)^{1/n}=|g'(0)|=r. \qquad \square$$

Proof of Theorem 5.44. By Theorem 1.49

$$T=\begin{bmatrix} T_1 & T_{12} & & & * \\ & T_2 & T_{23} & & \\ & & T_3 & \ddots & \\ & & & \ddots & T_{n-1\,n} \\ 0 & & & & T_n \end{bmatrix}=\begin{bmatrix} T_1 & * \\ 0 & B_{n-1}\end{bmatrix},$$

where $B_{n-1}=\begin{bmatrix} T_2 & & * \\ & \ddots & \\ 0 & & T_n\end{bmatrix}$ and $T_k\in\mathcal{B}_1(\Omega)$ $(1\leq k\leq n)$.

Since Ω is simply connected, $\sigma(T_1) \subset \sigma(T) = \overline{\Omega}$ and $\sigma(B_{n-1}) \subset \sigma(T) = \overline{\Omega}$. Thus $\sigma(T_1) = \overline{\Omega}$. Pick a $\lambda_0 \in \Omega$, if $\lambda_0 - B_{n-1}$ is invertible, i.e.,

$$\begin{bmatrix} X_{22} & \dots & X_{2n} \\ \dots & \dots & \\ X_{n2} & \cdots & X_{nn} \end{bmatrix} \begin{bmatrix} \lambda_0 - T_2 & & * \\ & \ddots & \\ 0 & & \lambda_0 - T_n \end{bmatrix} = \begin{bmatrix} 1 & & 0 \\ & \ddots & \\ 0 & & 1 \end{bmatrix}$$

for some operator $(X_{ij})_{i,j=2}^{n}$. Thus $X_{22}(\lambda_0 - T_2) = 1$, i.e., $(\lambda_0 - T_2)$ is bounded below. Since $T_2 \in \mathcal{B}_1(\Omega)$, $(\lambda_0 - T_2)$ is surjective. Thus $\lambda_0 \in \rho(T_2)$, a contradiction. Thus $\lambda_0 \in \sigma(B_{n-1})$ and $\sigma(B_{n-1}) \supset \overline{\Omega}$, i.e., $\sigma(B_{n-1}) = \overline{\Omega}$. By induction, we can prove that $\sigma(T_i) = \overline{\Omega}$ $(1 \le i \le n)$. Note that each T_i admits an upper triangular matrix representation

$$T_i = \begin{bmatrix} \lambda_0 & t_{12}^i & & * \\ & \lambda_0 & t_{23}^i & \\ & & \lambda_0 & \ddots \\ & 0 & & \ddots \end{bmatrix}$$

with respect to some ONB of the space. By Lemma 5.45

$$\lim_{n\to\infty} \left[\prod_{k=1}^{n} \left| \frac{t_{k\,k+1}^i}{t_{k\,k+1}^j} \right| \right]^{1/n} = 1 \qquad (i \ne j).$$

By the argument of Proposition 3.17 there are compact operators $\{K_i\}_{i=1}^n$, $\|K_i\| < \epsilon/2^i$ $(i = 1, 2, \cdots, n)$, such that $A_i = T_i + K_i \in \mathcal{B}_1(\Omega)$ and $\ker \tau_{A_i A_j} = \{0\}$ $(i \ne j)$. Since $\sigma_r(A_i) \cap \sigma_l(A_{i+1}) = \partial\Omega$, by Theorem 1.11 we can find compact operators $\{F_i\}_{i=1}^{n-1}$, $\|F_i\| < \epsilon/2^i$ $(i = 1, 2, \cdots, n-1)$, such that $F_i + T_{i\,i+1} \notin \operatorname{ran} \tau_{A_i A_{i+1}}$ $(i = 1, 2, \cdots, n-1)$. Thus we can find a compact operator K with $\|K\| < \epsilon$ such that

$$T + K = \begin{bmatrix} A_1 & T_{12}+F_1 & & & * \\ & A_2 & T_{23}+F_2 & & \\ & & A_3 & \ddots & \\ & & & \ddots & T_{n-1\,n}+F_{n-1} \\ & 0 & & & A_n \end{bmatrix}.$$

It is not difficult to show that $T + K \in (SI)$. □

If we loosen the restrictions on the domain Ω, some other conditions are needed to get a result similar to Theorem 5.44.

Proposition 5.46. *Let $B \in \mathcal{B}_n(\Omega)$ and let B have a unique (SI) decomposition. Then for each $\epsilon > 0$, there is a compact operator K with $\|K\| < \epsilon$, such that $B+K \in (SI)$.*

Proof. Since $B = \sum_{k=1}^{n} \dot{+} B_k$, $B \sim \bigoplus_{k=1}^{m} B_k$, where $B_k \in (SI)$. Thus without loss of generality we may assume that $B = \bigoplus_{k=1}^{m} B_k$. By Proposition 4.16 $\ker \tau_{B_i B_j} = \{0\}$ $(i \neq j)$. Here $B_k \in \mathcal{B}_{n_k}(\Omega)$ and $\sum_{i=1}^{m} n_i = n$. Let Ω_i be the component of $\rho_F(B_i) \cap \sigma(B_i)$ containing Ω. Thus $\Omega \subset \bigcap_{i=1}^{m} \Omega_i$. We always can find an index i_0 $(1 \leq i_0 \leq m)$ such that $\partial\Omega_{i_0} \cap \Omega_i \neq \varnothing$ or $\partial\Omega_{i_0} \cap \partial\Omega_i \neq \varnothing$ $(i \neq i_0)$. Without loss of generality, we may assume that $i_0 = 1$. Thus $\sigma_r(B_1) \cap \sigma_l(B_i) \neq \varnothing$ $(i \neq 1)$. By Theorem 1.11, we can find compact operators K_i, $\|K_i\| < \epsilon/2^i$, such that $K_i \notin \operatorname{ran} \tau_{B_1 B_i}$ $(2 \leq i \leq m)$. Thus

$$B+K = \begin{bmatrix} B_1 & K_2 & K_3 & \cdots & K_m \\ & B_2 & & & \\ & & B_3 & & \\ & & & \ddots & \\ & 0 & & & B_m \end{bmatrix},$$

where $K = \begin{bmatrix} 0 & K_2 & K_3 & \cdots & K_m \\ & & 0 & & \end{bmatrix}$ is compact and $\|K\| < \epsilon$. It is easily seen that $B + K \in (SI)$. □

By the same argument as Proposition 5.46, we can prove

Proposition 5.47. *Let $B \in \mathcal{B}_n(\Omega)$ and B have an n-(SI) decomposition. Then for each $\epsilon > 0$, there exists a compact operator K, $\|K\| < \epsilon$, such that $B + K \in (SI)$.*

5.6 Open problems

Problem 5.1. *Let B_1, B_2 be two essentially normal operators in $\mathcal{B}_n(\Omega)$ with $\Lambda(B_1) = \Lambda(B_2)$. Then for each $\epsilon > 0$ do there exist compact operators K_1 and K_2 such that $\|K_1\| < \epsilon$, $\|K_2\| < \epsilon$, $B_1 + K_1 \simeq_{u+K} B_2$ and $B_2 + K_2 \simeq_{u+K} B_1$?*

Problem 5.2. *Given B_1, $B_2 \in \mathcal{B}_n(\Omega)$, if there exists a compact operator K such that $B_1 + K = B_2$, then for each $\epsilon > 0$, does there exist a compact operator K_1, $\|K_1\| < \epsilon$, such that $B_1 + K_1 \sim B_2$?*

Problem 5.3. *Let T be an essentially normal operator with connected spectrum, and let $\epsilon > 0$. Does there exist a compact operator K with $\|K\| < \epsilon$, such that $T + K \in (SI)$?*

Problem 5.4. *Let $T \in \mathcal{B}_n(\Omega)$ with connected spectrum, and let $\epsilon > 0$. Does there exist a compact operator K with $\|K\| < \epsilon$ such that $T + K \in (SI)$?*

Problem 5.5. *Let $T \in \mathcal{L}(\mathcal{H})$ and let $\epsilon > 0$. Does there exist a compact operator K with $\|K\| < \epsilon$ such that $T + K \in (SI)$?*

5.7 Remarks

The concept of a $(\mathcal{U} + \mathcal{K})$-orbit of an operator was given by D.A. Herrero [6]. He tried to classify the essentially normal operators in terms of their $(\mathcal{U}+\mathcal{K})$-orbits. Al-Musallan, P.S.Guinand and L.Marcoux published nice work on this aspect (Al-Musallan [1], P.S.Guinand and L. Marcoux [1] [2], and L. Marcoux [1]). Theorem 5.1 is due to P.S. Guinand and L. Marcoux. In a personal communication with L. Marcoux, D.A. Herrero introduced the concept of a shift-like operator. An operator T is *shift-like* if T satisfies the following conditions.

(i) T is essentially normal with $\sigma(T) = \overline{D}$,

(ii) $\rho_F(T) = D$, $\text{ind}\,(T - \lambda) = -1$ and $\text{nul}\,(T - \lambda)^* = 0$ for $\lambda \in D$.

D.A. Herrero posed the following question: Suppose $T \in \mathcal{L}(\mathcal{H})$ is shift-like. Given $\epsilon > 0$, can we find W invertible so that $\|S - WTW^{-1}\| < \epsilon$ and $S - WTW^{-1}$ is compact?

L. Marcoux [1] described the closure of the $(\mathcal{U}+\mathcal{K})$-orbit of each shift-like operator and gave a confirmative answer to D.A. Herrero's question.

$A(\Omega, n)$ is a generalization of shift-like operators. Theorem 5.5 and Proposition 5.14 strengthen the main results of L.Marcoux [1]. Lemmas 5.11, 5.12 and Proposition 5.13 use the technique of L. Marcoux [1]. The other results of 5.1 are due to Y.Q. Ji, C.L. Jiang and Z.Y. Wang [1]. D.A. Herrero raised Question 5.16 and the same question for a general operator T. The results of 5.2, 5.3 of this chapter were mainly contributed by Y.Q. Ji, C.L. Jiang, S.Power, S.H. Sun and Z.Y. Wang (C.L. Jiang and Z.Y. Wang [1] [2], C.L. Jiang, S.H. Sun and Z.Y. Wang [1], Y.Q. Ji, C.L. Jiang and Z.Y. Wang [1], C.L. Jiang, S.Power and Z.Y. Wang [1]). The main results of 5.4 are due to K.Y. Guo and C.L. Jiang [1]. The results of 5.5 were proved by Y.Q. Ji, C.L. Jiang and Z.Y. Wang [2] [3]. The purpose of the work in these sections is to answer D.A. Herrero's question.

6. Strongly irreducible operators in nest algebras

In this chapter we will analyse the norm closure of the unitary orbit of the strongly irreducible operators in a nest algebra. Let $\mathcal{U}(\operatorname{alg}\mathcal{N} \cap (SI))$ denote the unitary orbit of the (SI) operators in the nest algebra $\operatorname{alg}\mathcal{N}$, i.e., $\mathcal{U}(\operatorname{alg}\mathcal{N} \cap (SI)) := \{T \in (SI) : \text{ there exist an } A \in \operatorname{alg}\mathcal{N} \text{ and a unitary operator } U \text{ such that } T = UAU^*\}$. By $\mathcal{U}(\operatorname{alg}\mathcal{N} \cap (SI))^-$ we shall denote the norm closure of $\mathcal{U}(\operatorname{alg}\mathcal{N} \cap (SI))$.

6.1 The $\mathcal{U}(\operatorname{alg}\mathcal{N} \cap (SI))^-$ of well-ordered nests

Denote $(QT)_C := \{T \in (QT) : \sigma(T) \text{ and } \sigma_W(T) \text{ are connected}\}$ and $(QT)_C^* := \{T : T^* \in (QT)_C\}$.

Theorem 6.1. (i) *If $\mathcal{N}$ is a well-ordered nest and all its atoms are finite dimensional, then $\mathcal{U}(\operatorname{alg}\mathcal{N} \cap (SI))^- = (QT)_C$.*

(ii) *If $\mathcal{N}^\perp$ is a well-ordered nest and all its atoms are finite dimensional, then $\mathcal{U}(\operatorname{alg}\mathcal{N} \cap (SI))^- = (QT)_C^*$, where $\mathcal{N}^\perp := \{N : N^\perp \in \mathcal{N}\}$.*

Note that D.A. Herrero has proved that if $\mathcal{N}$ is a well-ordered nest with finite dimensional atoms, then $\operatorname{alg}\mathcal{N} \subset (QT)$ (D.A. Herrero [6]). Thus, if $\mathcal{N}$ is refined to be a maximal nest $\mathcal{M}$, i.e., if $\mathcal{N} \subset \mathcal{M}$ and all atoms of $\mathcal{M}$ are one dimensional, and if the conclusion of Theorem 6.1 is true for $\mathcal{M}$, then we have

$$\mathcal{U}(\operatorname{alg}\mathcal{N} \cap (SI))^- = \mathcal{U}(\operatorname{alg}\mathcal{M} \cap (SI))^-.$$

Therefore, without loss of generality we may assume that the nest $\mathcal{N}$ is a maximal nest.

In order to prove Theorem 6.1, we need the following proposition.

Proposition 6.2. *Let $\mathcal{N}$ (or $\mathcal{N}^\perp$) be a well-ordered and maximal nest. If $T \in (QT)_C$ and $\sigma(T)$ consist of more than one point then there exists an operator $A \in \operatorname{alg}\mathcal{N} \cap (SI)$ such that $\Lambda(A) = \Lambda(T)$ and $T \in S(A)^-$.*

Proof. 1. Assume that $\mathcal{N}$ is a nest of type $\omega+1$, where ω is the first countable limit ordinal and $\mathcal{N} = \{0; \bigvee_{k=1}^{n}\{e_k\}\,(n=1,2,\cdots);\, \mathcal{H}\}$, where $\{e_k\}_{k=1}^{\infty}$ is an ONB of $\mathcal{H}$.

If $\sigma(T) = \sigma_{lre}(T)$ is a connected perfect set, then by Proposition 3.27, we can construct an operator $A \in (SI)(\mathcal{H})$ such that A admits an upper matrix representation

$$A = \begin{bmatrix} \lambda_1 & & * \\ & \lambda_2 & \\ 0 & & \ddots \end{bmatrix}$$

with respect to the ONB $\{e_k\}$, $\sigma(A) = \sigma_{lre}(A) = \sigma(T)$ and $\bigvee\{\ker(A-\lambda_k)^n : n \geq 1, k \geq 1\} = \mathcal{H}$, where $\{\lambda_k\}^- = \sigma(A)$. Thus $A \in \operatorname{alg}\mathcal{N}$. By the similarity orbit theorem $T \in S(A)^-$.

If $\rho_{s-F}(T)\cap\sigma(T) \neq \varnothing$, by Theorem 1.23, there is a compact operator K, $\|K\| < \epsilon$, such that $\sigma(T+K) = \sigma_W(T)$. Denote $T' = T+K$. Let $\{\Omega_k\}_{k=1}^{l}$ $(1 \leq l \leq \infty)$ be the components of $\rho_{s-F}(T')\cap\sigma(T')$. Denote $\Phi_k = (\overline{\Omega}_k)^\circ$ $(1 \leq k \leq l)$ and let $\{\lambda_k\}_{k=1}^{l_1}$ and $\{\mu_k\}_{k=1}^{\infty}$ be dense subsets of $\bigcup_{k=1}^{l}\Phi_k\backslash(\bigcup_{k=1}^{l}\Omega_k)$ and, respectively, $\sigma(T)\backslash(\bigcup_{k=1}^{l}\Phi_k)$.

Set $B_k = N_+(\Phi_k^*)^*$, where $N_+(\Phi_k^*)$ is the Bergmann operator on $A^2(\Phi_k^*)\,(1 \leq k \leq l)$. Then $B_k \in \mathcal{B}_1(\Phi_k)$ and $\sigma(B_k) = \overline{\Omega}_k$. Let V be the operator given in Example 3.22. Note that V is a quasinilpotent unicellular operator (therefore $V \in (SI)$) and V^k is not compact for all $k \geq 1$. Set $B_{\lambda_k} = \lambda_k + V^*$ $(1 \leq k \leq l_1)$ and $B_{\mu_j} = \mu_j + V^*$ $(1 \leq j < \infty)$.

Define

$$A_1 = \begin{bmatrix} B_1 \oplus \left(\bigoplus_{k=2}^{l} B_k^{(n_k)}\right) & 0 \\ 0 & \left(\bigoplus_{k=1}^{l_1} B_{\lambda_k}\right) \oplus \left(\bigoplus_{j=1}^{\infty} B_{\mu_j}\right) \end{bmatrix},$$

where $n_k = \operatorname{ind}(T-\lambda)$ $(\lambda \in \Omega_k,\ 1 \leq k \leq l)$.

Note that $A_1 \in (QT)$ and $\sigma_W(A_1) = \sigma(A_1)$ is connected. By Theorem 1.44, there is a compact operator K_1 such that $G = A_1 + K_1 \in \mathcal{B}_1(\Omega_1)$. Since B_1 and G are $\mathcal{B}_1(\Omega_1)$ operators, they have upper matrix representations

$$G = \begin{bmatrix} \lambda & g_1 & & * \\ & \lambda & g_2 & \\ & & \lambda & \ddots \\ 0 & & & \ddots \end{bmatrix}, \quad B_1 = \begin{bmatrix} \lambda & b_1 & & * \\ & \lambda & b_2 & \\ & & \lambda & \ddots \\ 0 & & & \ddots \end{bmatrix},$$

where $\lambda \in \Omega_1$. Let $r = \overline{\lim}_{n\to\infty}\left(\prod_{k=1}^{n}|g_k/b_k|\right)^{1/n}$.

By Proposition 3.17 and Theorem 1.11 there exists a compact operator K_2 such that

$$A_2 := G \oplus B_1^{(n_1-1)} + K_2 = \begin{bmatrix} G & D_1 & D_2 & \dots \\ & T_1 & & \\ & & T_2 & \\ & 0 & & \ddots \end{bmatrix} \quad \text{if } r \geq 1,$$

where $T_i \in \mathcal{B}_1(\Omega_1)$, $D_i \notin \operatorname{ran} \tau_{GT_i}$, $\ker \tau_{T_iG} = \{0\}$ $(i \geq 1)$ and $\ker \tau_{T_iT_j} = \{0\}$ $(i \neq j)$. By Corollary 3.14, $A_2 \in (SI)$. It is easily seen that $A_2 \in \mathcal{B}_{n_1}(\Omega_1)$.

$$A_2 := G \oplus B^{(n_1-1)} + K_2 = \begin{bmatrix} T_1 & & & D_1 \\ & T_2 & & D_2 \\ & & \ddots & \vdots \\ & 0 & & G \end{bmatrix} \quad \text{if } r < 1,$$

where $T_i \in \mathcal{B}_1(\Omega_1)$, $D_i \notin \operatorname{ran} \tau_{T_iG}$, $\ker \tau_{GT_i} = \{0\}$ $(i \geq 1)$ and $\ker \tau_{T_iT_j} = \{0\}$ $(i \neq j)$. By a similar argument to Proposition 3.13, $A_2 \in \mathcal{B}_{n_1}(\Omega_1) \cap (SI)$.

It is apparent that $\Lambda(A_2) = \Lambda(T)$ and by the similarity orbit theorem $T \in S(A_2)^-$.

Denote $\mathcal{K}_n = \ker(A_2 - \lambda)^n \ominus \ker(A_2 - \lambda)^{n-1}$ $(n = 1, 2, \cdots)$. Then $\mathcal{K} = \bigoplus_{n=1}^{\infty} \mathcal{K}_n$, $A_2 \in \mathcal{L}(\mathcal{K})$ and A_2 admits the representation

$$A_2 = \begin{bmatrix} \lambda & A_{12} & A_{13} & \dots \\ & \lambda & A_{23} & \dots \\ & & \lambda & \\ & 0 & & \ddots \end{bmatrix} \begin{matrix} \mathcal{K}_1 \\ \mathcal{K}_2 \\ \mathcal{K}_3 \\ \vdots \end{matrix}.$$

When $n_1 = \infty$, $A_2 \in \mathcal{B}_\infty(\Omega_1)$. Let $\{e_n^k\}_{n=1}^{\infty}$ be an ONB of the Hilbert space $\mathcal{K}_n$ $(k \geq 1)$ and let B be the right inverse of $(A_2 - \lambda)$. Set $N_1 = \bigvee\{e_1^1\}$ and let $x_1^2 = Be_1^1$. Since $(A_2 - \lambda)B = 1$, $x_1^2 \notin N_1$. Set $N_2 = \bigvee\{N_1, x_1^2\}$, $N_3 = \bigvee\{N_2, e_2^1\}$ and let $x_2^2 = Be_2^1$. Thus $x_2^2 \notin N_3$. In fact if $x_2^2 = \alpha_1 e_1^1 + \alpha_2 Be_1^1 + \alpha_3 e_2^1$, then $(A_2 - \lambda)x_2^2 = \alpha_2 e_1^1$, and we get a contradiction. Set $N_4 = \bigvee\{N_3, x_2^2\}$ and let $x_1^3 = B^2 e_1^1$. Then computation indicates that $x_1^3 \notin N_4$. Set $N_5 = \bigvee\{N_4, x_1^3\}$, $N_6 = \bigvee\{N_5, e_3^1\}$. Inductively, let $x_3^2 = Be_3^1$, $N_7 = \bigvee\{N_6, x_3^2\}$, $x_2^3 = B^2 e_2^1$, $N_8 = \bigvee\{N_7, x_2^3\}$, $x_1^4 = B^3 e_1^1$, $N_9 = \bigvee\{N_8, x_1^4\}, \cdots$. It is easy to see that each N_k is an invariant subspace of A_2 and $\dim(N_{k+1} \ominus N_k) = 1$ $(k \geq 1)$. From the construction, it is apparent that $\bigvee\{x_k^2 : k \geq 1\} = \bigvee\{Be_k^1 : k \geq 1\} = B\mathcal{K}_1 = \mathcal{K}_2$, $\bigvee\{x_k^3 : k \geq 1\} = \bigvee\{B^2 e_k^1 : k \geq 1\} = B^2\mathcal{K}_1 = \mathcal{K}_3$, and in general, $\bigvee\{x_k^n : k \geq 1\} = \bigvee\{B^{n-1} e_k^1 : k \geq 1\}$

$(n \geq 2)$. Thus $\bigvee\{N_k : k \geq 1\} = \mathcal{K}$. Let f_k denotes the unit vector of the one-dimensional subspace $N_k \ominus N_{k-1}$ ($k \geq 1$, here $N_0 = \{0\}$), then A_2 admits the matrix representation

$$A_2 = \begin{bmatrix} \lambda & & & * \\ & \lambda & & \\ & & \lambda & \\ & 0 & & \ddots \end{bmatrix}$$

with respect to the $ONB\, \{f_k\}_{k=1}^{\infty}$ of $\mathcal{K}$. It is obvious that there is a unitary operator U such that $A = UA_2U^* \in \operatorname{alg}\mathcal{N}$, $A \in (SI)$, $\Lambda(A) = \Lambda(T)$ and $T \in S(A)^-$.

By the same argument we can prove the conclusion when $n_1 < \infty$.

2. Assume that the order type of $\mathcal{N}$ is $\beta + m$, where β is a countable limit ordinal and m is finite. Then $\mathcal{H} = \left[\bigoplus\{\mathcal{H}_t : t \text{ is a limit ordinal, } \omega \leq t \leq \beta\}\right] \oplus \mathbb{C}^m$.

If $\sigma(T) = \sigma_{lre}(T)$ is a connected perfect set, choose a dense subset $\{\alpha_0,\, \alpha_t : \omega < t \leq \beta,\ t \text{ is limit ordinal}\}$ of $\sigma(T)$. According to Proposition 3.27, we can construct an (SI) operator A_ω satisfying

(i) $\sigma(A_\omega) = \sigma_{lre}(A_\omega) = \sigma(T)$, $\sigma_p(A_\omega) \cap \{\alpha_0,\, \alpha_t : \omega < t \leq \beta\} = \varnothing$;

(ii) there is a countable subset $\{\lambda_i\}_{i=1}^{\infty}$ of $\sigma_p(A_\omega)$ such that $\{\lambda_i\}^- = \sigma(A_\omega)$ and

$$\bigvee\{\ker(\lambda_i - A_\omega)^k : k \geq 1,\ i \geq 1\} = \mathcal{H};$$

(iii) $\sigma_p(A_\omega^*) = \varnothing$.

Define

$$A = \begin{bmatrix} A_\omega & C_{2\omega} & \cdots & C_t & \cdots & D \\ & A_{2\omega} & & & & \\ & & \ddots & & 0 & \\ & & & A_t & & \\ & & & & \ddots & \\ & 0 & & & & \lambda_0 + q_m \end{bmatrix} \begin{matrix} \mathcal{H}_\omega \\ \mathcal{H}_{2\omega} \\ \vdots \\ \mathcal{H}_t \\ \vdots \\ \mathbb{C}^m, \end{matrix}$$

where $A_{2\omega} = \alpha_{2\omega} + V^*, \cdots, A_t = \alpha_t + V^*$ $(\omega < t \leq \beta)$, $C_t \in \mathcal{L}(\mathcal{H}_t, \mathcal{H}_\omega)$, $C_t \notin \operatorname{ran}\tau_{A_\omega A_t}$, $D \in \mathcal{L}(\mathbb{C}^m, \mathcal{H}_\omega)$, $D \notin \operatorname{ran}\tau_{A_\omega \lambda_0 + q_m}$ and $\{\|C_t\| : \omega < t \leq \beta\}$ is bounded. Then it is easy to show that $\Lambda(A) = \Lambda(T)$, $A \in \operatorname{alg}\mathcal{N} \cap (SI)$ and $T \in S(A)^-$.

Let $\rho_{s-F}(T) \cap \sigma(T) \neq \varnothing$. In the first step, we have constructed an operator, denoted by A_ω, such that $\Lambda(A_\omega) = \Lambda(T)$ and $A_\omega \in \operatorname{alg}\mathcal{N}_\omega \cap (SI)$, where $\mathcal{N}_\omega := \{0;\ \bigvee_{k=1}^{n}\{e_{\omega k}\}\ (1 \leq n < \infty);\ \mathcal{H}_\omega\}$, i.e., $\mathcal{N}_\omega$ is the "restriction of $\mathcal{N}$ onto $\mathcal{H}_\omega$".

Choose pairwise distinct points $\{\lambda_t : t \text{ is a limit ordinal and } \omega < t < \beta\} \subset \sigma_{lre}(T)$ and for each t $(\omega < t < \beta)$ define

$$A_t = \lambda_t + V^* \in \mathcal{L}(\mathcal{H}_t).$$

Then $A_t \in (SI)(\mathcal{H}_t)$.

Define

$$A = \begin{bmatrix} A_\omega & C_{2\omega} & \dots & C_t & \dots & D \\ & A_{2\omega} & & & & \\ & & \ddots & & 0 & \\ & & & A_t & & \\ & & & & \ddots & \\ & 0 & & & & \lambda_0 + q_m \end{bmatrix} \begin{matrix} \mathcal{H}_\omega \\ \mathcal{H}_{2\omega} \\ \vdots \\ \mathcal{H}_t \\ \vdots \\ \mathbb{C}^m, \end{matrix}$$

where $\lambda_0 \in \sigma_{lre}(T)$ and $\lambda_0 \notin \{\lambda_t : \omega < t < \beta\}$, $C_t \in \mathcal{L}(\mathcal{H}_t, \mathcal{H}_\omega)$, $C_t \notin \operatorname{ran}\tau_{A_\omega A_t}$, $D \in \mathcal{L}(\mathbb{C}^m, \mathcal{H}_\omega)$, $D \notin \operatorname{ran}\tau_{A_\omega \lambda_0 + q_m}$ and $\{\|C_t\| : \omega < t < \beta\}$ is bounded. Note that $A_\omega \in \mathcal{B}_{n_1}(\Omega_1)$ and $\bigvee\{\ker(A_\omega - \lambda)^k : k \geq 1\} = \mathcal{H}_\omega$ for some $\lambda \in \Omega_1$. Thus it is routine to show that $A \in \operatorname{alg}\mathcal{N} \cap (SI)$, $\Lambda(A) = \Lambda(T')$ and $T' \in S(A)^-$.

3. $\mathcal{N}^\perp$ is a well-ordered nest. By 2 we can find an operator $A_1 \in \operatorname{alg}\mathcal{N}^\perp \cap (SI)$ such that $\Lambda(A_1) = \Lambda[T^*]$ and $T^* \in S(A_1)^-$. Thus $A := A_1^*$ satisfies the requirements of the proposition. □

Now we are in a position to prove Theorem 6.1.

Proof of Theorem 6.1. (i) If $\sigma(T) = \{\lambda_0\}$, for each $\epsilon > 0$ by Theorem 1.27 there exists an operator T_ϵ such that $\sigma(T_\epsilon)$ is a perfect set, $T_\epsilon \in (QT)_C$ and $\|T - T_\epsilon\| < \epsilon$. Thus it suffices to show that if $T \in (QT)_C$ and $\sigma(T)$ consists of more than one point then $T \in \mathcal{U}(\operatorname{alg}\mathcal{N} \cap (SI))^-$.

If $T \in (QT)_C$ then by Proposition 6.2, there exists an operator $A \in \operatorname{alg}\mathcal{N} \cap (SI)$ such that $\Lambda(A) = \Lambda(T)$ and $T \in S(A)^-$. Thus there is a sequence $\{X_n\}_{n=1}^\infty$ of invertible operators such that $B_n := X_n A X_n^{-1} \to T$ $(n \to \infty)$. For each n, there is a unitary operator U_n such that $C_n := U_n B_n U_n^* \in \operatorname{alg}\mathcal{N}$, i.e., $C_n \in \operatorname{alg}\mathcal{N} \cap (SI)$ and $U_n^* C_n U_n \to T$ $(n \to \infty)$. Therefore $T \in \mathcal{U}(\operatorname{alg}\mathcal{N} \cap (SI))^-$ and $(QT)_C \subset \mathcal{U}(\operatorname{alg}\mathcal{N} \cap (SI))^-$.

In order to show the converse inclusion, it suffices to prove that if $T \in (QT)$ with connected $\sigma(T)$ and disconnected $\sigma_W(T)$, then $T \notin \mathcal{U}(\operatorname{alg}\mathcal{N} \cap (SI))^-$. In fact, since $\sigma(T)$ is connected and $\sigma_W(T)$ is not, there is a component Ω of $\rho_F^\circ(T) \cap \sigma(T)$ such that $\sigma(T)\backslash\Omega$ and $\partial\Omega$ are disconnected. Thus there is a compact subset σ of Ω such that $\sigma(T)\backslash\sigma$ is disconnected. Let $T_n \in \mathcal{U}(\operatorname{alg}\mathcal{N} \cap (SI))$ and $T_n \to T$ $(n \to \infty)$ and let $\lambda \in \sigma$. By the stability property of semi-Fredholm operators and the compactness of σ there exists an $N \in \mathbb{N}$ such that $\operatorname{ind}(T_n - \lambda) = 0$ for $\lambda \in \sigma$ and $n > N$. Since $\sigma_p(T_n^*) \cap \rho_{s-F}^r(T_n^*) = \varnothing$, $\lambda \in \rho(T_n)$ $(n > N)$. This implies that $\sigma(T_n)$ is disconnected when $n > N$, a contradiction. By the same argument we can prove (ii). $\square$

6.2 The $\mathcal{U}(\operatorname{alg}\mathcal{N} \cap (SI))^-$ of continuous nests

In this section, we will prove the following two theorems.

Theorem 6.3. *Let $\mathcal{N}$ be a continuous nest. Then $\mathcal{U}(\operatorname{alg}\mathcal{N} \cap (SI))^- = \{T \in \mathcal{L}(\mathcal{H}) : \sigma(T)$ is connected$\}$.*

Theorem 6.4. *Let $\mathcal{N}$ be a nest with at least one infinite dimensional atom. Then $\mathcal{U}(\operatorname{alg}\mathcal{N} \cap (SI))^- = \{T \in \mathcal{L}(\mathcal{H}) : \sigma(T)$ is connected$\}$.*

Given an operator $T \in \mathcal{L}(\mathcal{H})$ with connected $\sigma(T)$ and given $\epsilon > 0$, by Theorem 1.27 there is an operator $T_\epsilon \in \mathcal{L}(\mathcal{H})$ with $\sigma(T_\epsilon)$ connected such that $\sigma_{lre}(T_\epsilon)$ is the closure of an analytic Cauchy domain and $\|T - T_\epsilon\| < \epsilon$. Thus for the nests of Theorems 6.3 and 6.4, it suffices to show that for each operator T with connected $\sigma(T)$ and whose $\sigma_{lre}(T)$ is the closure of an analytic Cauchy domain we can always find an (SI) operator A in $\operatorname{alg}\mathcal{N}$ such that $\|UAU^* - T\| < \epsilon$, where U is a unitary operator. In order to prove this, we need several lemmas.

Lemma 6.5. *Given a connected compact subset σ of $\mathbb{C}$ and pairwise disjoint connected open subsets Ω_k $(0 \le k \le l,\ 0 \le l \le \infty)$ of σ and given a sequence $\{n_k\}_{k=1}^{l}$ of numbers such that $\{n_k\}_{k=0}^{l} \subset \mathbb{N} \cup \{\infty\}$, $n_0 = \infty$ and $1 \le n_k \le \infty$ $(k \ge 1)$, then there exists an operator A in $\mathcal{B}_\infty(\Omega_0) \cap (SI)$ satisfying*

(i) $\sigma(A) = \sigma$, $\sigma_{lre}(A) = \sigma \backslash \bigcup_{k=0}^{l} \Omega_k$;

(ii) $\operatorname{ind}(A - \lambda) = \operatorname{nul}(A - \lambda) = n_k$ *(for all $\lambda \in \Omega_k$, $k = 0, 1, \cdots, l$).*

Proof. Denote $\Phi_k = (\overline{\Omega}_k)^\circ$, let $N_+(\Phi_k^*)$ be the Bergmann operator on $A^2(\Phi_k^*)$ and denote $A_0 = N_+(\Phi_0^*)^*$ and $A_k = N_+(\Phi_k^*)^{*(n_k)}$ $(k = 1, 2, \cdots, l)$. Thus $\sigma(A_0) = \overline{\Omega}_0$, $A_0 \in \mathcal{B}_1(\Phi_0) \cap (SI)$, $\sigma(A_k) = \overline{\Omega}_k$ and $A_k \in \mathcal{B}_{n_k}(\Phi_k)$ $(k = 1, 2, \cdots, l)$.

Let $\{\lambda_k\}_{k=1}^{\infty}$ be a dense subset of $\sigma \backslash \bigcup_{k=0}^{l} \Omega_k$. Set $T_k = \lambda_k + V^*$, where V is given in Example 3.22, and define

$$G = A_0 \oplus \left(\bigoplus_{k=1}^{l} A_k \right) \oplus \left(\bigoplus_{k=1}^{\infty} T_k \right).$$

Then G satisfies

(a) $\sigma(G) = \sigma_W(G) = \sigma$, $\sigma_{lre}(G) = \sigma \backslash \bigcup_{k=0}^{l} \Omega_k$;

(b) $\text{ind}(G-\lambda) = \text{nul}(G-\lambda) = 1$ for $\lambda \in \Omega_0$;

(c) $\text{ind}(G-\lambda) = \text{nul}(G-\lambda) = n_k$ for $\lambda \in \Omega_k$ $(k = 1, 2, \cdots, l)$.

By Theorem 1.44 for each $\epsilon > 0$, there exists a compact operator K with $\|K\| < \epsilon$ such that $G + K \in \mathcal{B}_1(\Omega_0)$. It is completely apparent that $G + K$ satisfies (a), (b) and (c).

Without loss of generality, we may assume that $0 \in \Omega_0$. By a similar argument to Proposition 6.2, we can find a compact operator K_1, $\|K_1\| < \epsilon$, such that $(G+K) \oplus A_0^{(\infty)} + K_1 \in \mathcal{B}_\infty(\Omega_0) \cap (SI)$. Denote $A = (G+K) \oplus A_0^{(\infty)} + K_1$; then A satisfies the requirements of the lemma. □

Lemma 6.6. *Let $\mathcal{N}$ be a continuous nest and $T \in \mathcal{L}(\mathcal{H})$ with connected spectrum $\sigma(T)$ and let $\sigma_{lre}(T)$ be the closure of an analytic Cauchy domain. Then there exists an operator $A \in \text{alg}\,\mathcal{N} \cap (SI)$ such that*

(i) $\Lambda(A) = \Lambda(T)$;

(ii) $\min\text{ind}(A-\lambda) = \begin{cases} 1, & \lambda \in \rho_{s-F}^{\circ}(A) \cap \sigma(A), \\ 0, & \text{ind}(A-\lambda) \neq 0 \end{cases} \quad (\lambda \in \rho_{s-F}(A))$;

(iii) $\rho_{s-F}^{s}(A) = \varnothing$.

Proof. Since $\sigma_{lre}(T)$ is a closure of an analytic Cauchy domain, $\sigma_{lre}(T)$ consists of finitely many components. Let $\{\sigma_k\}_{k=1}^{l_1}$ $(1 \le l_1 < \infty)$ and $\{\sigma_{-k}\}_{k=1}^{l_2}$ $(1 \le l_2 < \infty)$ be the components of $\sigma(T) \backslash \rho_{s-F}^{-}(T)$ and, respectively, $\sigma(T) \backslash \rho_{s-F}^{+}(T)$. Choose an open disc $\Omega_k \subset \sigma_k \cap \sigma_{lre}(T)$ $(-l_2 \le k \le l_1,\ k \neq 0)$ (if $[\sigma_{-k} \cap \sigma_{lre}(T)] \cap [\sigma_j \cap \sigma_{lre}(T)] \neq \varnothing$, then choose $\Omega_{-k} = \Omega_j$). According to Lemma 6.5

- Construct $A_k \in \mathcal{B}_\infty(\Omega_k) \cap (SI)$ $(1 \le k \le l_1)$ such that

$$\sigma(A_k) = \sigma_k,\ \sigma_{lre}(A_k) = \sigma_k \cap (\sigma_{lre}(T) \backslash \Omega_k),$$
$$\text{ind}(A_k - \lambda) = \text{nul}(A_k - \lambda) = \text{ind}(T-\lambda) \text{ for } \lambda \in \sigma_k \cap \rho_{s-F}^{+}(T),$$
$$\text{ind}(A_k - \lambda) = \text{nul}(A_k - \lambda) = 1 \text{ for } \lambda \in \sigma_k \cap \rho_{s-F}^{\circ}(T) \cap \sigma(T);$$

- Construct A_{-k} $(1 \le k \le l_2)$ such that $A_{-k}^* \in \mathcal{B}_\infty(\Omega_{-k}^*) \cap (SI)$,

$$\sigma(A_{-k}) = \sigma_{-k},\ \sigma_{lre}(A_{-k}) = \sigma_{-k} \cap (\sigma_{lre}(T) \backslash \Omega_{-k}),$$
$$\text{ind}(A_{-k} - \lambda) = -\text{nul}(A_{-k} - \lambda)^* = \text{ind}(T-\lambda) \text{ for } \lambda \in \sigma_{-k} \cap \rho_{s-F}^{-}(T),$$
$$\text{ind}(A_{-k} - \lambda) = -\text{nul}(A_{-k} - \lambda)^* = -1 \text{ for } \lambda \in \sigma_{-k} \cap \rho_{s-F}^{\circ}(T) \cap \sigma(T).$$

Set $A_+ = \bigoplus_{k=1}^{l_1} A_k \in \mathcal{L}(\mathcal{H}^{(l_1)})$, $A_- = \bigoplus_{k=1}^{l_2} A_{-k} \in \mathcal{L}(\mathcal{H}^{(l_2)})$. Define $Q \in \mathcal{L}(\mathcal{H}^{(l_2)}, \mathcal{H}^{(l_1)})$ as follows.

Assume that $Q = \begin{bmatrix} Q_{11} & \cdots & Q_{1\,l_2} \\ & \cdots \quad \cdots & \\ Q_{l_1 1} & \cdots & Q_{l_1 l_2} \end{bmatrix}$,

$$Q_{ij} \begin{cases} \notin \operatorname{ran} \tau_{A_i A_{-j}},\ Q_{ij} \text{ is compact}, & \text{if } \sigma(A_i) \cap \sigma(A_{-j}) \neq \varnothing, \\ = 0, & \text{otherwise.} \end{cases}$$

Define $\overline{A} = \begin{bmatrix} A_+ & Q \\ 0 & A_- \end{bmatrix} \begin{matrix} \mathcal{H}^{(l_1)} \\ \mathcal{H}^{(l_2)} \end{matrix}$. Assume that $P = \begin{bmatrix} P_{11} & P_{12} \\ P_{21} & P_{22} \end{bmatrix} \begin{matrix} \mathcal{H}^{(l_1)} \\ \mathcal{H}^{(l_2)} \end{matrix}$ is an idempotent commuting with $\overline{A}$. By Proposition 1.14 $\ker \tau_{A_- A_+} = \{0\}$, $\ker \tau_{A_k A_{k'}} = \{0\}$ ($1 \leq k,\ k' \leq l_1$ and $k \neq k'$) and $\ker \tau_{A_{-k} A_{-k'}} = \{0\}$ ($1 \leq k,\ k' \leq l_2$ and $k \neq k'$). Thus $P_{21} = 0$. Since P_{11} and P_{22} are idempotents and $A_k,\ A_{-k} \in (SI)$, $P_{11} = \bigoplus_{i=1}^{l_1} \epsilon_i$ and $P_{22} = \bigoplus_{i=1}^{l_2} \delta_i$, ϵ_i and δ_i are 0 or 1. Without loss of generality we may assume that $\epsilon_1 = 1$ (otherwise consider $1 - P$). Suppose that $P_{12} = \begin{bmatrix} E_{11} & \cdots & E_{1\,l_2} \\ & \cdots \quad \cdots & \\ E_{l_1 1} & \cdots & E_{l_1 l_2} \end{bmatrix}$. Since $\sigma(\overline{A}) \supset \sigma(T)$, $\Lambda_1 := \{-k :\ \sigma(A_{-k}) \cap \sigma(A_1) \neq \varnothing\} \neq \varnothing$. If $-k \in \Lambda_1$, computation shows that $E_{1k} A_{-k} = A_1 E_{1k} + Q_{1k} \delta_k$. Since $Q_{1k} \notin \operatorname{ran} \tau_{A_1 A_{-k}}$, $\delta_k = 0$. Similarly, $\epsilon_i = 0$ if $\sigma(A_i) \cap \sigma(A_{-k}) \neq \varnothing$. Since $\sigma(\overline{A}) = \sigma(T)$ and $l_1 + l_2 < \infty$, after a finite number of steps, $\epsilon_i = \delta_k = 0$ ($1 \leq i \leq l_1$, $1 \leq k \leq l_2$), i.e., $P = 0$ and $\overline{A} \in (SI)$. It is not difficult to see that $\Lambda(\overline{A}) = \Lambda(T)$, and

$$\min \operatorname{ind} (\overline{A} - \lambda) = \begin{cases} 1, & \lambda \in \rho^{\circ}_{s-F}(T) \cap \sigma(T), \\ 0, & \operatorname{ind} (T - \lambda) \neq 0. \end{cases}$$

By Proposition 2.34, $\rho^{s}_{s-F}(\overline{A}) = \varnothing$. Since $A_k \in \mathcal{B}_\infty(\Omega_k)$ and $A^*_{-k} \in \mathcal{B}_\infty(\Omega^*_{-k})$, there exist $\{\lambda_k\}_{k=1}^{\infty}$ and $\{\lambda_{-k}\}_{k=1}^{\infty}$ such that

(i) $\bigvee \{\ker (A_+ - \lambda_k) :\ k \geq 0\} = \mathcal{H}^{(l_1)}$, $\bigvee \{\ker (A_- - \lambda_k)^* :\ k \leq -1\} = \mathcal{H}^{(l_2)}$;

(ii) if $\mathcal{H}_1 = \ker (A_+ - \lambda_1)$, $\mathcal{H}_{n+1} = (\mathcal{H}_n + \ker (A_+ - \lambda_{n+1})) \ominus \mathcal{H}_n$ ($n \geq 1$), then $\dim \mathcal{H}_n = \infty$ ($n = 1, 2, \cdots$) and $\bigoplus_{n=1}^{\infty} \mathcal{H}_n = \mathcal{H}^{(l_1)}$;

(iii) if $\mathcal{H}_{-1} = \ker (A_- - \lambda_{-1})^*$, $\mathcal{H}_{-(n+1)} = (\mathcal{H}_{-n} + \ker (A_- - \lambda_{-(n+1)})^*) \ominus \mathcal{H}_{-n}$ ($n \geq 1$), then $\dim \mathcal{H}_{-n} = \infty$ ($n = 1, 2, \cdots$) and $\bigoplus_{n=1}^{\infty} \mathcal{H}_{-n} = \mathcal{H}^{(l_2)}$.

Let $\{t_n\}_{n=-\infty}^{\infty}$ be a two-sided sequence of numbers such that

$$0 = t_0 < t_1 < t_2 < \cdots < t_n < \cdots < \frac{1}{2} < \cdots < t_{-n} < \cdots < t_{-2} < t_{-1} = 1$$

and $\lim_{n \to \pm\infty} t_n = 1/2$. Assume that E is the spectral measure of $\mathcal{N}$ and let U be the unitary operator satisfying $U \mathcal{H}_n = \operatorname{ran} E[t_{n-1}, t_n]$ ($n = \pm 1, \pm 2, \cdots$). Thus $A := U \overline{A} U^* \in \operatorname{alg} \mathcal{N} \cap (SI)$ and A satisfies all the requirements of the lemma. □

Proof of Theorem 6.3. D.A. Herrero proved that if $\mathcal{N}$ and $\mathcal{N}^{\perp}$ are not well-ordered nests with finite dimensional atoms, then

$$\mathcal{U}(\operatorname{alg}\mathcal{N})^{-} = \{T \in \mathcal{L}(\mathcal{H}) : \ \sigma_0(T)\backslash\sigma_e(T)^{\wedge} = \varnothing\},$$

where $\sigma_e(T)^{\wedge}$ is the polynomially convex hull of $\sigma_e(T)$. Thus, considering the analysis before Lemma 6.5, it suffices to show that $\Delta := \{T \in \mathcal{L}(\mathcal{H}) : \sigma(T)$ is connected and $\sigma_{lre}(T)$ is the closure of an analytic Cauchy domain$\} \subset \mathcal{U}(\operatorname{alg}\mathcal{N} \cap (SI))^{-}$.

Assume that $T \in \Delta$. By Lemma 6.6 there exists an operator $A \in \operatorname{alg}\mathcal{N} \cap (SI)$ satisfying (i), (ii) and (iii) of Lemma 6.6. It follows from the similarity orbit theorem that for each $\epsilon > 0$, there is an invertible operator X such that $\|XAX^{-1} - T\| < \epsilon$. By the argument of Lemma 6.6, we can find a unitary operator U such that $A_\epsilon = UXAX^{-1}U^* \in \operatorname{alg}\mathcal{N} \cap (SI)$ and $\|U^*A_\epsilon U - T\| < \epsilon$. Since ϵ can be arbitrarily small $T \in \mathcal{U}(\operatorname{alg}\mathcal{N} \cap (SI))^{-}$. □

Proof of Theorem 6.4. Let $\mathcal{H} = \mathcal{H}_\alpha \oplus \mathcal{H}_0 \oplus \mathcal{H}_\beta$, where $\mathcal{H}_0$ is an atom of $\mathcal{N}$ and $\dim\mathcal{H}_0 = \infty$. It suffices to show that $\Delta = \{T \in \mathcal{L}(\mathcal{H}) : \sigma(T)$ is connected and $\sigma_{lre}(T)$ is the closure of an analytic Cauchy domain$\} \subset \mathcal{U}(\operatorname{alg}\mathcal{N} \cap (SI))^{-}$.

If $T \in \Delta$, by Lemma 6.6 there exists an operator $A \in (SI)(\mathcal{H})$ such that A admits the following representation

$$A = \begin{bmatrix} \lambda_1 & & & & & & * \\ & \lambda_2 & & & & & \\ & & \lambda_3 & & & & \\ & & & \ddots & & & \\ & & & & \lambda_{-3} & & \\ & 0 & & & & \lambda_{-2} & \\ & & & & & & \lambda_{-1} \end{bmatrix} \begin{matrix} \mathcal{H}_1 \\ \mathcal{H}_2 \\ \mathcal{H}_3 \\ \vdots \\ \mathcal{H}_{-3} \\ \mathcal{H}_{-2} \\ \mathcal{H}_{-1} \end{matrix}$$

with respect to the decomposition of the space $\mathcal{H} = \{\bigoplus \mathcal{H}_k : \ k = \pm 1, \pm 2, \cdots\}$, where $\dim\mathcal{H}_k = \infty$ $(k = \pm 1, \pm 2, \cdots)$, and $\Lambda(A) = \Lambda(T)$, $T \in S(A)^{-}$. Thus, for each $\epsilon > 0$, there is an invertible operator X such that $\|XAX^{-1} - T\| < \epsilon$. Observe that XAX^{-1} admits the same representation with respect to some decomposition of the space $\mathcal{H}$ as A has with respect to $\mathcal{H} = \bigoplus \mathcal{H}_k$. Thus, for simplicity we assume that $\|A - T\| < \epsilon$. Decompose $\mathcal{H}_1 = \mathcal{H}'_\alpha \oplus \mathcal{H}'_1$ and $\mathcal{H}_\beta = \mathcal{H}'_{-1} \oplus \mathcal{H}'_\beta$ so that $\dim\mathcal{H}'_\alpha = \dim\mathcal{H}_\alpha$ and $\dim\mathcal{H}'_\beta = \dim\mathcal{H}_\beta$. Find a unitary operator U so that $U\mathcal{H}_\alpha = \mathcal{H}'_\alpha$, $U\mathcal{H}_0 = \mathcal{H}'_1 \oplus \left(\bigoplus_{n=2}^{\infty}\mathcal{H}_n\right) \oplus \left(\bigoplus_{n=-\infty}^{-2}\mathcal{H}_n\right) \oplus \mathcal{H}'_{-1}$ and $U\mathcal{H}_\beta = \mathcal{H}'_\beta$.

Thus

$$U^*AU = \begin{bmatrix} \lambda_1 & & & & & & & \\ & \lambda_1 & & & & * & & \\ & & \lambda_2 & & & & & \\ & & & \ddots & & & & \\ & & & & \lambda_{-2} & & & \\ & 0 & & & & \lambda_{-1} & & \\ & & & & & & \lambda_{-1} \end{bmatrix} \begin{matrix} \mathcal{H}_\alpha \\ \mathcal{H}_0 \\ \vdots \\ \vdots \\ \\ \mathcal{H}_0 \\ \mathcal{H}_\beta. \end{matrix}$$

Thus $U^*AU \in \operatorname{alg}\mathcal{N} \cap (SI)$. Since ϵ is arbitrary, $T \in \mathcal{U}(\operatorname{alg}\mathcal{N} \cap (SI))^-$. □

6.3 The $\mathcal{U}(\operatorname{alg}\mathcal{N} \cap (SI))^-$ of non-well-ordered nests

Theorems 6.1, 6.3 and 6.4 imply that when $\mathcal{N}$ is some special nests, $\operatorname{alg}\mathcal{N}$ contains a large number of strongly irreducible operators. The main result in this section asserts that each nest algebra contains (SI) operators.

Theorem 6.7. *Let $\mathcal{N}$ be a maximal nest. Then $\mathcal{U}(\operatorname{alg}\mathcal{N} \cap (SI))^- = \{T \in \mathcal{L}(\mathcal{H}) : \sigma(T)$ is connected$\}$ if and only if $\mathcal{N}$ and $\mathcal{N}^\perp$ are not well-ordered.*

In order to prove Theorem 6.7, we need several lemmas.

Lemma 6.8. *Let σ be the closure of a connected Cauchy domain and Ω be an open disc in σ. Then there exists an operator $A \in \mathcal{L}(\mathcal{H}) \cap (SI)$ such that*

(i) $\sigma(A) = \sigma_{lre}(A) = \sigma$;

(ii) $\sigma_p(A) = \Omega$, $\operatorname{nul}(A-\lambda) = 1$ $(\lambda \in \Omega)$, *and* $\sigma_p(A^*) = \varnothing$;

(iii) *if* $\{\lambda_k\}_{k=1}^\infty \subset \Omega$ *is pairwise distinct and* $\lim_{k\to\infty} \lambda_k = \lambda_0 \in \Omega$, *then* $\bigvee\{\ker(A-\lambda_k) : k \geq 1\} = \mathcal{H}$;

(iv) $\|(A-\lambda)^{-1}\| \leq 2/\operatorname{dist}(\lambda, \sigma)$ *for* $\lambda \notin \sigma$.

Proof. Without loss of generality we may assume that Ω is the unit disc. Let S be the backward lateral shift, i.e., $S^* = T_z^* \in \mathcal{L}(\mathcal{H}_1)$, where $\mathcal{H}_1 = H^2$. Let M be a diagonal operator on $\mathcal{H}_1$ with $\sigma(M) = \sigma_{lre}(M) = \sigma$. Set $T = S^* \oplus M$. By Theorem 1.50 for each $\epsilon > 0$, there is a compact operator K, $\|K\| < \epsilon$, such that $A = T + K$ is quasisimilar to $T_z^* \in \mathcal{B}_1(\Omega)$. Choose ϵ small enough, so that A satisfies (i)–(iv). □

Lemma 6.9. *Let σ be the closure of a connected Cauchy domain and Ω be a connected open subset of σ. Then there exists an operator $W \in \mathcal{L}(\mathcal{H}) \cap (SI)$ satisfying*

(i) $\sigma(W) = \sigma_{lre}(W) = \sigma$;

(ii) $\sigma_p(W) \subset \Omega$, $\sigma_p(W^*) = \varnothing$;

(iii) *there exists* $\{\lambda_k\}_{k=1}^{\infty} \subset \Omega$ *such that* $\lim_{k\to\infty} \lambda_k = \lambda_0 \in \Omega$, $\mathrm{nul}\,(W - \lambda_k) = \infty$ $(k \geq 1)$ *and* $\bigvee\{\ker(W - \lambda_k) : k \geq 1\} = \mathcal{H}$.

Proof. Choose a sequence $\{D_n\}_{n=0}^{\infty}$ of open discs in Ω satisfying $D_n \backslash \overline{D}_m \neq \varnothing$ $(n \neq m, n \neq 0)$ and $D_0 \subset \bigcap_{n=1}^{\infty} D_n$.

Without loss of generality we may assume that D_0 is the unit disc and $D_1 = \alpha_1 + rD_0$. Let $S^* = T_z^* \in \mathcal{L}(\mathcal{H}_1)$, where $\mathcal{H}_1 = H^2$. Set $A_1 = \alpha_1 + rS^*$. Let $\mathcal{H} = \bigoplus_{n=1}^{\infty} \mathcal{H}_n$, where $\mathcal{H}_n = \mathcal{H}_1$ $(n \geq 2)$. For each $n \geq 2$, by Lemma 6.8, we can construct $A_n \in \mathcal{L}(\mathcal{H}_n) \cap (SI)$ satisfying

(a) $\sigma(A_n) = \sigma_{lre}(A_n) = \sigma$, $\sigma_p(A_n) = D_n$, $\sigma_p(A_n^*) = \varnothing$ and $\mathrm{nul}\,(A_n - \lambda) = 1$ for $\lambda \in D_n$;

(b) if $\{\mu_k\}_{k=1}^{\infty} \subset D_n$, pairwise distinct and $\lim\limits_{k\to\infty} \mu_k = \mu_0 \in D_n$, then $\bigvee\{\ker(A_n - \mu_k) : k \geq 1\} = \mathcal{H}_n$;

(c) $\|(A_n - \lambda)^{-1}\| \leq 2/\mathrm{dist}\,(\lambda, \sigma)$ for $\lambda \notin \sigma$.

It follows from $D_n \backslash \overline{D}_m \neq \varnothing$, (b) and Proposition 1.14 that $\ker \tau_{A_n A_m} = \{0\}$ $(n \neq m)$. Since $\sigma_r(A_1) \cap \sigma_l(A_n) \neq \varnothing$, by Theorem 1.11, we can find a compact operator $W_n \in \mathcal{L}(\mathcal{H}_n, \mathcal{H}_1)$, $\|W_n\| < 2^{-n}$, such that $W_n \notin \mathrm{ran}\,\tau_{A_1 A_n}$ $(n \geq 2)$.

Define

$$W = \begin{bmatrix} A_1 & W_2 & W_3 & \cdots \\ & A_2 & & 0 \\ & & A_3 & \\ & 0 & & \ddots \end{bmatrix} \in \mathcal{L}(\mathcal{H}).$$

By Corollary 3.14, $W \in (SI)(\mathcal{H})$. Let $\{\lambda_k\}_{k=1}^{\infty} \subset D_0$ be an arbitrary sequence such that $\lim_{k\to\infty} \lambda_k = \lambda_0 \in D_0$, pairwise distinct. Then $\bigvee\{\ker\big(\bigoplus_{n=2}^{\infty} A_n - \lambda_k\big) : k \geq 1\} = \bigoplus_{n=2}^{\infty} \mathcal{H}_n$ and $\bigvee\{\ker(A_1 - \lambda_k) : k \geq 1\} = \mathcal{H}_1$. Note that $\{\lambda_k\}_{k=1}^{\infty} \subset \rho_r(A_1)$, thus $\bigvee\{\ker(W - \lambda_n) : n \geq 1\} = \mathcal{H}$ and $\mathrm{nul}\,(W - \lambda_n) = \infty$ $(n = 0, 1, 2, \cdots)$. Since $\sigma_p(A_k) \subset D_k$ and $\sigma_p(A_k^*) = \varnothing$ $(k = 1, 2, \cdots)$, computation indicates that $\sigma_p(W) \subset \Omega$ and $\sigma_p(W^*) = \varnothing$. Observe that $W = \bigoplus_{n=1}^{\infty} A_n + K$, where K is a compact operator and $\|(A_n - \lambda)^{-1}\| < 2/\mathrm{dist}\,(\lambda, \sigma)$ for $\lambda \notin \sigma$ and for $n \geq 1$ we have $\sigma\big(\bigoplus_{n=1}^{\infty} A_n\big) = \sigma_{lre}\big(\bigoplus_{n=1}^{\infty} A_n\big) = \sigma$. Since $\sigma(W)$ is connected and $\sigma_p(W^*) = \varnothing$, $\sigma(W) = \sigma_{lre}(W) = \sigma$. □

Lemma 6.10. *Let* $T \in \mathcal{L}(\mathcal{H})$ *with connected spectrum* $\sigma(T)$ *and let* $\sigma_{lre}(T)$ *be the closure of an analytic Cauchy domain. Then there exists an operator* $A \in (SI)$ *satisfying*

(i) $\Lambda(A) = \Lambda(T)$;

(ii) $\min \operatorname{ind}(A-\lambda)=\begin{cases}0, & \operatorname{ind}(T-\lambda)\neq 0,\\ 1, & \lambda\in\rho_{s-F}^{\circ}(T)\cap\sigma(T);\end{cases}$

(iii) *A admits a representation* $A=\begin{bmatrix}A_1 & *\\ 0 & A_2\end{bmatrix}\begin{matrix}\mathcal{K}_1\\ \mathcal{K}_2\end{matrix}$ *and there is a subset* $\{\lambda_k : k = 0,\pm1,\pm2,\cdots\}$ *of complex numbers such that* $\operatorname{nul}(A_1-\lambda_k)=\infty$ $(k\geq 0)$, $\operatorname{nul}(A_2-\lambda_k)^*=\infty$ $(k<0)$, $\bigvee\{\ker(A_1-\lambda_k) : k\geq 0\}=\mathcal{K}_1$ *and* $\bigvee\{\ker(A_2-\lambda_k)^* : k<0\}=\mathcal{K}_2$, *where* $\mathcal{K}_1$, $\mathcal{K}_2$ *are infinite dimensional Hilbert spaces;*

(iv) *there is an open* disc $G\subset\sigma_{lre}(A)$ *such that* $G\cap\sigma_p(A_1)=G^*\cap\sigma_p(A_2^*)=\varnothing$.

Proof. Choose an open disc G_1 such that $\overline{G}_1\subset\sigma_{lre}(T)^{\circ}$. Denote $\sigma=\sigma(T)\backslash G_1$. Then σ is connected and $\sigma\cap\sigma_{lre}(T)$ is still the closure of an analytic Cauchy domain. Let $\{\sigma_k\}_{k=0}^{l_1}$ and $\{\sigma_{-k}\}_{k=1}^{l_2}$ be the components of $\sigma\backslash\rho_{s-F}^{-}(T)$ and, respectively, $\sigma\backslash\rho_{s-F}^{+}(T)$. For each k $(-l_2\leq k\leq l_1)$ choose an open disc Ω_k such that $\overline{\Omega}_k\subset[\sigma_k\cap\sigma_{lre}(T)]^{\circ}$ (if there is more than one k, $(\sigma_k\cap\sigma_{lre}(T))\cap(\sigma_{-j}\cap\sigma_{lre}(T))\neq\varnothing$ let Ω_{-j} equal one of the Ω_k's.) By Lemma 6.5 there is a B_k $(-l_2\leq k\leq l_1)$ such that

(i) If $k\geq 0$, $B_k\in\mathcal{B}_\infty(\Omega_k)\cap(SI)(\mathcal{H}_k)$, $\sigma(B_k)=\sigma_k$, $\sigma_{lre}(B_k)=\sigma_k\cap[\sigma_{lre}(T)\backslash\Omega_k]$, $\operatorname{ind}(B_k-\lambda)=\operatorname{nul}(B_k-\lambda)=\operatorname{ind}(T-\lambda)$ for $\lambda\in\sigma_k\cap\rho_{s-F}^{+}(T)$, $\operatorname{ind}(B_k-\lambda)=\operatorname{nul}(B_k-\lambda)=1$ for $\lambda\in\sigma_k\cap\rho_{s-F}^{\circ}(T)$;

(ii) If $k<0$, $B_k^*\in\mathcal{B}_\infty(\Omega_k^*)\cap(SI)(\mathcal{H}_k)$, $\sigma(B_k)=\sigma_k$, $\sigma_{lre}(B_k)=\sigma_k\cap[\sigma_{lre}(T)\backslash\Omega_k]$, $\operatorname{ind}(B_k-\lambda)=-\operatorname{nul}(B_k-\lambda)^*=\operatorname{ind}(T-\lambda)$ for $\lambda\in\sigma_k\cap\rho_{s-F}^{-}(T)$, $\operatorname{ind}(B_k-\lambda)=-\operatorname{nul}(B_k-1)^*=-1$ for $\lambda\in\sigma_k\cap\rho_{s-F}^{\circ}(T)$.

Choose open discs G and G_2 such that $\overline{G}\cup\overline{G}_2\subset G_1$ and $\overline{G}\cap\overline{G}_2=\varnothing$. By Lemma 6.9 we can construct an operator $W\in(SI)(\mathcal{K})$ satisfying

(i) $\sigma(W)=\sigma_{lre}(W)=\overline{G}_1$;

(ii) $\sigma_p(W)\subset G_2$, $\sigma_p(W^*)=\varnothing$;

(iii) there exists a sequence $\{\mu_k\}_{k=0}^{\infty}\subset G_2$ of distinct numbers such that $\lim_{k\to\infty}\mu_k=\mu_0$, $\operatorname{nul}(W-\mu_k)=\infty$ $(k\geq 1)$ and $\bigvee\{\ker(W-\mu_k) : k\geq 1\}=\mathcal{K}$.

For each k $(0\leq k\leq l_1)$, choose $R_k\in\mathcal{L}(\mathcal{H}_k,\mathcal{K})$ by

$$R_k\begin{cases}=0, & \text{if } \sigma(B_k)\cap\sigma(W)=\varnothing,\\ \notin\operatorname{ran}\tau_{WB_k} \text{ and } R_k \text{ is compact}, & \text{otherwise (Theorem 1.11)}.\end{cases}$$

Set $R=(R_0,R_1,\cdots,R_{l_1})$.

For each pair (i,j) $(0\leq i\leq l_1;\ 1\leq j\leq l_2)$ choose $Q_{ij}\in\mathcal{L}(\mathcal{H}_{-j},\mathcal{H}_i)$ by

$$Q_{ij}\begin{cases}=0, & \text{if } \sigma_i\cap\sigma_{-j}=\varnothing,\\ \notin\operatorname{ran}\tau_{B_iB_{-j}},\ Q_{ij} \text{ is compact}, & \text{if } \sigma_i\cap\sigma_{-j}\neq\varnothing.\end{cases}$$

Set

$$Q = \begin{bmatrix} Q_{01} & Q_{02} & \cdots & Q_{0l_2} \\ \vdots & \vdots & \ddots & \vdots \\ Q_{l_1 1} & Q_{l_1 2} & \cdots & Q_{l_1 l_2} \end{bmatrix} \in \mathcal{L}\left(\bigoplus_{k=1}^{l_2} \mathcal{H}_{-k}, \bigoplus_{k=0}^{l_1} \mathcal{H}_k\right).$$

Define

$$A = \begin{bmatrix} W & R & 0 \\ 0 & \bigoplus_{k=0}^{l_1} B_k & Q \\ 0 & 0 & \bigoplus_{k=1}^{l_2} B_{-k} \end{bmatrix} = \begin{bmatrix} A_1 & * \\ 0 & A_2 \end{bmatrix} \begin{matrix} \mathcal{K}_1 \\ \mathcal{K}_2, \end{matrix}$$

where $\mathcal{K}_1 = \mathcal{K} \oplus \left(\bigoplus_{k=0}^{l_1} \mathcal{H}_k\right)$, $\mathcal{K}_2 = \bigoplus_{k=1}^{l_2} \mathcal{H}_{-k}$, $A_1 = \begin{bmatrix} W & R \\ 0 & \bigoplus_{k=0}^{l_1} B_k \end{bmatrix}$ and $A_2 = \bigoplus_{k=1}^{l_2} B_{-k}$. It follows from the properties of W, B_k $(-l_2 \leq k \leq l_1)$ and Proposition 1.14 that $\ker \tau_{B_k B_{k'}} = \ker \tau_{B_{-k} B_{-k'}} = 0$ $(k \neq k')$, $\ker \tau_{\bigoplus_{k=1}^{l_2} B_{-k} \bigoplus_{k=0}^{l_1} B_k} = \ker \tau_{\bigoplus_{k=1}^{l_2} B_{-k} W} = \ker \tau_{\bigoplus_{k=0}^{l_1} B_k W} = \{0\}$. Since W and each B_k $(-l_2 \leq k \leq l_1)$ are strongly irreducible, by Lemma 4.24 $A \in (SI)$. From the construction of A, we can get (i) and (ii). Note that $\sigma\left(\bigoplus_{k=0}^{l_1} B_k\right) \cap \overline{G} \subset \sigma\left(\bigoplus_{k=0}^{l_1} B_k\right) \cap G_1 \subset \sigma \cap G_1 = \varnothing$ and $\sigma\left(\bigoplus_{k=1}^{l_2} B_{-k}\right) \cap \overline{G} \subset \sigma\left(\bigoplus_{k=1}^{l_2} B_{-k}\right) \cap G_1 \subset \sigma \cap G_1 = \varnothing$. Since $\sigma_p(W) \subset G_2$ and $\sigma_p(W^*) = \varnothing$, $\sigma_p(A_1) \cap G = \sigma_p(A_2^*) \cap G^* = \varnothing$. Since $\Omega_k \cap G_1 = \varnothing$ $(-l_2 \leq k \leq l_1)$, there are $\{\lambda_k\}_{k=1}^{\infty} \subset \sigma_p(A_1)$ and $\{\lambda_{-k}^*\}_{k=1}^{\infty} \subset \sigma_p(A_2^*)$ satisfy (iii). □

Lemma 6.11. *Let σ be the closure of a connected Cauchy domain and let $\{\sigma_k\}_{k=0}^{\infty}$ and $\{\Omega_k\}_{k=1}^{\infty}$ be two classes of subsets of σ° satisfying*

(i) *each σ_k is a connected Cauchy domain;*

(ii) *$\sigma_k \subset \sigma_{k+1}$ and $\sigma_{k+1} \backslash \overline{\sigma}_k$ is a connected Cauchy domain $(k = 0, 1, \cdots)$;*

(iii) *$\sigma = \left[\bigcup_{k=0}^{\infty} \sigma_k\right]^-$;*

(iv) *each Ω_k is an open disc and $\Omega_k \subset \sigma_{k+1} \backslash \overline{\sigma}_k$ $(k = 1, 2, \cdots)$.*

Then there exists an operator $T \in (SI)(\mathcal{H})$ satisfying

(a) *$\sigma(T) = \sigma_{lre}(T) = \sigma$, $\sigma_p(T) \subset \bigcup_{k=1}^{\infty} \Omega_k$ and $\sigma_p(T^*) = \varnothing$;*

(b) *there is a subset $\{\mu_n\}_{n=1}^{\infty}$ of $\sigma_p(T)$ such that $\mathrm{nul}(T - \mu_n) = \infty$ $(n = 1, 2, \cdots)$ and $\bigvee\{\ker(T - \mu_n) : n \geq 1\} = \mathcal{H}$;*

(c) *if $A \in \mathcal{L}(\mathcal{H})$ such that $\sigma(A) \cap \sigma^\circ = \varnothing$, then $\ker \tau_{AT} = \ker \tau_{TA} = \{0\}$.*

Proof. According to Lemma 6.9 we can construct an operator $T_k \in (SI)(\mathcal{H}_k)$ such that $\sigma(T_k) = \sigma_{lre}(T_k) = \sigma_k$, $\sigma_p(T_k) \subset \Omega_k$, $\sigma_p(T_k^*) = \varnothing$ and there is a sequence $\{\lambda_n^k\}_{n=0}^{\infty} \subset \Omega_k$ satisfying $\lim_{n\to\infty} \lambda_n^k = \lambda_0$, $\mathrm{nul}(T_k - \lambda_n^k) = \infty$ $(n = 1, 2, \cdots)$ and $\bigvee\{\ker(T_k - \lambda_k^n) : n \geq 1\} = \mathcal{H}_k$ $(k = 1, 2, \cdots)$. Since $\sigma_r(T_1) \cap \sigma_l(T_k) = \sigma_1 \cap \sigma_k \neq \varnothing$ $(k \geq 2)$, there is a compact operator $D_k \notin \mathrm{ran}\, \tau_{T_1 T_k}$, $\|D_k\| < 2^{-k}$ $(k \geq 2)$.

Set

$$T = \begin{bmatrix} T_1 & D_2 & D_3 & \dots \\ & T_2 & & \\ & & T_3 & \\ & 0 & & \ddots \end{bmatrix} \in \mathcal{L}(\mathcal{H}),$$

where $\mathcal{H} = \bigoplus_{k=1}^{\infty} \mathcal{H}_k$. Since $\{\Omega_k\}_{k=1}^{\infty}$ are pairwise disjoint, $\ker \tau_{T_i T_j} = \{0\}$ $(i \neq j)$. By Corollary 3.14 $T \in (SI)$. It follows from the construction of T that T satisfies (a) and (b). By Proposition 1.14 $\ker \tau_{AT} = \{0\}$. If there is an operator $X \in \mathcal{L}(\mathcal{H})$ such that $TX = XA$, let $X = \begin{bmatrix} X_1 \\ X_2 \\ \vdots \end{bmatrix}$, so we have $T_2 X_2 = X_2 A, \cdots, T_n X_n = X_n A$, $(n \geq 2)$. Since $\sigma(A) \cap \sigma^\circ = \varnothing$ and $\sigma(T_n) = \sigma_n \subset \sigma^\circ$, $\sigma(A) \cap \sigma(T_n) = \varnothing$. Thus $X_n = 0$ $(n \geq 2)$ and $T_1 X_1 = X_1 A$. For the same reason $X_1 = 0$ and $X = 0$, i.e., $\ker \tau_{TA} = \{0\}$. □

Lemma 6.12. *Let $n \in \mathbb{N}$ or $n = \infty$, let σ be a connected compact subset of $\mathbb{C}$ and Ω be a connected open subset of σ° such that $\sigma^\circ \backslash \overline{\Omega} \neq \varnothing$. Then there exists an operator $A \in (SI)(\mathcal{H})$ satisfying*

(i) $\sigma(A) = \sigma$, $\sigma_{lre}(A) = \sigma \backslash \Omega$, $\sigma_p(A^*) = \varnothing$;

(ii) $\operatorname{ind}(A - \lambda) = n$ *for* $\lambda \in \Omega$;

(iii) *there exists a subset $\{\lambda_k\}_{k=1}^{\infty}$ of σ such that* $\operatorname{nul}(A - \lambda_k) = \infty$ $(k \geq 1)$ *and* $\bigvee\{\ker(A - \lambda_k) : k \geq 1\} = \mathcal{H}$.

Proof. Let $\mathcal{H} = \mathcal{H}_1 \oplus \mathcal{H}_2$, $\dim \mathcal{H}_1 = \dim \mathcal{H}_2 = \infty$. Choose open discs G_1, G_2 such that $\overline{G}_2 \subset G_1 \subset \overline{G}_1 \subset \sigma^\circ \backslash \overline{\Omega}$. According to Lemma 6.5 we can construct an operator $A_1 \in \mathcal{B}_\infty(G_1) \cap (SI)(\mathcal{H}_1)$ satisfying $\sigma(A_1) = \sigma$, $\sigma_{lre}(A_1) = \sigma \backslash (G_1 \cup \Omega)$ and $\operatorname{ind}(A_1 - \lambda) = n$ for $\lambda \in \Omega$. By Lemma 6.9 we can find an operator $A_2 \in (SI)(\mathcal{H}_2)$ satisfying $\sigma(A_2) = \sigma_{lre}(A_2) = \overline{G}_1$, $\sigma_p(A_2) \subset G_2$, $\sigma_p(A_2^*) = \varnothing$ and there exists a sequence $\{\mu_i\}_{i=1}^{\infty} \subset G_2$ such that $\operatorname{nul}(A_2 - \mu_i) = \infty$ $(i \geq 1)$ and $\bigvee\{\ker(A_2 - \mu_i) : i \geq 1\} = \mathcal{H}_2$. By Proposition 1.14 $\ker \tau_{A_2 A_1} = \{0\}$. By Theorem 1.11, there is a compact operator $K \in \mathcal{L}(\mathcal{H}_2, \mathcal{H}_1)$ such that $K \notin \operatorname{ran} \tau_{A_1 A_2}$.

Define $A = \begin{bmatrix} A_1 & K \\ 0 & A_2 \end{bmatrix} \begin{matrix} \mathcal{H}_1 \\ \mathcal{H}_2 \end{matrix}$. By Corollary 3.14 $A \in (SI)(\mathcal{H})$ and satisfies (i), (ii) and (iii). □

Lemma 6.13. *Let $T \in \mathcal{L}(\mathcal{H})$ with connected spectrum $\sigma(T)$, and assume that $\sigma_{lre}(T)$ is the closure of an analytic Cauchy domain. Then there exists an operator $W \in (SI)(\mathcal{H})$ satisfying*

(i) $\Lambda(W) = \Lambda(T)$;

(ii) $\min\operatorname{ind}(W-\lambda) = \begin{cases} 0, & \text{if } \lambda \in \rho_{s-F}^{\pm}(W), \\ 1, & \text{if } \lambda \in \sigma(W) \cap \rho_{s-F}^{\circ}(W); \end{cases}$

(iii) $W = \begin{bmatrix} W_1 & * \\ 0 & W_2 \end{bmatrix} \begin{matrix} \mathcal{H}_1 \\ \mathcal{H}_2 \end{matrix}$, *where* $\dim \mathcal{H}_1 = \dim \mathcal{H}_2 = \infty$, *and there is a sequence* $\{\lambda_k : k = 0, \pm 1, \pm 2, \cdots\}$ *of numbers such that* $\bigvee\{\ker(W_1 - \lambda_k)^* : k \geq 0\} = \mathcal{H}_1$ *and* $\bigvee\{\ker(W_2 - \lambda_k) : k < 0\} = \mathcal{H}_2$;

(iv) *there is an open disc* $G \subset \sigma_{lre}(W)$ *such that* $G \cap \sigma_p(W_2) = G^* \cap \sigma_p(W_1^*) = \varnothing$.

Proof. Assume that

$$\{\Omega_{1i}\}_{i=1}^{l_1} \text{ are the components of } \rho_{s-F}^{-}(T),$$
$$\{\Omega_{2j}\}_{j=1}^{l_2} \text{ are the components of } \rho_{s-F}^{\circ}(T) \cap \sigma(T),$$
$$\{\Omega_{3k}\}_{k=1}^{l_3} \text{ are the components of } \rho_{s-F}^{+}(T).$$

Choose a connected Cauchy domain Φ_{ij} in $\sigma(T)$ ($i = 1, 2, 3$; $j = 1, 2, \cdots, j_i$) such that $\Phi_{ij} \supset \Omega_{ij}$, $\Phi_{ij} \backslash \overline{\Omega}_{ij}$ are connected Cauchy domains, $\{\overline{\Phi}_{ij}\}$ are pairwise disjoint and $\sigma(T) \backslash \bigcup \Phi_{ij}$ is the closure of an analytic Cauchy domain.

Choose an open disc $\sigma_0 \subset [\sigma(T) \backslash \bigcup \Phi_{ij}]^\circ$. Let $\{\sigma_k\}_{k=1}^{l_4}$ be the components of $\sigma(T) \backslash [\sigma_0^\circ \cup (\bigcup \Phi_{ij})]$. Choose an open disc G such that $\overline{G} \subset \sigma_0^\circ$. For each k $(0 \leq k \leq l_4)$ according to Lemma 6.11 we can construct an operator $E_k \in (SI)(\mathcal{H})$ satisfying

(i) $\sigma(E_k) = \sigma_{lre}(E_k) = \sigma_k$;

(ii) $\sigma_p(E_0) = \varnothing$ and there is a subset $\{\mu_n : n \geq 1\}$ of $\sigma_0 \backslash G$ such that $\operatorname{nul}(E_0 - \mu_n)^* = \infty$, $\bigvee\{\ker(E_0 - \mu_n)^* : n \geq 1\} = \mathcal{H}$ and $G^* \cap \sigma_p(E_0^*) = \varnothing$;

(iii) for each $k \geq 1$, $\sigma_p(E_k^*) = \varnothing$ and there is a subset $\{\mu_{kn} : n \geq 1\}$ of σ_k such that $\operatorname{nul}(E_k - \mu_{kn}) = \infty$, $\bigvee\{\ker(E_k - \mu_{kn}) : n \geq 1\} = \mathcal{H}$;

(iv) for each k and each operator F, if $\sigma(F) \cap \sigma_k^\circ = \varnothing$, then $\ker \tau_{E_k F} = \ker \tau_{F E_k} = \{0\}$.

According to Lemma 6.12, we construct the following (SI) operators.

- Construct $A_i \in (SI)(\mathcal{H})$ $(1 \leq i \leq l_1)$ such that $\sigma(A_i) = \overline{\Phi}_{1i}$, $\sigma_p(A_i) = \varnothing$, $\sigma_{lre}(A_i) = \overline{\Phi}_{1i} \backslash \Omega_{1i}$, $\operatorname{ind}(A_i - \lambda) = \operatorname{ind}(T - \lambda)$ for $\lambda \in \Omega_{1i}$ and there is a countable subset Λ_{1i} of $\sigma(A_i)$ such that $\operatorname{nul}(A_i - \lambda)^* = \infty$ $(\lambda \in \Lambda_{1i})$ and $\bigvee\{\ker(A_i - \lambda)^* : \lambda \in \Lambda_{1i}\} = \mathcal{H}$.
- Construct $B_k \in (SI)(\mathcal{H})$ $(1 \leq k \leq l_3)$ such that $\sigma(B_k) = \overline{\Phi}_{3k}$, $\sigma_p(B_k^*) = \varnothing$, $\sigma_{lre}(B_k) = \overline{\Phi}_{3k} \backslash \Omega_{3k}$, $\operatorname{ind}(B_k - \lambda) = \operatorname{ind}(T - \lambda)$ for $\lambda \in \Omega_{3k}$ and there is a countable

subset Λ_{3k} of $\sigma(B_k)$ such that $\mathrm{nul}\,(B_k - \lambda) = \infty$ $(\lambda \in \Lambda_{3k})$ and $\bigvee\{\ker(B_k - \lambda) : \lambda \in \Lambda_{3k}\} = \mathcal{H}$.

- Construct $C_j \in (SI)(\mathcal{H})$ $(1 \leq j \leq l_2)$ such that $\sigma(C_j) = \overline{\Phi}_{2j}$, $\sigma_p(C_j) = \varnothing$, $\sigma_{lre}(C_j) = \overline{\Phi}_{2j}\backslash\Omega_{2j}$, $\mathrm{ind}\,(C_j - \lambda) = -1$ for $\lambda \in \Omega_{2j}$ and there is a countable subset $\Lambda_{2j} \in \sigma(C_j)$ such that $\mathrm{nul}\,(C_j - \lambda)^* = \infty$ $(\lambda \in \Lambda_{3j})$ and $\bigvee\{\ker(C_j - \lambda)^* : \lambda \in \Lambda_{3j}\} = \mathcal{H}$.
- Construct $D_h \in (SI)(\mathcal{H})$ $(1 \leq h \leq l_2)$ such that $\sigma(D_h) = \overline{\Phi}_{2h}$, $\sigma_p(D_h^*) = \varnothing$, $\sigma_{lre}(D_h) = \overline{\Phi}_{2h}\backslash\Omega_{2h}$, $\mathrm{ind}\,(D_h - \lambda) = 1$ for $\lambda \in \Omega_{2h}$ and there is a countable subset Λ_{4h} of $\sigma(D_h)$ such that $\mathrm{nul}\,(D_h - \lambda) = \infty$ $(\lambda \in \Lambda_{4h})$ and $\bigvee\{\ker(D_h - \lambda) : \lambda \in \Lambda_{4h}\} = \mathcal{H}$.

By the definitions, it is easily seen that $\ker\tau_{A_iA_j} = \ker\tau_{B_iB_j} = \ker\tau_{C_iC_j} = \ker\tau_{D_iD_j} = \ker\tau_{E_iE_j} = \{0\}$ $(i \neq j)$.

Set $A = \bigoplus_{i=1}^{l_1} A_i \in \mathcal{L}(\mathcal{H}^{(l_1)})$, $B = \bigoplus_{k=1}^{l_3} B_k \in \mathcal{L}(\mathcal{H}^{(l_3)})$, $C = \bigoplus_{j=1}^{l_2} C_j$, $D = \bigoplus_{h=1}^{l_2} D_h \in \mathcal{L}(\mathcal{H}^{(l_2)})$ and $E = \bigoplus_{k=1}^{l_4} E_k \in \mathcal{L}(\mathcal{H}^{(l_4)})$.

Define $Q_i \in \mathcal{L}(\mathcal{H})$ $(1 \leq i \leq l_4)$ as follows

$$Q_i = \begin{cases} \text{compact and } \notin \mathrm{ran}\,\tau_{E_0E_i}, & \text{if } \sigma(E_i) \cap \sigma(E_0) \neq \varnothing, \\ 0, & \text{otherwise.} \end{cases}$$

Set $X_0 = (Q_1, Q_2, \cdots, Q_{l_4}) \in \mathcal{L}(\mathcal{H}^{(l_4)}, \mathcal{H})$.

Define $X_1 = (Q_{ij})_{l_1 \times l_4} \in \mathcal{L}(\mathcal{H}^{(l_4)}, \mathcal{H}^{(l_1)})$ as follows

$$Q_{ij} = \begin{cases} \text{compact and } \notin \mathrm{ran}\,\tau_{A_iE_j}, & \text{if } \sigma(A_i) \cap \sigma(E_j) \neq \varnothing, \\ 0, & \text{otherwise.} \end{cases}$$

$X_2 \in \mathcal{L}(\mathcal{H}^{(l_4)}, \mathcal{H}^{(l_2)})$, and $X_4 = \mathcal{L}(\mathcal{H}^{(l_4)}, \mathcal{H}^{(l_3)})$ are defined similarly.

$X_3 = (M_{ij})_{l_2 \times l_4} \in \mathcal{L}(\mathcal{H}^{(l_4)}, \mathcal{H}^{(l_2)})$ is defined as follows.

M_{ij} is compact and $M_{ij} + K \notin \mathrm{ran}\,\tau_{D_iE_j}$ for all $K \in \mathcal{K}(\mathcal{H})$ if $\sigma(D_i) \cap \sigma(E_j) = \overline{\Phi}_{1i} \cap \sigma_j \neq \varnothing$ (Corollary 1.12) and $M_{ij} = 0$ if $\sigma(D_i) \cap \sigma(E_j) = \varnothing$.

Define

$$W = \begin{bmatrix} E_0 & & & & & X_0 \\ & A & & 0 & & X_1 \\ & & C & & & X_2 \\ & & & D & & X_3 \\ & 0 & & & B & X_4 \\ & & & & & E \end{bmatrix} \begin{matrix} \mathcal{H} \\ \mathcal{H}^{(l_1)} \\ \mathcal{H}^{(l_2)} \\ \mathcal{H}^{(l_2)} \\ \mathcal{H}^{(l_3)} \\ \mathcal{H}^{(l_4)}. \end{matrix}$$

Assume that $P \in \mathcal{A}'(W)$ is idempotent. It follows from Proposition 1.14 and the properties of $\{E_k\}$ that P admits the following representation

$$P = \begin{bmatrix} P_1 & & & & & P_{16} \\ & P_2 & & 0 & & P_{26} \\ & & P_3 & & & P_{36} \\ & & P_{43} & P_4 & & P_{46} \\ & & & & P_5 & P_{56} \\ & 0 & & & & P_6 \end{bmatrix} \begin{matrix} \mathcal{H} \\ \mathcal{H}^{(l_1)} \\ \mathcal{H}^{(l_2)} \\ \mathcal{H}^{(l_2)} \\ \mathcal{H}^{(l_3)} \\ \mathcal{H}^{(l_4)}. \end{matrix}$$

Since $E_0 \in (SI)$ and since A, B, C, D, E are direct sums of (SI) operators with disjoint spectrum respectively, $P_1 = 0$ or 1, $P_2 = \bigoplus_{i=1}^{l_1} \delta_{2i}$, $P_3 = \bigoplus_{i=1}^{l_2} \delta_{3i}$, $P_4 = \bigoplus_{i=1}^{l_2} \delta_{4i}$, $P_5 = \bigoplus_{i=1}^{l_3} \delta_{5i}$ and $P_6 = \bigoplus_{i=1}^{l_4} \delta_{6i}$, where $\delta_{ji} = 0$ or 1. Without loss of generality, we can assume that $P_1 = 0$. By the argument of Lemma 4.24 we can get $P_2 = P_3 = P_5 = P_6 = 0$. Since $PW = WP$, $P_{43}X_2 + P_4X_3 + P_{46}E = DP_{46}$. Note that X_2 is compact, thus $P_{43}X_2$ is compact. For each j $(1 \leq j \leq l_2)$, there must exist an integer k such that $\sigma_{re}(D_j) \cap \sigma_{le}(E_k) = \overline{\Phi}_{1j} \cap \sigma_k \neq \varnothing$. Suppose that $P_{46} = (L_{ih})_{l_2 \times l_4}$. Then

$$D_j L_{jk} - L_{jk} E_k = \delta_{4j} M_{jk} + K,$$

where K is a compact operator. By the choice of M_{jk}, $\delta_{4j} = 0$. Thus $P_4 = 0$. Since $P^2 = P$, $P = 0$ and $W \in (SI)$.

Set $W_1 = \begin{bmatrix} E_0 & 0 & 0 \\ 0 & A & 0 \\ 0 & 0 & C \end{bmatrix}$, $W_2 = \begin{bmatrix} D & 0 & X_3 \\ 0 & B & X_4 \\ 0 & 0 & E \end{bmatrix}$. Then $W = \begin{bmatrix} W_1 & * \\ 0 & W_2 \end{bmatrix} \begin{matrix} \mathcal{H}_1 \\ \mathcal{H}_2 \end{matrix}$, where $\mathcal{H}_1 = \mathcal{H}^{(l_1+l_2+1)}$, $\mathcal{H}_2 = \mathcal{H}^{(l_2+l_3+l_4)}$. By the properties of $\{A_i\}$ and $\{C_i\}$ we have $\min\operatorname{ind}(W_1 - \lambda) = 0$ for $\lambda \in \rho_{s-F}(T) \cap \sigma(T)$ and

$$\operatorname{ind}(W_1 - \lambda) = \begin{cases} \operatorname{ind}(T - \lambda), & \lambda \in \rho_{s-F}^{-}(T), \\ -1, & \lambda \in \rho_{s-F}^{\circ}(T) \cap \sigma(T). \end{cases}$$

By the properties of $E_0, \{A_i\}$ and $\{C_i\}$ we can find a sequence $\{\lambda_k\}_{k=0}^{\infty}$ of numbers such that $\operatorname{nul}(W_1 - \lambda_k)^* = \infty$ $(k \geq 0)$ and $\bigvee\{\ker(W_1 - \lambda_k)^* : k \geq 0\} = \mathcal{H}_1$.

Similarly, by the properties of $\{E_i\}, \{B_i\}$ and $\{D_i\}$ we have $\min\operatorname{ind}(W_2 - \lambda) = 0$ for $\lambda \in \rho_{s-F}(T) \cap \sigma(T)$,

$$\operatorname{ind}(W_2 - \lambda) = \begin{cases} \operatorname{ind}(T - \lambda), & \lambda \in \rho_{s-F}^{+}(T), \\ 1, & \lambda \in \rho_{s-F}^{\circ}(T) \cap \sigma(T) \end{cases}$$

and there is a sequence $\{\lambda_k\}_{k=-1}^{-\infty}$ of numbers such that $\text{nul}\,(W_2 - \lambda_k) = \infty$ $(k \leq -1)$ and $\bigvee\{\ker\,(W_2 - \lambda_k) : k \leq -1\} = \mathcal{H}_2$.

It follows from $G \cap \left[(\bigcup_{k=1}^{l_4} \sigma_k) \cup (\bigcup\{\Phi_{ij} : i = 1, 2, 3;\ j = 1, 2, \cdots, l_i\})\right]$ and the properties of E_0 that we have $G \cap \sigma_p(W_2) = \varnothing$ and $G^* \cap \sigma_p(W_1^*) = \varnothing$. Thus W satisfies (iii) and (iv) of the lemma. It is easy to see that W satisfies (i) and (ii). Thus the proof of Lemma 6.13 is now complete. □

Now we are in a position to prove Theorem 6.7.

Proof of Theorem 6.7. In 6.1 we have proved that if $\mathcal{N}$ is well-ordered with finite dimensional atoms, then $\mathcal{U}(\text{alg}\,\mathcal{N} \cap (SI))^- = (QT)_C$. Thus we only need to show that if $\mathcal{N}$ is maximal and $\mathcal{N}$ and $\mathcal{N}^\perp$ are not well-ordered, then $\mathcal{U}(\text{alg}\,\mathcal{N} \cap (SI))^- = \{T \in \mathcal{L}(\mathcal{H}) : \sigma(T)$ is connected$\}$. By the same analysis as 6.2 it suffices to prove that $\Delta = \{T \in \mathcal{L}(\mathcal{H}) : \sigma(T)$ is connected and $\sigma_{lre}(T)$ is the closure of an analytic Cauchy domain$\} \subset \mathcal{U}(\text{alg}\,\mathcal{N} \cap (SI))^-$.

If $\mathcal{N}$ and $\mathcal{N}^\perp$ are not well-ordered, there are three possibilities.

Case A: There are $\{t_n\}_{n=-\infty}^{\infty} \subset [0,1]$ such that

$$0 = t_0 < t_1 < t_2 < \cdots < t_n < \cdots < t_{-n} < \cdots < t_{-2} < t_{-1} = 1,$$

$\lim_{n\to\infty} t_n = \lim_{n\to\infty} t_{-n}$ and $\dim M_{(t_{n-1}, t_n]} = \infty$ $(n = \pm 1, \pm 2, \cdots)$, where $M_{(t_{n-1}, t_n]} = E\big((t_{n-1}, t_n]\big)\mathcal{H}$ and E is the spectral measure associated with $\mathcal{N}$.

Case B: There are t_0, t_1, t_2, $t_3 \in [0,1]$, such that $0 < t_0 < t_1 < t_2 < t_3 < 1$ and

$$\mathcal{N}_0 := \{M_t : 0 \leq t \leq t_0\} \text{ is atomic,}$$
$$\mathcal{N}_1 := \{M_t \ominus M_{t_0} : t \leq t_1\} \text{ has the type } \omega + 1,$$
$$\mathcal{N}_2 := \{M_t \ominus M_{t_1} : t_1 \leq t \leq t_2\} \text{ is atomic,}$$
$$\mathcal{N}_3 := \{M_t \ominus M_{t_2} : t_2 \leq t \leq t_3\} \text{ has the type } 1 + \omega^*,$$
$$\mathcal{N}_4 := \{M_t \ominus M_{t_3} : t_3 \leq t \leq 1\} \text{ is atomic,}$$

where $M_t = M_{[0,t]} = E([0,t])\mathcal{H}$.

Case C: There are t_0, t_1, t_2, $t_3 \in [0,1]$ such that $0 < t_0 < t_1 < t_2 < t_3 < 1$ and

$$\mathcal{N}_0 := \{M_t : 0 \leq t \leq t_0\} \text{ is atomic,}$$
$$\mathcal{N}_1 := \{M_t \ominus M_{t_0} : t_0 \leq t \leq t_1\} \text{ has the type } 1 + \omega^*,$$
$$\mathcal{N}_2 := \{M_t \ominus M_{t_1} : t_1 \leq t \leq t_2\} \text{ is atomic,}$$
$$\mathcal{N}_3 := \{M_t \ominus M_{t_2} : t_2 \leq t \leq t_3\} \text{ has the type } \omega + 1,$$
$$\mathcal{N}_4 := \{M_t \ominus M_{t_3} : t_3 \leq t \leq 1\} \text{ is atomic.}$$

In Case A, according to Lemma 6.10, there exists an operator $A \in (SI)$ such that $\Lambda(A) = \Lambda(T)$, $\min\operatorname{ind}(A-\lambda) \leq \min\operatorname{ind}(T-\lambda)$ for $\lambda \in \rho_{s-F}(A)$ and $A = \begin{bmatrix} A_1 & A_{12} \\ 0 & A_2 \end{bmatrix} \begin{matrix} \mathcal{K}_1 \\ \mathcal{K}_2 \end{matrix}$, where

$$A_1 = \begin{bmatrix} \lambda_1 & & & * \\ & \lambda_2 & & \\ & & \lambda_3 & \\ & 0 & & \ddots \end{bmatrix} \begin{matrix} \mathcal{H}_1 \\ \mathcal{H}_2 \\ \mathcal{H}_3 \\ \vdots \end{matrix}, \quad A_2 = \begin{bmatrix} \ddots & & & * \\ & \lambda_{-3} & & \\ & & \lambda_{-2} & \\ & 0 & & \lambda_{-1} \end{bmatrix} \begin{matrix} \vdots \\ \mathcal{H}_{-3} \\ \mathcal{H}_{-2} \\ \mathcal{H}_{-1} \end{matrix},$$

$\mathcal{H}_n = \bigvee\{\ker(A_1-\lambda_k) : 1 \leq k \leq n\} \ominus \mathcal{H}_{n-1}$, $\mathcal{H}_{-n} = \bigvee\{\ker(A_2-\lambda_k) : -n \leq k \leq -1\} \ominus \mathcal{H}_{-n+1}$ $(n = 1, 2, \cdots)$, $\mathcal{H}_0 = \{0\}$, $\dim \mathcal{H}_n = \infty$ $(n = \pm 1, \pm 2, \cdots)$, $\mathcal{K}_1 = \bigoplus_{n=1}^{\infty} \mathcal{H}_n$ and $\mathcal{K}_2 = \bigoplus_{n=-1}^{-\infty} \mathcal{H}_n$, $\{\lambda_k : k = \pm 1, \pm 2, \cdots\}$ are given in Lemma 6.10 (iii).

By the similarity orbit theorem, $T \in S(A)^-$, i.e., for each $\epsilon > 0$, there exists an invertible operator X such that $\|XAX^{-1} - T\| < \epsilon$. It is easily seen that XAX^{-1} admits the same matrix representation with respect to another decomposition of the space:

$$XAX^{-1} = \begin{bmatrix} \lambda_1 & & & & & & & * \\ & \lambda_2 & & & & & & \\ & & \lambda_3 & & & & & \\ & & & \ddots & & & & \\ & & & & \ddots & & & \\ & & & & & \lambda_{-3} & & \\ & & 0 & & & & \lambda_{-2} & \\ & & & & & & & \lambda_{-1} \end{bmatrix} \begin{matrix} \mathcal{M}_1 \\ \mathcal{M}_2 \\ \mathcal{M}_3 \\ \vdots \\ \vdots \\ \mathcal{M}_{-3} \\ \mathcal{M}_{-2} \\ \mathcal{M}_{-1}, \end{matrix}$$

where $\dim \mathcal{M}_n = \infty$ $(n = \pm 1, \pm 2, \cdots)$.

Choose a unitary operator U so that $U\mathcal{M}_n = M_{(t_{n-1}, t_n]}$ $(n = \pm 1, \pm 2, \cdots)$. Then $UXAX^{-1}U^* \in \operatorname{alg}\mathcal{N} \cap (SI)$, i.e., $T \in \mathcal{U}(\operatorname{alg}\mathcal{N} \cap (SI))^-$.

If B is the case, for simplicity we only prove the conclusion of the theorem when $t_0 = 0$ and $t_3 = 1$. Denote the operator A in Case A by A_1 which satisfies (i), (ii), (iii) and (iv) of Lemma 6.10. Let $\{f_\alpha\}_{\alpha \in \Lambda}$ be the unit vectors of the atoms of $\mathcal{N}_2$, $\bigvee\{f_\alpha : \alpha \in \Lambda\} = M_{t_2} \ominus M_{t_1}$. Assume that G is the open disc contained in $\sigma_{lre}(A)$ given in Lemma 6.10 (iv), then choose $c_\alpha \in G$ $(\alpha \in \Lambda)$ such that $\{c_\alpha\}$ is pairwise distinct and define $A_3 = \sum c_\alpha f_\alpha \otimes f_\alpha$. By the construction of A_1 in

Lemma 6.10, $G \subset \sigma_{lre}(A_1)$. Thus for each α there is a unit vector $g_\alpha \in \mathcal{K}_1$ such that $g_\alpha \notin \operatorname{ran}(A_1 - c_\alpha)$. Let $\{d_\alpha\}_{\alpha\in\Lambda}$ be positive numbers satisfying $\sum_{\alpha\in\Lambda} d_\alpha = 1$. Set $K = \sum_{\alpha\in\Lambda} d_\alpha g_\alpha \otimes f_\alpha$ and

$$A = \begin{bmatrix} A_1 & K & A_{12} \\ 0 & A_3 & 0 \\ 0 & 0 & A_2 \end{bmatrix} \begin{matrix} \mathcal{K}_1 \\ M_{t_2} \ominus M_{t_1} \\ \mathcal{K}_2. \end{matrix}$$

Then it is easily seen that $\Lambda(A) = \Lambda(T)$ and $\min\operatorname{ind}(A - \lambda) \leq \min\operatorname{ind}(T - \lambda)$ for $\lambda \in \rho_{s-F}(T)$. By Lemma 6.10 (iii), (iv) we have $\ker \tau_{A_3 A_1} = \ker \tau_{A_2 A_3} = \{0\}$.

Assume that P is an idempotent commuting with A and

$$P = \begin{bmatrix} P_{11} & P_{12} & P_{13} \\ P_{21} & P_{22} & P_{23} \\ P_{31} & P_{32} & P_{33} \end{bmatrix} \begin{matrix} \mathcal{K}_1 \\ M_{t_2} \ominus M_{t_1} \\ \mathcal{K}_2. \end{matrix}$$

Then by Proposition 1.14, $P = \begin{bmatrix} P_{11} & P_{12} & P_{13} \\ 0 & P_{22} & 0 \\ 0 & 0 & P_{33} \end{bmatrix}$. Observe that $P' = \begin{bmatrix} P_{11} & 0 & P_{13} \\ 0 & 0 & 0 \\ 0 & 0 & P_{33} \end{bmatrix}$ is an idempotent commuting with $\begin{bmatrix} A_1 & 0 & A_{12} \\ 0 & 0 & 0 \\ 0 & 0 & A_2 \end{bmatrix}$ and $A' = \begin{bmatrix} A_1 & A_{12} \\ 0 & A_2 \end{bmatrix} \in (SI)$, thus $P' = 0$ or 1. Without loss of generality we can assume that $P' = 0$, or $P = \begin{bmatrix} 0 & P_{12} & 0 \\ 0 & P_{22} & 0 \\ 0 & 0 & 0 \end{bmatrix}$. Since $PA = AP$, $P_{12}A_3 = A_1 P_{12} + K P_2$. It follows from $P_{22}A_3 = A_3 P_{22}$ and the pairwise distinction of the c_α's that $P_{22} = \bigoplus_{\alpha\in\Lambda} \delta_\alpha$, where $\delta_\alpha = 0$ or 1. Thus for each $\alpha \in \Lambda$

$$(A_1 P_{12} - P_{12} A_3) f_\alpha = A_1 P_{12} f_\alpha - c_\alpha P_{12} f_\alpha = -\delta_\alpha d_\alpha g_\alpha.$$

Since $g_\alpha \notin \operatorname{ran}(A_1 - c_\alpha)$, $\delta_\alpha = 0$. Therefore $P = 0$ and $A \in (SI)$. By the similarity orbit theorem, $T \in S(A)^-$, i.e., for each $\epsilon > 0$ there exists an invertible operator X such that $\|XAX^{-1} - T\| < \epsilon$. By Lemma 6.10 (iii) A_1 and A_2^* admit upper triangular matrix representations

$$A_1 = \begin{bmatrix} \lambda_0 & & & * \\ & \lambda_1 & & \\ & & \lambda_2 & \\ 0 & & & \ddots \end{bmatrix} \begin{matrix} e_0^1 \\ e_1^1 \\ e_2^1, \\ \vdots \end{matrix} \quad A_2 = \begin{bmatrix} \ddots & & & * \\ & \lambda_{-3} & & \\ & & \lambda_{-2} & \\ 0 & & & \lambda_{-1} \end{bmatrix} \begin{matrix} \vdots \\ e_3^2 \\ e_2^2 \\ e_1^2 \end{matrix}$$

with respect to some $ONB\,\{e_n^1\}_{n=0}^{\infty}$ of $\mathcal{K}_1$ and, respectively, $ONB\,\{e_n^2\}_{n=1}^{\infty}$ of $\mathcal{K}_2$.

Set

$$\mathcal{M} = \left\{ \begin{array}{l} \bigvee_{i=1}^{n} \{e_i^1\}\,(n=0,1,2,\cdots);\ \bigvee_{i=1}^{\infty} \{e_i^1\} \oplus N\,(N \in \mathcal{N}_2); \\ \bigvee_{i=1}^{\infty} \{e_i^1\} \oplus (M_{t_2} \ominus M_{t_1}) \oplus \bigvee_{j=n}^{\infty} \{e_j^2\}\,(n=0,1,2,\cdots) \end{array} \right\}.$$

Then $\mathcal{M}$ is a maximal atomic nest, and unitarily equivalent to $\mathcal{N}$. Thus, there exists a unitary operator U such that $UXAX^{-1}U^* \in \operatorname{alg}\mathcal{N}$. Therefore $T \in \mathcal{U}(\operatorname{alg}\mathcal{N} \cap (SI))^-$.

For Case C, we only prove the conclusion of the theorem when $t_1 = t_2$.

According to Lemma 6.13 we get an operator $W \in (SI)$ satisfying (i)–(iv) of Lemma 6.13. Let $W = \begin{bmatrix} W_1 & W_{12} \\ 0 & W_2 \end{bmatrix} \begin{matrix} \mathcal{H}_1 \\ \mathcal{H}_2 \end{matrix}$.

Let $N_{-\infty} = \bigcap\{M_{t_n} : -\infty < n < \infty\}$, $N_\infty = \bigvee\{M_{t_n} : -\infty < n < \infty\}$. Let $\mathcal{N}_- = \{M_t \in \mathcal{N} : 0 \le t \le t_0\}$, $\mathcal{N}_+ = \{M_t \ominus M_{t_3} : t_3 \le t \le 1\}$. Let $\{f_\alpha\}_{\alpha\in\Lambda_1}$ and $\{g_\beta\}_{\beta\in\Lambda_2}$ be the unit vectors of the atoms of $\mathcal{N}_-$ and, respectively, $\mathcal{N}_+$. Define $B_1 = \sum_{\alpha\in\Lambda_1} c_\alpha f_\alpha \otimes f_\alpha$ and $B_2 = \sum_{\beta\in\Lambda_2} d_\beta g_\beta \otimes g_\beta$, where $\{c_\alpha,\, \alpha \in \Lambda_1;\ d_\beta,\, \beta \in \Lambda_2\} \subset G \subset \sigma_{lre}(W)$ are pairwise distinct and G is given in Lemma 6.13 (iv). In a similar way to Case B, construct operators $E_1 \in \mathcal{L}(\mathcal{H}_1 \oplus \mathcal{H}_2, \bigvee\{f_\alpha : \alpha \in \Lambda_1\})$ and $E_2 \in \mathcal{L}(\bigvee\{g_\beta : \beta \in \Lambda_2\}, \mathcal{H}_1 \oplus \mathcal{H}_2)$ such that $E_1^* f_\alpha \notin \operatorname{ran}(W_1 - c_\alpha)^*$, $E_2\, g_\beta \notin \operatorname{ran}(W_2 - d_\beta)$ $(\alpha \in \Lambda_1,\, \beta \in \Lambda_2)$.

Set

$$A = \begin{bmatrix} B_1 & E_1 & 0 \\ 0 & W & E_2 \\ 0 & 0 & B_2 \end{bmatrix} \begin{matrix} \bigvee\{f_\alpha : \alpha \in \Lambda_1\} \\ \mathcal{H}_1 \oplus \mathcal{H}_2 \\ \bigvee\{g_\beta : \beta \in \Lambda_2\}. \end{matrix}$$

By the same argument as Case B, $A \in (SI)$ and $T \in S(A)^-$. Thus for each $\epsilon > 0$, $\|XAX^{-1} - T\| < \epsilon$ for some invertible operator X. Note that by (i), (ii) and (iii) of Lemma 6.13

$$W = \begin{bmatrix} \ddots & & & & & * \\ & \lambda_{-2} & & & & \\ & & \lambda_{-1} & & & \\ & & & \lambda_0 & & \\ & & & & \lambda_1 & \\ & 0 & & & & \lambda_2 \\ & & & & & & \ddots \end{bmatrix} \begin{matrix} \vdots \\ e_{-2} \\ e_{-1} \\ e_0 \\ e_1 \\ e_2 \\ \vdots \end{matrix}$$

with respect to some $ONB\,\{e_n\}_{n=-\infty}^{\infty}$ of $\mathcal{H}_1 \oplus \mathcal{H}_2$. Thus by the argument of Case B, there is a unitary operator U such that $UXAX^{-1}U^* \in \operatorname{alg}\mathcal{N}$ and therefore $T \in \mathcal{U}(\operatorname{alg}\mathcal{N} \cap (SI))^-$. The proof of Theorem 6.7 is now complete. $\square$

An operator Banach algebra $\mathcal{B}$ is a *strongly irreducible algebra* if the commutant $\mathcal{A}'(\mathcal{B})$ of $\mathcal{B}$ contains no non-trivial idempotents. By the results of this chapter we have

Theorem 6.14. *Each nest algebra is strongly irreducible.*

Using Proposition 4.16, Theorem 4.17 we get

Theorem 6.15. *Let $T \in \mathcal{B}_n(\Omega)$. Then*

(i) *when $n = 1$, $\mathcal{A}'(T)$ is strongly irreducible;*

(ii) *if $\mathcal{A}'(T)$ is semi-simple, $\mathcal{A}'(T)$ can be uniquely expressed as a direct sum of finitely many strongly irreducible algebras.*

6.4 Open problems

Problem 6.1. *If $\mathcal{U}_a(\operatorname{alg}\mathcal{N} \cap (SI)) = \{A \in \mathcal{L}(\mathcal{H}) :$ for each $\epsilon > 0$, there exist a compact operator K_ϵ and a unitary operator U_ϵ such that $U_\epsilon A U_\epsilon^* - K_\epsilon \in \operatorname{alg}\mathcal{N} \cap (SI)$ and $\|K_\epsilon\| < \epsilon\}$, then what is $\mathcal{U}_a(\operatorname{alg}\mathcal{N} \cap (SI))$?*

Problem 6.2. *What kind of Banach algebra has a unique (SI) decomposition?*

6.5 Remarks

The reader is referred to the excellent treatise of K.R. Davidson [1] for everything related to nest algebra. The work in this chapter was motivated by two aspects. In answering one of W.B. Arveson's questions, D.A. Herrero [6], [9] described the sets $\mathcal{U}_a(\operatorname{alg}\mathcal{N})$ and $\mathcal{U}(\operatorname{alg}\mathcal{N})^-$ for each nest $\mathcal{N}$. K.R. Davidson suggested the authors find out what the two sets $\mathcal{U}_a(\operatorname{alg}\mathcal{N} \cap (SI))$ and $\mathcal{U}(\operatorname{alg}\mathcal{N} \cap (SI))^-$ are. On the other hand, W.B. Arveson used a continuous nest algebra to prove Voiculescu's non-commutative Weyl–von Neumann theorem (W.B. Arveson [2]). Z.J. Jiang suggested the authors prove $SI(\mathcal{H})^- = \{T : \sigma(T) \text{ is connected}\}$ again in terms of continuous nests (6.2). The results of this chapter are due to Y.Q. Ji, C.L. Jiang and Z.Y. Wang [2] [3] [5] [6].

7. Some other results about strongly irreducible operators

7.1 The sums of strongly irreducible operators

H. Radjavi [1] proved a very interesting result about irreducible operators. It asserts that each operator on a Hilbert space can be expressed as the sum of two irreducible operators. In this section we will consider a more delicate problem, i.e., is every operator the sum of two (SI) operators? This is a more difficult problem. Up to now, we have only given confirmative answers for some special classes of operators, but we have proved that each operator is the sum of three (SI) operators.

Theorem 7.1. *Each operator on a Hilbert space $\mathcal{H}$ can be expressed as the sum of three strongly irreducible operators.*

In order to prove Theorem 7.1, we need several lemmas.

Lemma 7.2. *Let $T \in \mathcal{L}(\mathcal{H})$; then T is unitarily equivalent to the operator T' of the form*

$$T' = \begin{bmatrix} A_1 & B_1 & & 0 \\ C_1 & A_2 & B_2 & \\ & C_2 & A_3 & \ddots \\ 0 & & \ddots & \ddots \end{bmatrix} \begin{matrix} \mathcal{H}_1 \\ \mathcal{H}_2 \\ \mathcal{H}_3 \\ \vdots \end{matrix} \tag{7.1}$$

where each $\mathcal{H}_n$ is a finite dimensional space.

We will omit the proof of Lemma 7.2. The reader can refer to I.D. Berg and K.R. Davidson [1, P.131] or C.K. Fong and P.Y. Wu [1, Lemma 2.2].

A notable fact is that if $(t_{ij})_{i,j=1}^{\infty}$ is the matrix representation of T on an infinite dimensional Hilbert space $\mathcal{H}$ with respect to an ONB $\{e_n\}_{n=1}^{\infty}$ of $\mathcal{H}$, then the upper triangular part of T may not be bounded on $\mathcal{H}$, i.e., if $A = (t'_{ij})_{i,j=1}^{\infty}$ is the matrix representation of a linear operator A with respect to $\{e_n\}_{n=1}^{\infty}$, where $t'_{ij} = t_{ij}$ if $i \le j$ and $t'_{ij} = 0$ if $i > j$, then it may happen that $A \notin \mathcal{L}(\mathcal{H})$. For example, $T = \bigoplus_{n=2}^{\infty} T_n$, where

$$T_n = i\frac{\log\log n}{\log n}\begin{bmatrix} 0 & 1 & 1/2 & \cdots & \frac{1}{n-1} \\ -1 & 0 & 1 & \cdots & \frac{1}{n-2} \\ & \cdots & \cdots & & \\ -\frac{1}{n-1} & -\frac{1}{n-2} & -\frac{1}{n-3} & \cdots & 0 \end{bmatrix} \in \mathcal{L}(\mathbb{C}^n), \quad \mathcal{H}=\bigoplus_{n=2}^{\infty}\mathbb{C}^n,$$

then $T \in \mathcal{L}(\mathcal{H})$. But its upper triangular part is not bounded (K. R. Davidson [1, P.34]). But we have

Lemma 7.3. *Let $T \in \mathcal{L}(\mathcal{H})$; then there exists an ONB $\{e_n\}_{n=1}^{\infty}$ of $\mathcal{H}$, such that the matrix representation (t_{ij}) of T with respect to $\{e_n\}$ has a bounded upper triangular part.*

Proof. By Lemma 7.2, $T \simeq T'$, T' admits the representation (7.1). Since $\|C_k\| \leq \|T\|$ $(k = 1, 2, \cdots)$, by calculation it is easily seen that $\begin{bmatrix} 0 & & & 0 \\ C_1 & 0 & & \\ & C_2 & 0 & \\ 0 & & \ddots & \ddots \end{bmatrix}\begin{matrix}\mathcal{H}_1\\ \mathcal{H}_2\\ \mathcal{H}_3\\ \vdots\end{matrix}$ is bounded. Thus $T'' = \begin{bmatrix} A_1 & B_1 & & 0 \\ & A_2 & B_2 & \\ & & A_3 & \ddots \\ 0 & & & \ddots \end{bmatrix}\begin{matrix}\mathcal{H}_1\\ \mathcal{H}_2\\ \mathcal{H}_3\\ \vdots\end{matrix}$ is bounded. Let $U \in \mathcal{L}(\mathcal{H})$ be a unitary operator such that $T' = U^*TU$. Denote $\mathcal{K}_n = U\mathcal{H}_n$ $(n = 1, 2, \cdots)$; then computation shows that

$$T = \begin{bmatrix} A_1' & B_1' & & 0 \\ C_1' & A_2' & B_2' & \\ & C_2' & A_3' & \ddots \\ 0 & & \ddots & \ddots \end{bmatrix}\begin{matrix}\mathcal{K}_1\\ \mathcal{K}_2\\ \mathcal{K}_3\\ \vdots\end{matrix},$$

where $A_n' = U|_{\mathcal{H}_n}A_n$, $B_n' = U|_{\mathcal{H}_n}B_n$ and $C_n' = U|_{\mathcal{H}_{n+1}}C_n$ $(n = 1, 2, \cdots)$. Thus

$$\begin{bmatrix} A_1' & B_1' & & 0 \\ & A_2' & B_2' & \\ & & A_3 & \ddots \\ 0 & & & \ddots \end{bmatrix}\begin{matrix}\mathcal{K}_1\\ \mathcal{K}_2\\ \mathcal{K}_3\\ \vdots\end{matrix} = UT''$$

is bounded. Since each $\mathcal{K}_n$ is a finite dimensional Hilbert space, there is an ONB of $\mathcal{K}_n$ such that each A_n' admits an upper triangular matrix representation with respect to this basis. Thus, we get an $ONB\,\{e_n\}_{n=1}^{\infty}$ of $\mathcal{H}$ such that UT'' has an upper triangular matrix representation with respect to it and T has a bounded upper triangular part with respect to the $ONB\,\{e_n\}_{n=1}^{\infty}$. □

Lemma 7.4. *Assume that* $0 < r < 1$, $T = S^* + X$, *where* S *is the unilateral shift on the Hilbert space* $\mathcal{H}$ *and* $X \in \mathcal{L}(\mathcal{H})$ *with* $\|X\| \le 1 - r$. *Then for each* λ, $|\lambda| < 1$, $T - \lambda$ *is surjective and* $\operatorname{nul}(T - \lambda) = 1$.

Proof. Note that for each λ, $|\lambda| < 1$, $\|S(X-\lambda)\| \le \|X\| + |\lambda| < (1-r)+r = 1$. Thus $1 + S(X - \lambda)$ is invertible. Therefore $T - \lambda = S^*[1 + S(X - \lambda)]$ is surjective, and $(T-\lambda)^* = [1+S(X-\lambda)]^* S$ is injective. Thus $\operatorname{ind}(T-\lambda) = \operatorname{ind} S^* + \operatorname{ind}[1+S(X-\lambda)] = 1$ and $\operatorname{nul}(T - \lambda) = 1$. □

Lemma 7.5. *Assume that* $X \in \mathcal{L}(\mathcal{H})$ *admits a block-upper triangular matrix representation*

$$X = \begin{bmatrix} X_{11} & X_{12} & & * \\ & X_{22} & X_{23} & \\ & & X_{33} & \ddots \\ & 0 & & \ddots \end{bmatrix}$$

with respect to the decomposition of the space $\mathcal{H} = \bigoplus_{n=1}^{\infty} \mathcal{H}_n$, *where* $\sup_n\{\dim \mathcal{H}_n\} < \infty$. *Then for each* r $(0 < r < 1)$, *there exists a* $\delta > 0$ *such that* $S^* + \delta_1 X \in \mathcal{B}_1(D_r)$ *for all* δ_1, $0 < \delta_1 \le \delta$, *where* $D_r = \{\lambda \in \mathbb{C} : |\lambda| < r\}$.

Proof. Let q_n $(n \ge 1)$ denote the $n \times n$ Jordan nilpotent

$$q_n = \begin{bmatrix} 0 & 1 & & & 0 \\ & 0 & 1 & & \\ & & 0 & \ddots & \\ & & & \ddots & 1 \\ & 0 & & & 0 \end{bmatrix}$$

and $n_j = \dim \mathcal{H}_j$. Since $\{n_j : j \ge 1\}$ is bounded, we can find a $\delta > 0$ such that $\delta < (1 - r)/\|X\|$ and $\sigma(q_{n_j} + \delta_1 X_{jj}) \subset D_r$ for all $j \ge 1$ and $\delta_1 \le \delta$. Thus

$$T := S^* + \delta_1 X = \begin{bmatrix} q_{n_1} + \delta_1 X_{11} & & * \\ & q_{n_2} + \delta_1 X_{22} & \\ & 0 & \ddots \end{bmatrix} \begin{matrix} \mathcal{H}_1 \\ \mathcal{H}_2 \\ \vdots \end{matrix}$$

admits an upper triangular matrix representation with respect to an ONB of $\mathcal{H}$, and the diagonal $d(T)\subset D_r$. Thus $\bigvee\{\ker(T-\lambda)^k : \lambda\in D_r,\ k\geq 1\}=\mathcal{H}$. By Lemma 7.4, $D_r\subset \rho_r(T)$. It is easy to see that $T\in\mathcal{B}_1(D_r)$. □

Corollary 7.6. *Assume that $X\in\mathcal{L}(\mathcal{H})$ is a triangular operator; then for each r, $0<r<1$, there exists a $\delta>0$ such that $S^*+\delta_1 X\in\mathcal{B}_1(D_r)$ provided $0<\delta_1\leq\delta$.*

In the next lemma, we will consider the unilateral shift as the operator T_z on H^2.

Lemma 7.7. *Assume that ϕ is a function analytic in $\overline{D}$; then for each r, $0<r<1$, there exists a $\delta>0$ such that $S^*+\delta_1\phi(S)\in\mathcal{B}_1(D_r)$ provided $0<\delta_1\leq\delta$.*

Proof. We may assume that ϕ is analytic in D_{r_1} $(r_1>1)$. Fix a r_2, $1<r_2<r_1$ and let $M=\max\{|\phi(z)| : |z|\leq r_2\}$. Choose a δ, $0<\delta<(1-r)/M$. Set $T=S^*+\delta_1\phi(S)$, $0<\delta_1\leq\delta$. By Lemma 7.4, $T-\lambda$ is surjective and $\operatorname{nul}(T-\lambda)=1$ for all $\lambda\in D_r$. Thus it suffices to show that when δ is sufficiently small, $\bigvee\{\ker T^k : k\geq 1\}=H^2$. Let $Y=1+\delta_1 S\phi(S)$; since $\|\delta_1 S\phi(S)\|<1$, Y is invertible, and

$$T(Y^{-1}S)=(S^*+\delta_1\phi(S))Y^{-1}S=S^*[1+\delta_1 S\phi(S)]Y^{-1}S=1.$$

Let $e\in H^2$ and $e(t)\equiv 1$; then $e\in\ker S^*$ and

$$T^kY^{-1}(SY^{-1})^{k-1}e=T^k(Y^{-1}S)^{k-1}Y^{-1}e=TY^{-1}e=S^*e=0.$$

This implies that $Y^{-1}(SY^{-1})^{k-1}e\in\ker T^k$ $(k\geq 1)$. Thus it is necessary to show that $\bigvee\{(SY^{-1})^k e : k\geq 0\}=H^2$. Observe that $SY^{-1}=\psi(S)$, where $\psi(z)=z/[1+\delta_1 z\phi(z)]$. Therefore we need prove that $\bigvee\{\psi^k : k\geq 0\}=H^2$. We first show that ψ is a univalent analytic function in $\overline{D}$. In fact, if $\psi(z_1)=\psi(z_2)$ for z_1, $z_2\in\overline{D}$, then

$$z_1+\delta_1 z_1 z_2\phi(z_2)=z_2+\delta_1 z_1 z_2\phi(z_1).$$

Thus

$$z_1-z_2=\delta_1 z_1 z_2(\phi(z_1)-\phi(z_2)).$$

Denote $L=\sup\{|\phi(\omega_1)-\phi(\omega_2)|/|\omega_1-\omega_2| : \omega_1,\omega_2\in D_{r_2},\ \omega_1\neq\omega_2\}$. It is obvious that $L<\infty$. Thus, $1/\delta_1=|z_1z_2|\,|\phi(z_1)-\phi(z_2)|/|z_1-z_2|\leq L$. This implies that if we choose $\delta<1/L$, ψ is a univalent analytic function in $\overline{D}$. By the property of univalent analytic function in $\overline{D}$, $\bigvee\{\psi^k : k\geq 0\}=\mathcal{H}$ (J. Dazord [1]). □

Proof of Theorem 7.1. Given $T \in \mathcal{L}(\mathcal{H})$, by Lemma 7.3 $T = A + B$, where $A, B \in \mathcal{L}(\mathcal{H})$, A, B^* are triangular operators with respect to a suitable ONB of $\mathcal{H}$, i.e.,

$$A = \begin{bmatrix} a_{11} & a_{12} & & * \\ & a_{22} & a_{23} & \\ & & a_{33} & \ddots \\ & 0 & & \ddots \end{bmatrix}, \quad B = \begin{bmatrix} 0 & & & 0 \\ b_{21} & 0 & & \\ b_{31} & b_{32} & 0 & \\ \vdots & & \ddots & \ddots \end{bmatrix}.$$

Fix a number r, $0 < r < 1$. By Corollary 7.6 and Lemma 7.7 there is a $\delta > 0$ such that $S^* + \delta^2 A$, $S^* + \delta B^*$ and $S^* + \delta S$ are in $\mathcal{B}_1(D_r)$. Thus

$$T = (1/\delta^2)(S^* + \delta^2 A) + (1/\delta)(S + \delta B) + (-1/\delta^2)(S^* + \delta S). \qquad \square$$

To consider the sum of two (SI) operators, we first analyse the case of a finite dimensional space.

Proposition 7.8. *Every operator on a finite dimensional Hilbert space is the sum of two strongly irreducible operators.*

Proof. Let $T \in \mathcal{L}(\mathbb{C}^n)$, and $A = T - (1/n)\mathrm{tr}\, T$, where $\mathrm{tr}\, T$ denotes the sum of the diagonal elements of the matrix representation of T with respect to any ONB of the space.

If $\mathrm{rank}\, A \leq 1$, then $A \simeq \begin{bmatrix} 0 & a & 0 & \dots & 0 \\ & 0 & 0 & \dots & 0 \\ & & 0 & & \\ & & & \ddots & \\ & 0 & & & 0 \end{bmatrix}$. Let $b = (1/2n)\mathrm{tr}\, T$, $c \neq 0$ and $c \neq a$. Thus

$$T \simeq \begin{bmatrix} b & c & & 0 \\ & b & c & \\ & & b & \ddots \\ & & & \ddots & c \\ & 0 & & & b \end{bmatrix} + \begin{bmatrix} b & a-c & & & 0 \\ & b & -c & & \\ & & b & -c & \\ & & & b & \ddots \\ & 0 & & & \ddots & -c \\ & & & & & b \end{bmatrix} = T_1 + T_2.$$

Since T_1 and T_2 are similar to q_n, T_1 and T_2 are strongly irreducible.

If rank $A > 1$, then classical matrix theory indicates that

$$A \sim \begin{bmatrix} 0 & a_{12} & \cdots & a_{1\,n-1} & a_{1n} \\ a_{21} & 0 & & & a_{2n} \\ \vdots & & \ddots & & \vdots \\ a_{n-1\,1} & & & 0 & a_{n-1\,n} \\ a_{n1} & a_{n2} & \cdots & a_{n\,n+1} & 0 \end{bmatrix} = \begin{bmatrix} 0 & a_{12} & \cdots & a_{1n} \\ & 0 & & a_{2n} \\ & & \ddots & \vdots \\ 0 & & & 0 \end{bmatrix} + \begin{bmatrix} 0 & & & \\ a_{21} & 0 & & 0 \\ \vdots & & \ddots & \\ a_{n1} & \cdots & \cdots & 0 \end{bmatrix},$$

where $a_{ij} \neq 0$ $(i \neq j)$ (M.-D. Choi, C. Lausce and H. Radjavi [1]). It is easily seen that T is the sum of two (SI) operators. □

Theorem 7.9. *Assume that $\mathcal{H}$ is an infinite dimensional Hilbert space and $T \in \mathcal{L}(\mathcal{H})$ admits the matrix representation $(a_{ij})_{i,j=1}^{\infty}$ with respect to some ONB such that $a_{ij} = 0$ for $i - j > n \geq 0$, where n is a fixed integer. Then T is the sum of two strongly irreducible operators T_1 and T_2. Furthermore, if T is compact, then there is a $\delta > 0$ such that for each $z_0 \in \mathbb{C}$ and arbitrary $R > 1/\delta$, T_1, T_2 can be chosen so that $\sigma(T_1) = \{z \in \mathbb{C} : |z - z_0| \leq R\}$ and $\sigma(T_2) = \{z \in \mathbb{C} : |z + z_0| \leq R\}$.*

Proof. If $n = 0$, T is a triangular operator. By Corollary 7.6, for r, $0 < r < 1$, there exists a $\delta > 0$ such that $S^* + \delta_1 T \in \mathcal{B}_1(D_r)$ provided that $\delta_1 \leq \delta$. Thus $T = (1/\delta_1)(S^* + \delta_1 T) + (-1/\delta_1)S^*$ and $(1/\delta_1)(S^* + \delta_1 T)$ and $(-1/\delta_1)S^*$ are (SI) operators.

Furthermore, if $T \in \mathcal{K}(\mathcal{H})$, $\sigma(S^* + \delta_1 T) = \overline{D} \cup \sigma_0(S^* + \delta_1 T)$. Since $S^* + \delta_1 T \in (SI)$, $\sigma(S^* + \delta_1 T)$ is connected. Thus $\sigma_0(S^* + \delta_1 T) = \varnothing$ and $\sigma((1/\delta_1)(S^* + \delta_1 T)) = \overline{D}_{1/\delta_1}$.

Let $\delta_2 = 1/R$ and set $T_1 = (1/\delta_2)(S^* + \delta_2 T - \delta_2 z_0)$ and $T_2 = (-1/\delta_2)(S^* - \delta_2 z_0)$. Thus $T = T_1 + T_2$, $T_1, T_2 \in (SI)(\mathcal{H})$, $\sigma(T_1) = \{z \in \mathbb{C} : |z - z_0| \leq R\}$ and $\sigma(T_2) = \{z \in \mathbb{C} : |z + z_0| \leq R\}$.

If $n \geq 1$, $T = A + B$, where

$$A = \begin{bmatrix} A_1 & & * \\ & A_2 & \\ 0 & & \ddots \end{bmatrix}, \quad B = \begin{bmatrix} 0 & & & \\ & B_1 & & 0 \\ & & B_2 & \\ & 0 & & \ddots \end{bmatrix}$$

A_i, B_i are $2n \times 2n$ matrices $(i = 1, 2, \cdots)$ and the 0 in the block diagonal of B is an $n \times n$ 0-matrix. Lemma 7.5 indicates that for each r, $0 < r < 1$, there is a $\delta > 0$ such that $S^* + \delta_1 A$ and $S^* - \delta_1 B \in \mathcal{B}_1(D_r)$ when $\delta_1 < \delta$. Thus $T = (1/\delta_1)(S^* + \delta_1 A) + (-1/\delta_1)(S^* - \delta_1 B)$ and $(1/\delta_1)(S^* + \delta_1 B)$, $(-1/\delta_1)(S^* - \delta_1 B) \in (SI)(\mathcal{H})$.

If T is compact, by the argument of the case of $n = 0$, $\sigma((1/\delta_1)(S^* + \delta_1 A)) = \sigma((-1/\delta_1)(S^* - \delta_1 B)) = \overline{D}_{1/\delta_1}$.

Let $\delta_2 = 1/R$ and set $T_1 = (1/\delta_2)(S^* + \delta_2 A - \delta_2 z_0)$, $T_2 = (-1/\delta_2)(S^* - \delta_2 B - \delta_2 z_0)$. Thus $T = T_1 + T_2$, $T_1, T_2 \in (SI)(\mathcal{H})$ and $\sigma(T_1) = \{z \in \mathbb{C} : |z - z_0| \leq R\}$, $\sigma(T_2) = \{z \in \mathbb{C} : |z + z_0| \leq R\}$. □

Corollary 7.10. *Assume that $T \in \mathcal{L}(\mathcal{H})$ admits a matrix representation $(a_{ij})_{i,j=-\infty}^{\infty}$ with respect to a two sided ONB $\{e_n\}_{n=-\infty}^{\infty}$ of $\mathcal{H}$ satisfying $a_{ij} = 0$ if $|i \pm j| > n \geq 0$, where n is a fixed integer. Then T is the sum of two strongly irreducible operators.*

Proof. Rearrange $\{e_k\}_{k=-\infty}^{\infty}$ as $\{e_0, e_1, e_{-1}, e_2, e_{-2}, \cdots\}$, i.e., consider the *ONB* $\{f_k\}_{k=1}^{\infty}$ of $\mathcal{H}$, where

$$f_{2k} = e_k \quad (k \geq 1), \qquad f_{2k+1} = e_{-k} \quad (k \geq 0).$$

Thus the matrix $(b_{ij})_{i,j=1}^{\infty}$ of T with respect to the *ONB* $\{f_k\}$ satisfies that $b_{ij} = 0$ if $i - j > 2n + 1$. By Theorem 7.9, T is the sum of two (SI) operators. □

Corollary 7.11. *Each bilateral weighted shift is the sum of two strongly irreducible operators.*

If T is a triangular operator, or an analytic Toeplitz operator, then by Theorem 7.9, T is the sum of two (SI) operators. The following proposition is a stronger result.

Proposition 7.12. *Each triangular operator (or analytic Toeplitz operator) is the sum of two strongly irreducible triangular operators (or analytic Toeplitz operators, respectively).*

Proof. If T admits the upper triangular matrix representation $T = (a_{ij})_{i,j=1}^{\infty}$ with respect to an *ONB* $\{e_k\}_{k=1}^{\infty}$ of $\mathcal{H}$ and S is the lateral shift (with respect to the *ONB* $\{e_k\}$), then fix an r, $0 < r < 1$. By Corollary 7.6, there is a $\delta > 0$ such that $S^* + \delta T \in \mathcal{B}_1(D_r)$. Thus $T = (1/\delta)(S^* + \delta T) + (-1/\delta)S^*$ satisfies the conclusion of the proposition.

The same argument can be used if T is an analytic Toeplitz operator. □

Proposition 7.13. *Assume that $T = A + B$, where A is a triangular operator, and B admits the matrix representation $B = (b_{ij})_{i,j=1}^{\infty}$ with respect to some ONB such that $\sum_{i,j} |b_{ij}| < \infty$. Then T is the sum of two strongly irreducible operators.*

Proof. Note that $S^* + \delta_1 A \in \mathcal{B}_1(D_r)$ for some r, $0 < r < 1$, and δ_1, $0 < \delta_1 \leq \delta$. It follows from $\sum_{i,j} |b_{ij}| < \infty$ and S.-C. Lin [1] or B. Chan [1] that when δ_1 is small enough, $S + \delta_1 B^* \sim S$. Thus $T = (1/\delta)(S^* + \delta A) + (-1/\delta)(S^* - \delta_1 B)$ satisfies the conclusion of the proposition. □

Theorem 7.14. *Each compact operator is the sum of two strongly irreducible operators.*

Proof. By a result of R.G. Douglas and C.M. Pearcy [1],

$$T = \begin{bmatrix} T_1 & T_{12} & & * \\ & T_2 & T_{23} & \\ & & T_3 & \ddots \\ & 0 & & \ddots \end{bmatrix} \begin{matrix} \mathcal{H}_1 \\ \mathcal{H}_2 \\ \mathcal{H}_3 \\ \vdots \end{matrix},$$

where T_n is a cyclic operator on an infinite dimensional Hilbert space $\mathcal{H}_n$, $n = 1, 2, \cdots$. Thus, each T_n admits a matrix representation $T_n = (a_{ij}^n)_{i,j=1}^{\infty}$ with respect to a suitable ONB such that $a_{ij}^n = 0$ $(i - j > 1)$. Applying Theorem 7.9 to each T_n, $T_n = A_n + B_n$, where $A_n \in \mathcal{B}_1(D^{(n)})$, $B_n \in \mathcal{B}_1(G^{(n)})$, $D^{(n)} = \{z \in \mathbb{C} : |z - 1/2^n| < R\}$, $G^{(n)} = \{z \in \mathbb{C} : |z + 1/2^n| < R\}$ for a fixed R.

Since $\sigma_{re}(A_n) \cap \sigma_{le}(A_{n+1}) \neq \emptyset$, by Theorem 1.13 $\mathcal{M}_n = \operatorname{ran} \tau_{A_n A_{n+1}}$ is not dense in $\mathcal{L}(\mathcal{H}_{n+1}, \mathcal{H}_n)$ $(n = 1, 2, \cdots)$. Similarly, $\mathcal{N}_n = \operatorname{ran} \tau_{B_n B_{n+1}}$ is not dense in $\mathcal{L}(\mathcal{H}_{n+1}, \mathcal{H}_n)$. Thus

$$\mathcal{M}_n \cup \mathcal{N}_n \neq \mathcal{L}(\mathcal{H}_{n+1}, \mathcal{H}_n) \qquad (n = 1, 2, \cdots).$$

Choose $C_{n\,n+1} \in \mathcal{L}(\mathcal{H}_{n+1}, \mathcal{H}_n) \backslash (\mathcal{M}_n \cup \mathcal{N}_n)$ $(n = 1, 2, \cdots)$ where $\epsilon_n := \|C_{n\,n+1}\|$ will be determined later.

Define $A_{n\,n+1}$, $B_{n\,n+1} \in \mathcal{L}(\mathcal{H}_{n+1}, \mathcal{H}_n)$ $(n = 1, 2, \cdots)$ as follows

$$\begin{aligned}
&A_{n\,n+1} = B_{n\,n+1} = \frac{1}{2} T_{n\,n+1} && \text{if } T_{n\,n+1} \notin \mathcal{M}_n \cup \mathcal{N}_n; \\
&A_{n\,n+1} = T_{n\,n+1} + C_{n\,n+1},\ B_{n\,n+1} = -C_{n\,n+1} && \text{if } T_{n\,n+1} \in \mathcal{M}_n \backslash \mathcal{N}_n; \\
&A_{n\,n+1} = -C_{n\,n+1},\ B_{n\,n+1} = T_{n\,n+1} + C_{n\,n+1} && \text{if } T_{n\,n+1} \in \mathcal{N}_n \backslash \mathcal{M}_n; \\
&A_{n\,n+1} = \frac{1}{2} T_{n\,n+1} + C_{n\,n+1},\ B_{n\,n+1} = \frac{1}{2} T_{n\,n+1} - C_{n\,n+1} && \text{if } T_{n\,n+1} \in \mathcal{M}_n \cap \mathcal{N}_n.
\end{aligned}$$

Thus $A_{n\,n+1} \notin \mathcal{M}_n$, $B_{n\,n+1} \notin \mathcal{N}_n$ and $A_{n\,n+1} + B_{n\,n+1} = T_{n\,n+1}$ $(n = 1, 2, \cdots)$.

Define

$$A=\begin{bmatrix} A_1 & A_{12} & T_{13} & & * \\ & A_2 & A_{23} & T_{24} & \\ & & A_3 & A_{34} & \ddots \\ & & & A_4 & \ddots \\ & 0 & & & \ddots \end{bmatrix}\begin{matrix}\mathcal{H}_1\\ \mathcal{H}_2\\ \mathcal{H}_3\\ \mathcal{H}_4\\ \vdots\end{matrix}, \quad B=\begin{bmatrix} B_1 & B_{12} & & 0 \\ & B_2 & B_{23} & \\ & & B_3 & \ddots \\ & 0 & & \ddots \end{bmatrix}\begin{matrix}\mathcal{H}_1\\ \mathcal{H}_2\\ \mathcal{H}_3\\ \vdots\end{matrix}.$$

Thus $T = A + B$. Observe that for each n, $\bigoplus_{k=n}^{\infty} A_k$ is a compact perturbation of $\bigoplus_{k=n}^{\infty}(RS^* - 1/2^k)$. Thus

$$\sigma\left(\bigoplus_{k=n}^{\infty} A_k\right) = \sigma\left(\bigoplus_{k=n}^{\infty}\left(RS^* - \frac{1}{2^k}\right)\right) \cup \sigma_0\left(\bigoplus_{k=n}^{\infty} A_k\right).$$

Since each $RS^*-1/2^k$ is a cosubnormal operator, $\sigma(\bigoplus_{k=n}^{\infty}(RS^*-1/2^k)=[\bigcup_{k=n}^{\infty} D^{(k)}]^-$. Since $\sigma_0\big(\bigoplus_{k=n}^{\infty} A_k\big) \subset \bigcup_{k=n}^{\infty}\sigma_0(A_k)$ and since $A_k \in (SI)$, $\sigma_0\big(\bigoplus_{k=n}^{\infty} A_k\big) = \varnothing$ and $\sigma\big(\bigoplus_{k=n}^{\infty} A_k\big) = \big[\bigcup_{k=n}^{\infty} D^{(k)}\big]^-$.

Set

$$X_n=\begin{bmatrix} A_1 & A_{12} & & & * \\ & A_2 & A_{23} & & \\ & & A_3 & \ddots & \\ & & & \ddots & A_{n-1\,n} \\ & 0 & & & A_n \end{bmatrix}\begin{matrix}\mathcal{H}_1\\ \mathcal{H}_2\\ \mathcal{H}_3\\ \vdots\\ \mathcal{H}_n\end{matrix}, \quad Y_n=\begin{bmatrix} A_{n+1} & A_{n+1\,n+2} & \cdots \\ & A_{n+2} & \\ 0 & & \ddots \end{bmatrix}\begin{matrix}\mathcal{H}_{n+1}\\ \mathcal{H}_{n+2}\\ \vdots\end{matrix}$$

and

$$Y_n'=\begin{bmatrix} A_{n+1} & A'_{n+1\,n+2} & & 0 \\ & A_{n+2} & A'_{n+2\,n+3} & \\ & & A_{n+3} & \ddots \\ & 0 & & \ddots \end{bmatrix},$$

where

$$A'_{n+k,\,n+k+1}=\begin{cases} 0, & \text{if } A_{n+k,\,n+k+1} = \frac{1}{2}T_{n+k,\,n+k+1}, \\ C_{n+k,\,n+k+1}, & \text{if } A_{n+k,\,n+k+1} = T_{n+k,\,n+k+1} + C_{n+k,\,n+k+1} \\ & \text{or } A_{n+k,\,n+k+1} = \frac{1}{2}T_{n+k,\,n+k+1} + C_{n+k,\,n+k+1}, \\ -C_{n+k,\,n+k+1}, & \text{if } A_{n+k,\,n+k+1} = -C_{n+k,\,n+k+1}. \end{cases}$$

Thus Y'_n is a compact perturbation of Y_n. Denote $\Omega_n = \bigcap_{k=1}^n D^{(k)}$. Since $A_k \in \mathcal{B}_1(D^{(k)})$ $(1 \leq k \leq n)$, computation shows that $X_n \in \mathcal{B}_n(\Omega_n)$. By the upper semi-continuity of the spectrum, we can choose $\{\epsilon_k : k = 1, 2, \cdots\}$ so that

$$\sigma(Y'_n) \subset \left\{z \in \mathbb{C} : \operatorname{dist}\left(z, \bigcup_{k=n+1}^{\infty} D^{(k)}\right) < 2^{-(n+2)}\right\},$$

$$\sigma(Y_n) \subset \left\{z \in \mathbb{C} : \operatorname{dist}\left(z, \bigcup_{k=n+1}^{\infty} D^{(k)}\right) < 2^{-(n+2)}\right\} \cup E_n,$$

where E_n is an almost countable set and $E_n \subset \sigma_0(Y_n)$. Thus $\Gamma_n = \Omega_n \backslash \sigma(Y_n)$ is not empty. By Proposition 1.14, $\ker \tau_{Y_n X_n} = \{0\}$. Thus it follows from Proposition 3.15 that $A \in (SI)$. By the same argument we can prove that $B \in (SI)$. □

To analyse the small compact perturbation, we need the following useful lemma. The reader can find its proof in D.A. Herrero [1].

Lemma 7.15. *Each operator $T \in \mathcal{L}(\mathcal{H})$ can be expressed as $T = A + B$, where A, B are in the norm closure of the nilpotent operators on $\mathcal{H}$.*

Theorem 7.16. *Given $T \in \mathcal{L}(\mathcal{H})$ and $\epsilon > 0$, there exists a compact operator K with $\|K\| < \epsilon$ such that $T + K$ is the sum of two strongly irreducible operators.*

Proof. By Lemma 7.15 $T = A + B$, where A, B are the limits of nilpotent operators. It follows from the upper continuity of the spectrum and the stability property of semi-Fredholm operators that $\sigma(A)$ and $\sigma(B)$ are connected and $\operatorname{ind}(A - \lambda) = 0$ for $\lambda \in \rho_{s-F}(A)$ and $\operatorname{ind}(B - \lambda) = 0$ for $\lambda \in \rho_{s-F}(B)$. By Theorem 5.29, there are compact operators K_1 and K_2 such that $A + K_1 \in (SI)(\mathcal{H})$, $B + K_2 \in (SI)(\mathcal{H})$ and $\|K_1 + K_2\| < \epsilon$. Let $K = K_1 + K_2$, thus $T + K = (A + K_1) + (B + K_2)$ and the conclusion of the theorem is true. □

7.2 Some applications of strongly irreducible operators

To consider the relation between (SI) operators and the invariant subspace problem, we have

Theorem 7.17. *If the strongly irreducible operators in each infinite dimensional subspace $\mathcal{X}$ of $\mathcal{L}(H^2 \oplus \mathcal{H})$ do not form a dense open set, then each operator $A \in \mathcal{L}(\mathcal{H})$ has a non-trivial invariant subspace.*

Proof. If the condition of Theorem 7.17 is satisfied and A is an operator on $\mathcal{H}$ which has no non-trivial invariant subspace, then for each $x \in \mathcal{H}$, $x \neq 0$, x is a cyclic vector of A. Without loss of generality, we may assume that $\|A\| < 1$ and $\sigma(A)^\wedge \subset D$, where $\sigma(A)^\wedge$ denotes the polynomial defined by the convex hull of $\sigma(A)$. Define an operator U: $\mathcal{H} \to \mathcal{L}(H^2 \oplus \mathcal{H})$ by

$$Ux = \begin{bmatrix} (x, e_0)T_z^* & 0 \\ x \otimes e_0 & (x, e_0)A \end{bmatrix},$$

where $e_0 \in \ker T_z^*$ and $\|e_0\| = 1$. Denote $\mathcal{X} = \operatorname{ran} U \subset \mathcal{L}(H^2 \oplus \mathcal{H})$.

Claim. (i) U is a linear mapping, $\|x\| \leq \|Ux\| \leq 2\|x\|$ therefore $\operatorname{ran} U$ is a (closed) subspace of $\mathcal{L}(H^2 \oplus \mathcal{H})$;
(ii) $Ux \in (SI)$ if and only if $(e_0, x) \neq 0$;
(iii) let $\mathcal{X}_1$ denote $\mathcal{X} \cap (SI)$; then $\mathcal{X}_1$ is a dense open subset of $\mathcal{X}$.

In fact, for $x, y \in \mathcal{H}$ and $\alpha \in \mathbb{C}$,

$$U(\alpha x + y) = \begin{bmatrix} (\alpha x + y, e_0)T_z^* & 0 \\ (\alpha x + y, e_0) \otimes e_0 & (\alpha x + y, e_0)A \end{bmatrix}$$

$$= \alpha \begin{bmatrix} (x, e_0)T_z^* & 0 \\ x \otimes e_0 & (x, e_0)A \end{bmatrix} + \begin{bmatrix} (y, e_0)T_z^* & 0 \\ y \otimes e_0 & (y, e_0)A \end{bmatrix}$$

$$= \alpha Ux + Uy.$$

Note that $\|Ux\| \geq \|x \otimes e_0\| = \|x\|$. Thus $\|Ux\| \leq \max\{\|(x, e_0)T_z^*\|, \|(x, e_0)\| + \|(x, e_0)A\|\} \leq 2\|x\|$, i.e., (i) is true.

If $(x, e_0) \neq 0$,

$$Ux = (x, e_0) \begin{bmatrix} T_z^* & 0 \\ \dfrac{x \otimes e_0}{(x, e_0)} & A \end{bmatrix} \in (SI)$$

by Proposition 2.32. This proves (ii).

Set $\mathcal{M} = H^2 \ominus \bigvee\{e_0\}$ and $\mathcal{X}_0 = U(\mathcal{M})$. By (i) $\mathcal{X}_0$ is a closed subspace of $\mathcal{X}$. Since $Ux = \begin{bmatrix} 0 & 0 \\ x \otimes e_0 & 0 \end{bmatrix}$ is strongly reducible for $x \in \mathcal{M}$, $\mathcal{X}_0$ is a non-trivial subspace of $\mathcal{X}$. Thus $\mathcal{X}_1 = \mathcal{X} \backslash \mathcal{X}_0$ is a dense open subset of $\mathcal{X}$. □

Now we consider the relation between (SI) operators and the stable invariant subspaces of operators.

An invariant subspace $\mathcal{M}$ of $T \in \mathcal{L}(\mathcal{H})$ is called a *stable invariant subspace* of T, if for each $\epsilon > 0$, there exists a $\delta > 0$ such that if $T' \in \mathcal{L}(\mathcal{H})$ and $\|T - T'\| < \delta$ there is an invariant subspace $\mathcal{M}'$ of T' such that $\|P_{\mathcal{M}} - P_{\mathcal{M}'}\| < \epsilon$, where $P_{\mathcal{M}}$ and $P_{\mathcal{M}'}$ denote the orthogonal projections onto $\mathcal{M}$ and $\mathcal{M}'$ respectively.

The concept of stable invariant subspace was given by S. Campbell and J. Daughtry [1] when they were considering the stable solution of matrix equations of second degree. J.B. Conway and P.R. Halmos [1] and G.T. Adams [1] described the stable invariant subspaces of matrices on finite dimensional Hilbert spaces. For infinite dimensional Hilbert spaces we have

Theorem 7.18. *Given $T \in \mathcal{L}(\mathcal{H})$ with connected spectrum $\sigma(T)$, then T has no non-trivial stable reducing invariant subspace.*

A subspace $\mathcal{M}$ of $\mathcal{H}$ is a *stable reducing invariant subspace* of T if $\mathcal{M}$ is a stable invariant subspace of T and there is a topologically complementary subspace $\mathcal{N}$ of $\mathcal{M}$ such that $\mathcal{N}$ is also a stable invariant subspace of T.

In order to prove Theorem 7.18, we need the following lemmas.

Lemma 7.19. *Let $\mathcal{M}$ be a stable invariant subspace of $T \in \mathcal{L}(\mathcal{H})$. Then for each invertible operator $X \in \mathcal{L}(\mathcal{H})$, $X(\mathcal{M})$ is a stable invariant subspace of XTX^{-1}.*

Proof. For each $\epsilon > 0$, there is a $\delta > 0$ such that when $\|T - T'\| < \epsilon$, there is an invariant subspace $\mathcal{M}'$ of T' such that

$$\|P_{\mathcal{M}} - P_{\mathcal{M}'}\| < \epsilon(\|X\|\,\|X^{-1}\|)^{-1}.$$

Note that $\operatorname{ran} XP_{\mathcal{M}'}X^{-1} = X(\mathcal{M}')$, $\operatorname{ran} XP_{\mathcal{M}}X^{-1} = X(\mathcal{M})$, $XP_{\mathcal{M}'}X^{-1}$, $XP_{\mathcal{M}}X^{-1}$ are idempotent operators and $X(\mathcal{M}')$, $X(\mathcal{M})$ are invariant subspaces of $XT'X^{-1}$ and XTX^{-1} respectively. Then by a result of T. Kato [1, Theorem 6.35],

$$\begin{aligned}\|P_{X(\mathcal{M}')} - P_{X(\mathcal{M})}\| &\le \|XP_{\mathcal{M}'}X^{-1} - XP_{\mathcal{M}}X^{-1}\| \\ &\le \|X\|\,\|X^{-1}\|\,\|P_{\mathcal{M}'} - P_{\mathcal{M}}\| < \epsilon.\end{aligned}$$

Thus $\mathcal{M}$ is a stable invariant subspace of T. □

Lemma 7.20. *Let $\mathcal{M}$ be a stable invariant subspace of $T \in \mathcal{L}(\mathcal{H})$. If there is a non-trivial complementary subspace $\mathcal{M}_1$ of $\mathcal{M}$ and $\mathcal{M}_1$ is also a stable invariant subspace of T, then there exists a number $\delta > 0$ such that $\|T - T'\| < \delta$ implies that T' is a strongly reducible operator and T' has two invariant subspaces $\mathcal{M}'$ and $\mathcal{M}_1'$ satisfying*

(i) $\dim \mathcal{M} = \dim \mathcal{M}'$, $\dim \mathcal{M}_1' = \dim \mathcal{M}_1$;

(ii) $M' \cap \mathcal{M}_1' = \{0\}$, $M' \dot{+} \mathcal{M}_1' = \mathcal{H}$.

Proof. Since $\mathcal{M}$ and $\mathcal{M}_1$ are complementary invariant subspaces of T, there is an invertible operator $X \in \mathcal{L}(\mathcal{H})$ such that $X(\mathcal{M})$ is orthogonal to $X(\mathcal{M}_1)$ and $X(\mathcal{M}) \oplus X(\mathcal{M}_1) = \mathcal{H}$ and $X(\mathcal{M}), X(\mathcal{M}_1)$ are invariant subspaces of XTX^{-1}. By Lemma 7.19, $X(\mathcal{M})$ and $X(\mathcal{M}_1)$ are stable invariant subspaces of T. Thus there is a $\delta_0 > 0$ such that T' has two invariant subspaces $\mathcal{M}'$ and $\mathcal{M}_1'$ satisfying $\|P_{\mathcal{M}'} - P_{X(\mathcal{M})}\| < 1/4$ and $\|P_{\mathcal{M}_1'} - P_{X(\mathcal{M}_1)}\| < 1/4$ provided that $T' \in \mathcal{L}(\mathcal{H})$ and $\|T' - T\| < \delta_0$.

Note that

$$\|P_{\mathcal{M}'} - P_{\mathcal{M}_1'}\| \geq \|P_{X(\mathcal{M})} - P_{X(\mathcal{M}_1)}\| - \|P_{X(\mathcal{M})} - P_{\mathcal{M}'}\| - \|P_{X(\mathcal{M}_1)} - P_{\mathcal{M}_1'}\| > \frac{1}{2}.$$

Thus $\mathcal{M}' \cap \mathcal{M}_1' = \{0\}$. Since $\|P_{\mathcal{M}'} - P_{X(\mathcal{M})}\| < 1/4$ and $\|P_{\mathcal{M}_1'} - P_{X(\mathcal{M}_1)}\| < 1/4$, $\dim M' = \dim X(\mathcal{M}) = \dim \mathcal{M}$ and $\dim M_1' = \dim X(\mathcal{M}_1) = \dim \mathcal{M}_1$.

Set $P' = P_{\mathcal{M}'} + P_{\mathcal{M}_1'}$. Thus

$$\begin{aligned}\|P' - 1\| &= \|P_{\mathcal{M}'} - P_{X(\mathcal{M})} + P_{\mathcal{M}_1'} - P_{X(\mathcal{M}_1)}\| \\ &\leq \|P_{\mathcal{M}'} - P_{X(\mathcal{M})}\| + \|P_{\mathcal{M}_1'} - P_{X(\mathcal{M}_1)}\|.\end{aligned}$$

Therefore P' is invertible and $\operatorname{ran} P' = \mathcal{H}$, i.e., $\mathcal{M}' \dot{+} \mathcal{M}_1' = \mathcal{H}$. □

Proof of Theorem 7.18. By Theorem 3.30, $T \in (SI)(\mathcal{H})^-$, i.e., there is a sequence $\{T_n\}_{n=1}^{\infty}$ of (SI) operators such that $\lim_{n\to\infty} \|T_n - T\| = 0$. Thus the theorem is a conclusion of Lemma 7.20. □

To compare with Theorem 7.18 we have

Theorem 7.21. *Assume that $T \in \mathcal{L}(\mathcal{H})$, $\sigma(T) = \sigma_1 \cup \sigma_2$, where σ_1, σ_2 are disjoint non-empty clopen subsets of $\sigma(T)$. Then T has a non-trivial stable reducing invariant subspace.*

The following lemma is needed in the proof of Theorem 7.21.

Lemma 7.22. *Let T and $\sigma(T) = \sigma_1 \cup \sigma_2$ be given as in Theorem 7.21, $E_i = E(T, \sigma_i)$ $(i = 1, 2)$, let $\mathcal{M}$ be an invariant subspace of T and $\mathcal{M} = E_1(\mathcal{M}) \dot{+} E_2(\mathcal{M})$. Then the following are equivalent*

(i) *$\mathcal{M}$ is a stable invariant subspace of T;*

(ii) *$E_i(\mathcal{M})$ is a stable invariant subspace of T $(i = 1, 2)$;*

(iii) *$E_i(\mathcal{M})$ is a stable invariant subspace of* $T|_{\text{ran } E_i}$ $(i = 1, 2)$.

Proof. Let Ω_1, Ω_2 be two analytic Cauchy domains such that $\Omega_1 \supset \sigma_1$, $\Omega_2 \supset \sigma_2$, $\overline{\Omega}_1 \cap \sigma_2 = \overline{\Omega}_2 \cap \sigma_1 = \varnothing$ and $\partial\Omega_1 \cup \partial\Omega_2 \subset \rho(T)$. Thus $E_j = \frac{1}{2\pi i} \int_{\partial\Omega_j} (\lambda - T)^{-1}\, d\lambda$ $(j = 1, 2)$.

(i)⇒ (ii). It follows from $\partial\Omega_1 \cup \partial\Omega_2 \subset \rho(T)$ and the upper continuity of the spectrum that if $T_n \to T$ $(n \to \infty)$ then $(T_n - \lambda)^{-1} \to (T - \lambda)^{-1}$ $(n \to \infty)$ uniformly on $\partial\Omega_1 \cup \partial\Omega_2$. Thus for each $\epsilon > 0$, there is a $\delta > 0$, such that $\sigma(T') = \sigma_1' \cup \sigma_2'$, $\sigma_1' \subset \Omega_1$, $\sigma_2' \subset \Omega_2$, $\|E_1 - E_1'\| + \|E_2 - E_2'\| < \epsilon/2$ and $\|P_{\mathcal{M}'} - P_{\mathcal{M}}\| < \epsilon/(2\|E_1\|)$ for some invariant subspace M_1' of T' provided that $\|T' - T\| < \delta$, where $E_j' = \frac{1}{2\pi i} \int_{\partial\Omega_j} (\lambda - T')^{-1}\, d\lambda$ $(j = 1, 2)$. Therefore

$$
\begin{aligned}
\|P_{E_1'(\mathcal{M}')} - P_{E_1(\mathcal{M})}\| &\le \|E_1'|_{\mathcal{M}'} - E_1|_{\mathcal{M}}\| = \|E_1' P_{\mathcal{M}'} - E_1 P_{\mathcal{M}}\| \\
&= \|E_1' P_{\mathcal{M}'} - E_1 P_{\mathcal{M}'}\| + \|E_1 P_{\mathcal{M}'} - E_1 P_{\mathcal{M}}\| \\
&\le \|E_1' - E_1\| + \|E_1\| \|P_{\mathcal{M}'} - P_{\mathcal{M}}\| < \epsilon,
\end{aligned}
$$

and $E_1(\mathcal{M})$ is a stable invariant subspace of T. Similarly, we can prove that $E_2(\mathcal{M})$ is a stable invariant subspace of T.

(ii)⇒(iii). Since $E_1(\mathcal{M}), E_2(\mathcal{M})$ are stable invariant subspaces of T, for each $\delta > 0$ by the argument of (i)⇒(ii), there exists a $\delta_1 > 0$ such that $\|T - T'\| < \delta_1$ implies that $\|P_{E_j(\mathcal{M})} - P_{E_j'(\mathcal{M})}\| < \delta$, where $E_j' = \frac{1}{2\pi i} \int_{\partial\Omega_i} (T' - \lambda)^{-1}\, d\lambda$ $(j = 1, 2)$ and $\sigma(T') = \sigma_1' \cup \sigma_2'$, $\sigma_1' \subset \Omega_1$, $\sigma_2' \subset \Omega_2$.

Set

$$
\left.
\begin{aligned}
T_j' &= \frac{1}{2\pi i} \int_{\partial\Omega_j} T'(T' - \lambda)^{-1}\, d\lambda = T'|_{\text{ran } E_j'}, \\
T_j &= \frac{1}{2\pi i} \int_{\partial\Omega_j} T(T - \lambda)^{-1}\, d\lambda = T|_{\text{ran } E_j}
\end{aligned}
\right\} \quad (j = 1, 2).
$$

Thus $T' = T_1' + T_2'$ and $T = T_1 + T_2$. Choose $\delta < \epsilon$ and δ_1 sufficient small so that $\|T_1' - T_1\| < \delta$ and $\|T_2' - T_2\| < \delta$. Note that $E_i(\mathcal{M})$ and $E_i'(\mathcal{M})$ are invariant subspaces of T_i and T_i' respectively, $\|P_{E_i(\mathcal{M})} - P_{E_i'(\mathcal{M})}\| < \epsilon$ implies that $E_i(\mathcal{M})$ is a stable invariant subspace of T_i $(i = 1, 2)$.

(iii)⇒(i) is obvious. □

Proof of Theorem 7.21. Let Ω_1, Ω_2, E_1 and E_2 be given as in Lemma 7.22. Note that $\mathcal{H}$ is a stable invariant subspace of T, thus by Lemma 7.22 $E_1(\mathcal{H}), E_2(\mathcal{H})$ are non-trivial stable invariant subspaces of T. □

If $\mathcal{H}$ is a finite dimensional Hilbert space $\mathcal{H}$, we have

Proposition 7.23. *Let $T \in \mathcal{L}(\mathbb{C}^n)$. Then the following are equivalent.*

(i) *Each invariant subspace of T is stable invariant.*

(ii) *T has a unique strongly irreducible decomposition.*

(iii) *$\mathcal{A}'(T)$ is abelian.*

Proof. (ii)⇒(iii). A notable fact is that $A \in \mathcal{L}(\mathbb{C}^n)$ is a (SI) operator if and only if $A \sim \lambda + q_n$. Since $\mathcal{A}'(\lambda + q_n)$, the commutant of $\lambda + q_n$, is abelian, $\mathcal{A}'(A)$ is abelian. By the Jordan decomposition theorem $T = \sum_{k=1}^{m} \dot{+} T_k$, where $T_k \in (SI)$. It follows from Proposition 4.16 that $\mathcal{A}'(T) = \sum \dot{+} \mathcal{A}'(T_k)$. Since each $\mathcal{A}'(T_k)$ is abelian, $\mathcal{A}'(T)$ is abelian.

(iii)⇒(ii) is a direct consequence of Theorem 4.19.

(ii)⇒(i). Assume that T has a unique (SI) decomposition $T = \sum_{k=1}^{m} \dot{+} T_k$, where each $T_k \in (SI)$. By Proposition 4.16 $\ker \tau_{T_k T'_k} = \{0\}$ $(k \neq k')$. Suppose that $T_k \sim (\lambda_k + q_{n_k})$. Since $\ker \tau_{T_k T'_k} = \{0\}$ $(k \neq k')$, $\lambda_k \neq \lambda_{k'}$ $(k \neq k')$. By C. Apostol, L.A. Fialkow, D.A. Herrero and D. Voiculescu [1, Lemma A1.24] each invariant subspace of $\lambda_k + q_{n_k}$ is stable invariant. But each invariant subspace $\mathcal{M}$ of T is the orthogonal direct sum of invariant subspaces of $(\lambda_k + q_{n_k})$'s. By Lemma 7.22 $\mathcal{M}$ is a stable invariant subspace of T.

(i)⇒(ii). Assume that $T \sim \bigoplus_{n=1}^{m} (\lambda_k + q_{n_k}) = T'$. If $\lambda_1 = \lambda_2$, by the same argument as Lemma 7.22 we can prove that $\mathbb{C}^{k_1} \oplus 0$ and $0 \oplus \mathbb{C}^{k_2}$ are stable invariant subspaces of $(\lambda_1 + q_{k_1}) \oplus (\lambda_2 + q_{k_2})$. Therefore, $(\lambda_1 + q_{k_1}) \oplus (\lambda_2 + q_{k_2})$ has stable reducing invariant subspaces. This is contradictory to the conclusion of Theorem 7.18, thus $\lambda_1 \neq \lambda_2$. By the same argument, $\lambda_k \neq \lambda_{k'}$ $(k \neq k')$. By Proposition 1.14, $\ker \tau_{\lambda_k + q_{n_k}, \lambda_{k'} + q_{n_{k'}}} = \{0\}$. Thus by Proposition 4.16, T has a unique (SI) decomposition. □

7.3 Open problems

Problem 7.1. *Is each operator a sum of two (SI) operators?*

Problem 7.2. *Does there exist an infinite dimensional subspace $\mathcal{X}$ of $\mathcal{L}(\mathcal{H})$, such that $\mathcal{X} \cap (SI)(\mathcal{H})$ is a dense open subset of $\mathcal{X}$?*

Problem 7.3. *What is the character of the stable invariant subspaces of an (SI) operator?*

7.4 Remarks

The results of 7.1 are basically due to C.L. Jiang and P.Y. Wu [1]. 7.2 contains some unpublished results of Y.Q. Ji, C.L. Jiang and Z.Y. Wang.

REFERENCES

Adams, G. T.

[1] A non-linear characterization of stable invariant subspaces, Integral Equations and Operator Theory 6 (1983), 473–487.

Adams, R. A.

[1] Sobolev spaces, Academic Press, New York–San Francisco–London, 1975.

Apostol, C., Douglas, R.G. and Foiaş, C.

[1] Quasisimilar models for nilpotent operators, Trans. Amer. Math. Soc. 224 (1976), 407–415.

Apostol, C., Fialkow, L. A., Herrero, D. A. and Voiculescu, D.

[1] Approximation of Hilbert space operator. II, Research Notes in Math. 102. (Longman, Harlow, Essex, 1984).

Agler, J., Franks, E. and Herrero, D. A.

[1] Spectral pictures of operators quasisimilar to the unilateral shift, J. Reine und Angewandte Mathematik 422 (1991), 1–20.

Apostol, C. and Morrel, B. B.

[1] On uniform approximation of operators by simple models, Indiana Univ. Math. J. 26 (1977), 427–442.

Allan, G. R.

[1] On one-sided inverses in Banach algebras of holomorphic vector-valued functions, J. London Math. Soc. 42 (1967), 463–470.

[2] Holomorphic vector-valued functions on a domain of holomorphy, J. London Math. Soc. 42 (1967), 509–513.

Al-Musallan.

[1] An upper estimate for the distance to the essentially G_1 operators, Dissertation, Arizona state university, 1989.

Arveson, W. B.

[1] An invitation to C^*-algebras, Grad. Texts in Math. 39. (New York–Heidelberg–Berlin: Springer–Verlag, 1976).

[2] Ten lectures on operator algebras, CBMS series 55, Amer. Math. Soc. Providence, 1983.

Baker, I. N., Deddens, J. A. and Ullman, J. L.

[1] A theorem on entire functions with applications to Toeplitz operators, Duke Math. J. 41 (1974), 703–745.

Bart, H., Gohberg, I. C. and Kaashoek, M. A.

[1] stable factorizations of monic matrix polynomials and stable invariant subspaces, Integral Equations and Operator Theory 1 (1978), 496–517.

Behnke, H.

[1] Structure of certain non-normal operators, J. Math. Mech. 18, 103–107.

Berberian, S. K.

[1] Approximate proper vectors, Proc. Amer. Math. Soc. 13 (1962), 111–114.

Berg, I. D. and Davidson, K. R.

[1] Almost commuting matrices and a quantitative version of Brown–Douglas–Fillmore theorem, Acta. Math. 166 (1991), 121–161.

Berger, C. A. and Shaw, B. I.

[1] Intertwining, analytic structure, and the trace norm estimate, Proceedings of a conference on operator theory, Halifax, Nova Scotia 1973, Lect. Notes in Math. 345. (Berlin–Heidelberg–New York: Springer–Verlag, 1973), 1–6.

Bonsule, F. F. and Duncan, J.

[1] Complete normed algebras, Springer–Verlag, New York and Berlin, 1973.

Campbell, S. and Daughtry, J.

[1] The stable solutions of quadratic matrix equations, Proc. Amer. Math. Soc. 74 (1979), 19–23.

Cao, G. F. (See Jiang, C. L.)

Chan, B.

[1] Perturbation of shifts, J. Math. Appl. 88 (1982), 572–590.

Choi, M.-D., Lausce, C. and Radjavi, H.

[1] On commutators and invariant subspaces, Linear Multilinear Algebra 9 (1981), 329–340.

Conway, J. B.

[1] A course in functional analysis, Grad. Texts in Math. 95 (New York–Heidelberg–Berlin: Springer–Verlag, 1990).

Conway, J. B. and Halmos, P. R.

[1] Finite-dimensional points of continuity of Lat. Lin. Alg. Appl. 31 (1980), 93–102.

Conway, J. B. and Morrel, B. B.

[1] Roots and logarithms of bounded operators on Hilbert space, J. Funct. Anal. 70 (1987), 179–193.

Cowen, C. C.

[1] The commutant of an analytic Toeplitz operator, Trans. Amer. Math. Soc. 239 (1978), 1–31.

Cowen, M. J. and Douglas, R. G.

[1] Complex geometry and operator theory, Bull. Amer. Math. Soc. 83 (1977), 131–133.

Daughtry, J. (See also Campbell, S.)

[1] Isolated solutions of quadratic matrix equations, J. Lin. Alg. Appl. 21 (1978), 89–94.

Davidson, K. R. (See also Berg, I. D.)

[1] Nest algebra, Research Notes in Math. 191 (Longman, Harlow, Essex, 1988).

Davidson, K. R. and Herrero, D. A.

[1] The Jordan form of a bitriangular operator, J. Funct. Anal. 94 (1990), 27–73.

Davis, C. and Rosenthal, P.

[1] Solving linear operator equations, Can. J. Math. 26 (1974), 1384–1389.

Dazord, J.

[1] Invariant subspaces for functions of the shift, Operator Theory Adv. Appl. 14 (1984), 109–113.

Deddens, J. A. (See Baker, I. N.)

Deddens, J. A. and Wong, T. K.

[1] The commutant of analytic Toeplitz operators, Trans. Amer. Math. Soc. 184 (1973), 261–273.

Dixmier, J.

[1] Les algebres d'opérateurs dans l'espace Hilbertien, Ganthier–Villars, Paris, 1969.

Dixmier, J. and Foiaş, C.

[1] Sur le spectre ponctuel d'un opérateur, Colloq. Math. János Bolyai. 5. Hilbert space operators and operator algebras, Tihang (Hungary) 1970. (Amsterdam–London: North Holland Publ. Co, 1972) 127–133.

Douglas, R. G. (See also Apostol, C.; Cowen, M. J.)

[1] Banach algebras techniques in operator theory. (New York and London: Academic Press, 1972).

Douglas, R. G. and Pearcy, C. M.

[1] A note on quasitriangular operators, Duke Math. J. 37 (1970), 177–188.

[2] Invariant subspaces of non-quasitriangular operator, Lect. Notes in Math. 345. (Berlin–Heidelberg–New York: Springer–Verlag, 1973), 13–57.

Duncan, J. (See Bonsule, F. F.)

Dunford, N. and Schwartz, J. T.

[1] Linear operators. Part II: Spectral theory. Self-adjoint operators in Hilbert space. (New York–London: Interscience, 1963).

Engloš, M.

[1] Density of algebras generated by Toeplitz operators on Bergman spaces, Arkiv för Matematic, 30 (1992), 227–243.

Fialkow, L. A. (See also Apostol, C.)

[1] A note on the range of the operator $X \mapsto AX - XB$, Illinois J. Math. 25 (1981), 112–124.

Fisher, S. D.

[1] Function theory on planar Domains, Wiley–Interscience Series, John Wiley and Sons, New York (1983).

Foiaş, C. (See Apostol, C.; Dixmier, J.)

Fong, C. K. and Jiang, C. L.

[1] Normal operators similar to irreducible operators, Acta. Math. Sinica, New Series 10 (1994), 192–135.

[2] On irreducible operators, Northeastern Math. J. 8 (1992), 385–390.

[3] Approximation by Jordan type operators, Houston J. Math. 19 (1993), 51-62.

Fong, C. K. and Wu, P. Y.

[1] Diagonal operators: dilation, sum and product, Acta Sci. Math. (Szeged) 57 (1993), 125–138.

Franks, E. (See Agler, J.)

Gilfeather, F.

[1] Strong reducibility of operators, Ind. Univ. Math. J. 22 (1972), 393–397.

[2] On Suzuki structure of non-self-adjoint operator on Hilbert space, Acta Sci. Math. (Szeged) 32, 239–249.

Gohberg, I. C. (See Bart, H.)

Gray, L. J.

[1] Jordan representation for a class of nilpotent operators, Indiana Univ. Math. J. 26 (1977), 57–64.

Guinand, P. S. and Marcoux, L.

[1] Between the unitary and similarity orbits of normal operators, Pacific J. Math. 159 (1993), 299–335.

[2] On the $(\mathcal{U} + \mathcal{K})$-orbits of certain weighted shifts, Integral Equations and Operator Theory 17 (1993), 516–543.

Guo, K.Y. and Jiang, C.L.

[1] Hyponormal strongly irreducible operators and Berezin perturbation, Acta Sci. Math. (Szeged) (to appear).

Halmos, P. R. (See also Conway, J. B.)

[1] A Hilbert space problem book. (Princeton, New Jersey: D. Van Nostrand, 1967).

[2] Irreducible operators, Mich. Math. J. 15 (1968), 215–223.

Hartogs, H. and Rosenthal, A.

[1] über fdgen analytischer funktionen, Math. Ann. 104 (1931), 606–610.

Herrero, D. A. (See also Apostol, C.; Agler, J.; Davidson, K. R.)

[1] Approximation of Hilbert space operators I 2ed ed., Research Notes in Math. 224. (Longman, Harlow, Essex, 1990).

[2] Spectral pictures of operators in the Cowen–Douglas class $\mathcal{B}_n(\Omega)$ and its closure, J. Operator Theory 10 (1987), 213–222.

[3] On multicyclic operators, Integral Equations and Operator Theory 1 (1978), 57–102.

[4] The Fredholm structure of an n-multicyclic operator, Indiana Univ. Math. J. 36 (1987), 549–566.

[5] Quasisimilar operators with different spectra, Acta Sci. Math. (Szeged) 41 (1979), 101–118.

[6] Compact perturbations of nest algebras, index obstructions and a problem of Arveson, J. Funct. Anal. 55 (1984), 78–109.

[7] A unicellular universal quasinilpotent operator, Proc. Amer. Math. Soc. 110 (1990), 649–652.

[8] A trace obstruction to approximation by block-diagonal operators, Amer. J. Math. 108 (1986), 451–484.

[9] Compact perturbations of continuous nest algebras, J. London Math. Soc. 27 (1983), 339–344.

[10] Operator algebras of finite strict multiplicity, Indiana Univ. Math. J. 22 (1972), 13–24.

[11] Operator algebras of finite strict multiplicity. II, Indiana Univ. Math. J. 27 (1978), 9–18.

Herrero, D. A. and Jiang, C. L.

[1] Limits of strongly irreducible operators and the Riesz decomposition theorem, Mich. Math. J. 37 (1990), 283–291.

Herrero, D. A., Taylor, T. J. and Wang, Z. Y.

[1] Variation of the point spectrum under compact perturbations, Topics in operator theory, Constantin Apostol Memorial Issue, OT: Advances and Applications 32. (Basel–Bsoton–Stutlgart: Birkhäuser–Verlag, 1988), 113–158.

Ji, Y. Q., Jiang, C. L. and Wang, Z. Y.

[1] The $(\mathcal{U}+\mathcal{K})$-orbit of essentially normal operators and the compact perturbation of strongly irreducible operators, Mathematics and Its Applications 356 (1996) 307–314.

[2] Strongly irreducible operators in Nest algebras, Integral Equations and Operator Theory 28 (1997) 28–44.

[3] Strongly irreducible operators in Nest algebras with well-ordered nest, Michigan Math. J. 44 (1997), 85–98.

[4] Essentially normal+small compact = strongly irreducible, Chinese Ann. Math. (To appear).

Jiang, C. L. (See also Fong, C. K.; Guo, K.Y.; Herrero, D. A.; Ji, Y. Q.)

[1] Similarity, reducibility and approximation of Cowen–Douglas operators, J. Operator Theory 32 (1994), 77–89.

[2] Strongly irreducible operator and Cowen–Douglas operators, Northeastern Math. J. 1 (1991), 1–3.

[3] Approximation of direct sums of strongly irreducible operators, Northeastern Math. J. 1 (1989), 1–3.

Jiang, C. L., Cao, G. F. and Sun, S. L.

[1] The complete irreducibility of analytic Toeplitz operators, Mathematics and Its Application 356 (1996), 315–321.

Jiang, C. L., Power, S. and Wang, Z. Y.

[1] Biquasitriangular+small compact = strongly irreducible, (submitted).

Jiang, C. L., Sun, S. H. and Wang, Z. Y.

[1] Essentially normal operator+compact operator = strongly irreducible operator, Trans. Amer. Math. Sco. 349 (1997), 217–233.

Jiang, C. L. and Wang, Z. Y.

[1] The spectral picture and the closure of the similarity orbit of strongly irreducible operators, Integral Equations and Operator Theory 24 (1996), 81–105.

[2] A class of strongly irreducible operators with nice property, J. Operator Theory 36 (1996), 3–19.

Jiang, C. L. and Wu, P. Y.

[1] A class of operators which are similar to irreducible operators (Preprint).

[2] Sums of strongly irreducible operators (Preprint).

Jiang, Z. J.

[1] Topics in operator theory, Seminar reports in functional analysis (Chinese, lecture notes), Jilin University, 1979.

[2] On the structure of linear operators (Chinese), the report on the national symposium on operator theory, Jiu Jiang (1987).

Jiang, Z. J. and Sun, S. L.

[1] On completely irreducible operators (Chinese), Acta Sci. Natur. Univ. Jilin. 4(1992), 20–29.

Kaashoek, M. A. (See Bart, H.)

Kato, T.

[1] Perturbation theory for linear operators. (New York: Springer–Verlag, 1966).

Kelley, R. L.

[1] Weighted shifts on Hilbert space, Dissertation, Univ. of Michigan, 1966.

Lambert, A.

[1] Strictly cyclic operator algebra, Pacific J. Math. 38 (1971), 717–726.

[2] Strictly cyclic weighted shift, Proc. Amer. Math. Soc. 29 (1971), 331–336.

Lausce, C. (See Choi, M.-D.)

Lin, S.-C.

[1] Perturbations of isometries on Hilbert space, Indiana Univ. Math. J. 21 (1971), 363–372.

Lumer, G. and Rosenblum, M.

[1] Linear operator equations, Proc. Amer. Math. Soc. 10 (1959), 32–41.

Marcoux, L. (See also Guinand, P. S.)

[1] The closure of the $(\mathcal{U}+\mathcal{K})$-orbit of shift-like operators, Indiana Univ. Math. J. 41 (1992), 1211–1223.

Morrel, B. B. (See Apostol, C.; Conway, J. B.)

Pearcy, C. M. (See Douglas, R. G.)

Power, S. (See Jiang, C. L.)

Radjavi, H. (See also Choi, M.-D.)

[1] Every operator is the sum of two irreducible ones, Proc. Amer. Math. Soc. 21 (1969), 251–252.

Radjavi, H. and Rosenthal, P.

[1] Invariant subspaces, Ergebnisse der Math. und ihrer Grenz. 77. (New York–Heidelberg–Berlin: Springer–Verlag, 1973).

Rickart, C.

[1] General theory of Banach algebra. (Princeton, New Jersey: D. Van Nostrand, 1960).

Rosenblum, M. (See also Lumer, G.)

[1] On the operator equation $BX - XA = Q$, Duke Math. J. 23 (1956), 263–269.

Rosenthal, A. (See Hartogs)

Rosenthal, P. (See Davis, C.; Radjavi, H.)

Rudin, W.

[1] Real and complex analysis, 2ed ed., McGraw–Hill Book Company, New York, 1974.

Schwartz, J. T. (See Dunford, N.)

Shaw, B. I. (See Berger, C. A.)

Shilds, A. L.

[1] Weighted shift operators and analytic function theory, Math. Surveys 13. (Providence, Rhode Island: Amer. Math. Soc., 1974) 49–128.

Shubin, M. A.

[1] On holomorphic families of subspaces of a Banach space, Mat. Issled. (Kishinev) 5, vyp. 4 (18) (1970), 153–165; English Transl., Integral Equation and Operator Theory 2 (1979), 407–420.

Sun, S. H. (See Jiang, C. L.)

Sun, S. L. (See Jiang, C. L.; Jiang, Z. J.)

Suzuki, N.

[1] The algebraic structure of non-self-adjoint operators, Acta Sci. Math. (Szeged) 27, 173–184.

Taylor, T. J. (See Herrero, D. A.)

Thomson, T. E.

[1] Intersection of commutants of analytic Toeplitz operators, Proc. Amer. Math. Soc. 52 (1975) 1305–1310.

Ullman, J. L. (See Baker, I. N.)

Voiculescu, D. (See also Apostol, C.)

[1] A non-commutative Weyl–von Neumann theorem, Rev. Roum. Math. Pures et Appl. 21 (1976) 97–113.

[2] Norm-limits of algebraic operators, Rev. Roum. Math. Pures et Appl. 19 (1974), 371–378.

Wang, Z. Y. (See also Herrero, D. A.; Ji, Y. Q.; Jiang, C. L.;)

[1] The multiplication operators on Sobolev space, Proceedings of The First China Postdoctoral Academic Congress, BeiJing (1993), 1167–1170.

[2] Compact perturbation of the algebra of a tensor product of nests, J. Operator Theory 27 (1992), 251–282.

[3] On semi-strictly cyclic operators, East China Normal Univ. J. 3(1981), 26–34.

Wong, T. K. (See Deddens, J. A.)

Wu, P. Y. (See Fong, C. K.; Jiang, C. L.)